Introduction to Space-Time Wireless Communications

Wireless designers constantly seek to improve the spectrum efficiency/capacity, link reliability, and coverage of wireless networks. Space-time wireless technology that uses multiple antennas along with appropriate signaling and receiver techniques offers a powerful tool for improving wireless performance. Some aspects of this technology have already been incorporated into 3G mobile and fixed wireless standards. More advanced space-time techniques are planned for future mobile networks, wireless LANs and WANs.

The authors present the basics of space-time wireless propagation, the space-time channel, diversity and capacity performance, space-time coding, space-time receivers, interference cancellation for single carrier modulation, and extensions of OFDM and DS-spread spectrum modulation. They also cover space-time multi-user communications and system design tradeoffs.

This book is an introduction to this rapidly growing field for graduate students in wireless communications and for wireless designers in industry. Homework problems and other supporting material are available on a companion website.

Arogyaswami Paulraj is a pioneer of space-time wireless communications technology. He received his Ph.D. from the Indian Institute of Technology and is a Professor of Electrical Engineering at Stanford University, where he supervises the Smart Antennas Research Group. He is the author of nearly 300 research papers and holds 18 patents. He has held several positions in Indian industry, leading programs in military sonars and high-speed computing before moving to Stanford University. He founded Iospan Wireless to develop MIMO space-time technology for fixed wireless access. He is a Fellow of the IEEE and a member of the Indian National Academy of Engineering.

Dhananjay A. Gore was a graduate student in the Smart Antennas Research Group and received his Ph.D. in electrical engineering from Stanford University in March 2003. Dr. Gore is currently with Qualcomm Inc., San Diego, CA.

Rohit Nabar was a graduate student in the Smart Antennas Research Group and received his Ph.D. from Stanford University in February 2003. Between graduation and September 2004 he was a postdoctoral researcher at ETH, Zurich. He is currently a lecturer in the Communications and Signal Processing Research Group at the Department of Electrical and Electronic Engineering, Imperial College, London.

Introduction to Space-Time Wireless Communications

Arogyaswami Paulraj

Stanford University

Rohit Nabar

ETH, Zurich

Dhananjay Gore

Stanford University

CAMBRIDGE
UNIVERSITY PRESS

CAMBRIDGE UNIVERSITY PRESS
Cambridge, New York, Melbourne, Madrid, Cape Town, Singapore, São Paulo

Cambridge University Press
The Edinburgh Building, Cambridge CB2 8RU, UK

Published in the United States of America by Cambridge University Press, New York

www.cambridge.org
Information on this title: www.cambridge.org/9780521826150

First published 2003
Reprinted 2005 (with corrections), 2006
This digitally printed version 2008

A catalogue record for this publication is available from the British Library

ISBN 978-0-521-82615-0 hardback
ISBN 978-0-521-06593-1 paperback

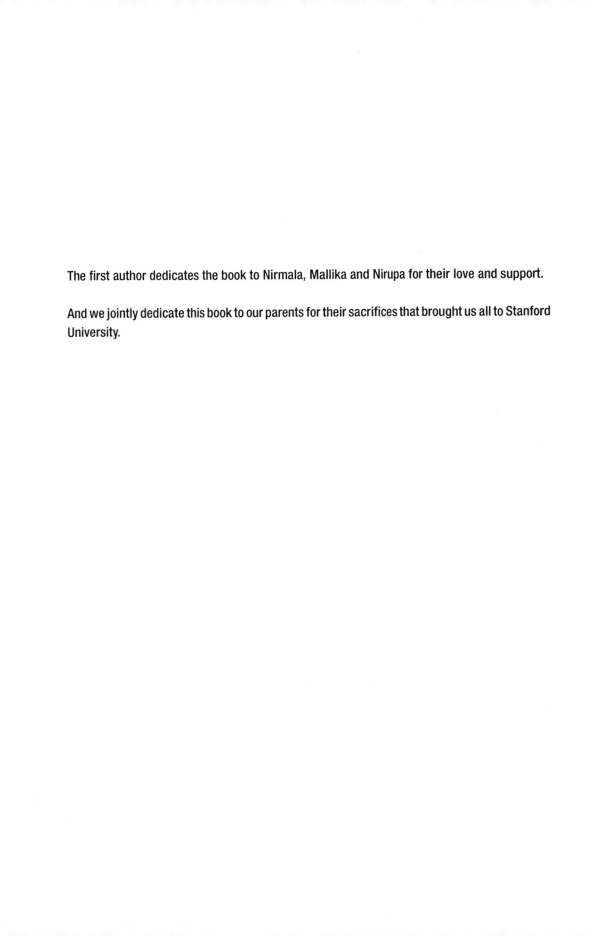

The first author dedicates the book to Nirmala, Mallika and Nirupa for their love and support.

And we jointly dedicate this book to our parents for their sacrifices that brought us all to Stanford University.

Contents

Figures

Tables

Preface

Use of multiple antennas in wireless links with appropriate space-time (ST) coding/ modulation and demodulation/decoding is rapidly becoming the new frontier of wireless communications. Recent years have seen the field mature substantially, both in theory and practice. Recent advances in theory include the solid understanding of capacity and other performance limits of ST wireless links, ST propagation and channel models, and also ST modulation/coding and receiver design. A growing awareness of the huge performance gains possible with ST techniques has spurred efforts to integrate this technology into practical systems. One example of this integration is the transmit diversity technique currently incorporated into different 2.5G and 3G standards. In the standards arena, recent efforts have focused on introducing spatial multiplexing concepts that require multiple antennas at both ends of the link (MIMO) into the UMTS standard for mobile wireless, the IEEE 802.16 standard for fixed and nomadic wireless and the IEEE 802.11 standard for wireless LANs. In addition, a number of proprietary products have been built to exploit ST technology, with ideas spanning from simple beamforming systems to the more complex spatial multiplexing concepts.

This book is an introduction to the theory of ST wireless communications. This area of technology has grown so large in the past few years that this book cannot cover all aspects in moderate detail. Rather, our aim has been to provide a coherent overview of the key advances in this field emphasizing basic theory and intuition. We have attempted to keep the presentation as simple as possible without sacrificing accuracy. ST theory is full of subtlety and nuances and the reader is guided to references for greater detail. A companion web page (http://publishing.cambridge.org/resources/0521826152) will provide a growing compilation of proofs, exercises, classroom slides, references and errata. The reader is encouraged to use this web site to improve the effectiveness of this book as a teaching tool. This book was written in the hope of being useful to graduate students and engineers in industry who wish to gain a basic understanding of this new field. A companion source for details on ST coding can be found in *Space-Time Block Coding for Wireless Communications* (by E. Larsson and P. Stoica) concurrently published by Cambridge University Press.

This work was supported in part by grants from the National Science Foundation (Grant Nos. CCR-0241919 and CCR-021921), Intel Corp. and Sprint Corp.

Stanford University's Smart Antennas Research Group (SARG) has been fortunate in attracting brilliant students and visitors, and the book is, in a sense, a distillation of what they have discovered through the efforts of others and themselves. The first author – Prof. Paulraj – acknowledges the many who have made contributions to SARG and therefore indirectly to this book. First, he thanks Prof. Thomas Kailath whose work in the area of directions-of-arrival estimation provided a foundation for SARG's work in ST wireless communications. Special acknowledgement is also due to Prof. David Gesbert and Dr Constantinos Papadias who contributed to an initial effort on a book project in 1998 which was, however, overwhelmed by a startup company project. Acknowledgement is also due to a number of companies (too numerous to list) and to the Federal Government who have supported SARG over the years. Special mention is due to Dr Bill Sanders at the Army Research Office and Dr John Cozzens at NSF, both of whom believed in and supported work on ST technology several years before it reached its current "superstar" status. The generous recent support of SARG by Sprint Corp. (sponsor Khurram Sheikh) and Intel Corp. (sponsor Dr E. Tsui) is most gratefully acknowledged. Prof. Paulraj also acknowledges the work of former colleagues at SARG who contributed their great talents to this field. These are (in a roughly chronological order) – Dr Chih-Yuan Chang, Dr Derek Gerlach, Dr Ayman Naguib, Dr Shilpa Talwar, Prof. Allejan van der Veen, Dr Kjell Gustaffson, Dr Constantinos Papadias, Dr Michaela van der Veen, Prof. K. Giridhar, Dr Boon Ng, Rupert Stuezle, S. Ratnavel, Prof. David Gesbert, Dr Jenwei Liang, Dr Erik Lindskog, Dr Mats Cederval, Prof. Umpathi Reddy, Dr Joachim Sorelius, Dr Suhas Diggavi, Dr S. Kuwahara, Dr T. Maeda, Dr Junheo Kim, Jens Kamman, Tushar Moorthy, Wonil Roh, Dr Sumeet Sandhu, Dr Sriram Mudulodu, Prof. Robert Heath, Prof. Helmut Bölcskei, Prof. K. V. S. Hari, Sebastian Peroor, Prof. Petre Stoica, Dr Osama Ata, Dr Hemanth Sampath, Daniel Baum, Dr Claude Oesteges, Prof. Thomas Strohmer, Dr Alexei Gorokhov, and Prof. Huzur Saran. Thanks are also due for many insights learnt from working closely with brilliant engineers such as Dr Vinko Erceg, Dr Jose Tellado, Frank McCarthy and Dr Rajeev Krishnamoorthy at Iospan Wireless (technology since acquired by Intel Corp.), who successfully built the first chip sets for a MIMO-OFDM wireless system. They are the true pioneers of commercial MIMO wireless.

The authors acknowledge the help during this book project from many at Stanford. Our current PhD students and visitors at SARG – Ozgur Oyman, Eric Stauffer, Andrew Brzezinski, Oghenkome Oteri, Eunchul Yoon, Majid Emami, Swaroop Sampath and Dr Alexei Gorokhov for checking the math and for other support. Prof. Thomas Kailath took time to read the first few chapters and made several useful suggestions. Mallika Paulraj, whose mastery of English well exceeded our own, helped improve the readability of the book.

Finally we wish to acknowledge the advice and comments from a number of reviewers, many anonymous and a few known. We thank them all. We are happy to acknowledge: Dr Constantinos Papadias, who coordinated a multi-person team review effort at Lucent Bell Labs, Prof. Helmut Bölcskei, Prof. Jorgen Bach Anderson and Prof. Eric Larsson for their careful reading and comments. Thanks are also due to Maureen Storey for handling the copy editing in a timely manner. We finally acknowledge the superb support from Dr Phil Meyler at Cambridge University Press, whose prompt and efficient handling of this publication project has been remarkable.

Abbreviations

3G	third generation
ADD	antenna division duplexing
AMPS	Advanced Mobile Phone Service
AOA	angle-of-arrival
AOD	angle-of-departure
AWGN	additive white Gaussian noise
BER	bit error rate
BPSK	binary phase shift keying
CCI	co-channel interference
CDF	cumulative distribution function
CDMA	code division multiple access
COFDM	coded orthogonal frequency division multiplexing
CP	cyclic pre-fix
CW	continuous wave
D-BLAST	diagonal Bell Labs layered space-time
DE	diagonal encoding
DFE	decision feedback equalizer
DPC	dirty paper coding
DS	direct sequence
EM	electromagnetic
ESPRIT	estimation of signal parameters via rotational invariance techniques
EXIT	extrinsic information transfer
FDD	frequency division duplexing
FEC	forward error correction
FFT	fast Fourier transform
FH	frequency hopping
FIR	finite impulse response
GDD	generalized delay diversity
GDFE	generalized decision feedback equalizer
GSM	global system for mobile
HE	horizontal encoding

HO	homogeneous channels
ICI	interchip interference
IFFT	inverse fast Fourier transform
IID	independent identically distributed
IIR	infinite impulse response
IMTS	improved mobile telephone service
ISI	intersymbol interference
LHS	left-hand side
LOS	line-of-sight
LP	Lindskog–Paulraj
MAI	multiple access interference
MF	matched filter
MFB	matched-filter bound
MIMO	multiple input multiple output
MIMO-BC	MIMO broadcast channel
MIMO-MAC	MIMO multiple access channel
MIMO-MU	multiple input multiple output multiuser
MIMO-SU	multiple input multiple output single user
MISO	multiple input single output
ML	maximum likelihood
MLSE	maximum likelihood sequence estimation
MLSR	maximal-length shift register
MMSE	minimum mean square error
MRC	maximum ratio combining
MSI	multistream interference
MUSIC	multiple signal classification
OFDM	orthogonal frequency division multiplexing
OSTBC	orthogonal space-time block code/codes/coding
OSUC	ordered successive cancellation
PAM	pulse amplitude modulation
PAR	peak-to-average ratio
PDF	probability density function
PEP	pairwise error probability
PER	packet error rate
PSK	phase shift keying
QAM	quadrature amplitude modulation
QoS	quality of service
QPSK	quadrature phase shift keying
RF	radio frequency
RHS	right-hand side
RMS	root mean square

ROC	region of convergence
SC	single carrier
SDD	standard delay diversity
SDMA	space division multiple access
SER	symbol error rate
SIMO	single input multiple output
SINR	signal to interference and noise ratio
SIR	signal to interference ratio
SISO	single input single output
SM	spatial multiplexing
SNR	signal to noise ratio
SS	spread spectrum
ST	space-time
STBC	space-time block code/codes/coding
STTC	space-time trellis code/codes/coding
SUC	successive cancellation
SUI	Stanford University interim
SVD	singular value decomposition
TDD	time division duplexing
TDM	time division multiplexing
TDMA	time division multiple access
UMTS	universal mobile telecommunications system
US	uncorrelated scattering
VE	vertical encoding
WSS	wide sense stationarity
WSSUS	wide sense stationary uncorrelated scattering
XIXO	(single or multiple) input (single or multiple) output
XPC	cross-polarization coupling
XPD	cross-polarization discrimination
ZF	zero forcing
ZMCSCG	zero mean circularly symmetric complex Gaussian

Symbols

$\approx$	approximately equal to
$\star$	convolution operator
$\otimes$	Kronecker product
$\odot$	Hadamard product
$\mathbf{0}_m$	$m \times m$ all zeros matrix
$\mathbf{0}_{m,n}$	$m \times n$ all zeros matrix
$\mathbf{1}_{D,L}$	$1 \times L$ row vector with $[\mathbf{1}_{D,L}]_{1,i} = \begin{cases} 1 & \text{if } i = D \\ 0 & \text{if } i \neq D \end{cases}$
$\|a\|$	magnitude of the scalar a
$\mathbf{A}^*$	elementwise conjugate of $\mathbf{A}$
$\mathbf{A}^\dagger$	Moore–Penrose inverse (pseudoinverse) of $\mathbf{A}$
$[\mathbf{A}]_{i,j}$	ijth element of matrix $\mathbf{A}$
$\|\mathbf{A}\|_F^2$	squared Frobenius norm of $\mathbf{A}$
$\mathbf{A}^H$	conjugate transpose of $\mathbf{A}$
$\mathbf{A}^T$	transpose of $\mathbf{A}$
$c(\mathcal{X})$	cardinality of the set $\mathcal{X}$
$\delta(x)$	Dirac delta (unit impulse) function
$\delta[x]$	Kronecker delta function, defined as
	$\delta[x] = \begin{cases} 1 & \text{if } x = 0 \\ 0 & \text{if } x \neq 0, x \in \mathcal{Z} \end{cases}$
$\det(\mathbf{A})$	determinant of $\mathbf{A}$
$\text{diag}\{a_1, a_2, \ldots, a_n\}$	$n \times n$ diagonal matrix with $[\text{diag}\{a_1, a_2, \ldots, a_n\}]_{i,i} = a_i$
$\mathcal{E}$	expectation operator
$f(x)$	PDF of the random variable X
$f(x_1, x_2, \ldots, x_N)$	joint PDF of the random variables $X_1, X_2, \ldots, X_N$
$F(x)$	CDF of the random variable X
$F(x_1, x_2, \ldots, x_N)$	joint CDF of the random variables $X_1, X_2, \ldots, X_N$
$\mathbf{I}_m$	$m \times m$ identity matrix
$\min(a_1, a_2, \ldots, a_n)$	minimum of $a_1, a_2, \ldots, a_n$
$Q(x)$	Q-function, defined as $Q(x) = (1/\sqrt{2\pi}) \int_x^\infty e^{-t^2/2} dt$

$r(\mathbf{A})$	rank of the matrix $\mathbf{A}$
$\mathcal{R}$	real field
$\Re\{\mathbf{A}\}, \Im\{\mathbf{A}\}$	real and imaginary parts of $\mathbf{A}$, respectively
$\mathrm{Tr}(\mathbf{A})$	trace of $\mathbf{A}$
$u(x)$	unit step function, defined as $u(x) = \begin{cases} 1 & \text{if } x \geq 0, x \in \mathcal{R} \\ 0 & \text{if } x < 0, x \in \mathcal{R} \end{cases}$
$\mathrm{vec}(\mathbf{A})$	stacks $\mathbf{A}$ into a vector columnwise[1]
$(x)_+$	defined as $(x)_+ = \begin{cases} x & \text{if } x \geq 0, x \in \mathcal{R} \\ 0 & \text{if } x < 0, x \in \mathcal{R} \end{cases}$
$\mathcal{Z}$	integer field

[1] If $\mathbf{A} = [\mathbf{a}_1 \ \mathbf{a}_2 \ \cdots \ \mathbf{a}_n]$ is $m \times n$, then $\mathrm{vec}(\mathbf{A}) = [\mathbf{a}_1^T \ \mathbf{a}_2^T \ \cdots \ \mathbf{a}_n^T]^T$ is $mn \times 1$.

1 Introduction

The radio age began over a 100 years ago with the invention of the radiotelegraph by Guglielmo Marconi and the wireless industry is now set for rapid growth as we enter a new century and a new millennium. The rapid progress in radio technology is creating new and improved services at lower costs, which results in increases in air-time usage and the number of subscribers. Wireless revenues are currently growing between 20% and 30% per year, and these broad trends are likely to continue for several years.

Multiple access wireless communications is being deployed for both fixed and mobile applications. In fixed applications, the wireless networks provide voice or data for fixed subscribers. Mobile networks offering voice and data services can be divided into two classes: high mobility, to serve high speed vehicle-borne users, and low mobility, to serve pedestrian users. Wireless system designers are faced with a number of challenges. These include the limited availability of the radio frequency spectrum and a complex time-varying wireless environment (fading and multipath). In addition, meeting the increasing demand for higher data rates, better quality of service (QoS), fewer dropped calls, higher network capacity and user coverage calls for innovative techniques that improve spectral efficiency and link reliability. The use of multiple antennas at the receiver and/or transmitter in a wireless system, popularly known as space-time (ST) wireless or multiantenna communications or smart antennas is an emerging technology that promises significant improvements in these measures. This book is an introduction to the theory of ST wireless communications.

1.1 History of radio, antennas and array signal processing

The origins of radio date back to 1861 when Maxwell, while at King's College in London, proposed a mathematical theory of electromagnetic (EM) waves. A practical demonstration of the existence of such waves was performed by Hertz in 1887 at the University of Karlsruhe, using stationary (standing) waves. Following this, improvements in the generation and reception of EM waves were pursued by many researchers in Europe. In 1890, Branly in Paris developed a "coherer" that could detect the presence of EM waves using iron filings in a glass bottle. The coherer was further refined by

1

Righi at the University of Bologna and Lodge in England. Other contributions came from Popov in Russia, who is credited with devising the first radio antenna during his attempts to detect EM radiation from lightning.

In the summer of 1895, Marconi, at the age of 21, was inspired by the lectures on radio waves by Righi at the University of Bologna and he built and demonstrated the first radio telegraph. He used Hertz's spark transmitter, Lodge's coherer and added antennas to assemble his instrument. In 1898, Marconi improved the telegraph by adding a four-circuit tuning device, allowing simultaneous use of two radio circuits. That year, his signal bridged the English Channel, 52 km wide, between Wimereux and Dover. His other technical developments around this time included the magnetic detector, which was an improvement over the less efficient coherer, the rotary spark and the use of directive antennas to increase the signal level and to reduce interference in duplex receiver circuits. In the next few years, Marconi integrated many new technologies into his increasingly sophisticated radio equipment, including the diode valve developed by Fleming, the crystal detector, continuous wave (CW) transmission developed by Poulsen, Fessenden and Alexanderson, and the triode valve or audion developed by Forrest.

Civilian use of wireless technology began with the installation of the first 2 MHz land mobile radiotelephone system in 1921 by the Detroit Police Department for police car dispatch. The advantages of mobile communications were quickly realized, but its wider use was limited by the lack of channels in the low frequency band. Gradually, higher frequency bands were used, opening up the use of more channels. A key advance was made in 1933, when Armstrong invented frequency modulation (FM), which made possible high quality radio communications. In 1946, a Personal Correspondence System introduced by Bell Systems began service and operated at 150 MHz with speech channels 120 kHz apart. As demand for public wireless services began to grow, the Improved Mobile Telephone Service (IMTS) using FM technology was developed by AT&T. These were the first mobile systems to connect with the public telephone network using a fixed number of radio channels in a single geographic area. Extending such technology to a large number of users with full duplex channels needed excessive bandwidth. A solution was found in the cellular concept (known as cellularization) conceived by Ring at Bell Laboratories in 1947. This concept required dividing the service area into smaller cells, and using a subset of the total available radio channels in each cell. AT&T proposed the first high capacity analog cellular telephone system called the Advanced Mobile Phone Service (AMPS) in 1970. Mobile cellular systems have evolved rapidly since then, incorporating digital communication technology and serve nearly one billion subscribers worldwide today. While the Global System for Mobile (GSM) standard developed in Europe has gathered the largest market share, cellular networks in the USA have used the IS-136 (using time division multiple access or TDMA) and IS-95 (using Code Division Multiple Access or CDMA) standards. With increasing use of wireless internet in the late 1990s, the demand for higher spectral efficiency and data rates has led to the development of the so called Third Generation (3G)

Table 1.1. *Performance goals for antennas in wireless communications*

Antenna design	AOA estimation	Link performance
Gain	Error variance	Coverage
Bandwidth	Bias	Quality
Radiation pattern	Resolution	Interference reduction
Size		Spectral efficiency

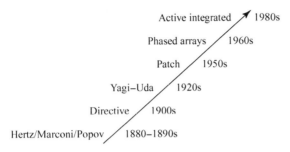

Figure 1.1: Developments in antenna (EM) technology.

wireless technologies. 3G standardization failed to achieve a single common world-wide standard and now offers UMTS (wideband CDMA) and 1XRTT as the primary standards. Limitations in the radio frequency (RF) spectrum necessitate the use of innovative techniques to meet the increased demand in data rate and QoS.

The use of multiple antennas at the transmitter and/or receiver in a wireless communication link opens a new dimension – space, which if leveraged correctly can improve performance substantially. Table 1.1 details the three main areas of study in the field of radio antennas and their applications. The first covers the electromagnetic design of the antennas and antenna arrays. The goals here are to meet design requirements for gain, polarization, beamwidth, sidelobe level, efficiency and radiation pattern. The second area is the angle-of-arrival (AOA) estimation and, as the name indicates, focuses on estimating arrival angles of wavefronts impinging on the antenna array with minimum error and high resolution. The third area of technology that this book focuses on is the use of antenna arrays to improve spectral efficiency, coverage and quality of wireless links.

A timeline of the key developments in the field of antenna design is given in Fig. 1.1. The original antenna design work came from Marconi and Popov among others in the early 1900s. Marconi soon developed directional antennas for his cross-Atlantic links. Antenna design improved in frequency of operation and bandwidth in the early part of the twentieth century. An important breakthrough was the Yagi–Uda arrays that offered high bandwidth and gain. Another important development was the patch antenna that offers low profile and cost. The use of antennas in arrays began in World War II, mainly

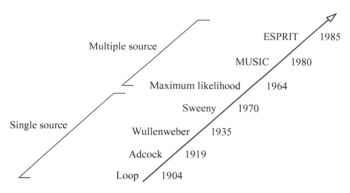

Figure 1.2: Developments in AOA estimation.

for radar applications. Array design brought many new issues to the fore, such as gain, beamwidth, sidelobe level, and beamsteering.

The area of AOA estimation had its beginnings in World War I when loop antennas were used to estimate signal direction (see Fig. 1.2 for a timeline of AOA technology). Adcock antennas were a significant advance and were used in World War II. Wullenweber arrays were developed in 1938 for lower frequencies and where accuracy was important, and are used in aircraft localization to this day. These techniques addressed the single source signal wavefront case. If there are multiple sources in the same frequency channel or multipath arrivals from a single source, new techniques are needed. The problem of AOA estimation in the multisource case was properly addressed in the 1970s and 1980s. Capon's method [Capon *et al.*, 1967], a well-known technique, offered reasonable resolution performance although it suffered from bias even in asymptotically large data cases. The multiple signal classification (MUSIC) technique proposed by Schmidt in 1981 was a major breakthrough. MUSIC is asymptotically unbiased and offers improved resolution performance. Later a method called estimation of signal parameters via rotational invariance techniques (ESPRIT) that has the remarkable advantage of not needing exact characterization of the array manifold and yet achieves optimal performance was proposed [Paulraj *et al.*, 1986; Roy *et al.*, 1986].

The third area of antenna applications in wireless communications is link enhancement (see Fig. 1.3). The use of multiple receive antennas for diversity goes back to Marconi and the early radio pioneers. So does the realization that steerable receive antenna arrays can be used to mitigate co-channel interference in radio systems. The use of antenna arrays was an active reseach area during and after World War II in radar systems. More sophisticated applications of adaptive signal processing at the wireless receiver for improving diversity and interference reduction had to wait until the 1970s for the arrival of digital signal processors at which point these techniques were vigorously developed for military applications. The early 1990s saw new proposals for using antennas to increase capacity of wireless links. Roy and Ottersten in 1996 proposed the use of base-station antennas to support multiple co-channel users. Paulraj and Kailath in

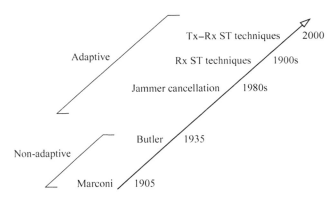

Figure 1.3: Developments in antenna technology for link performance enhancement.

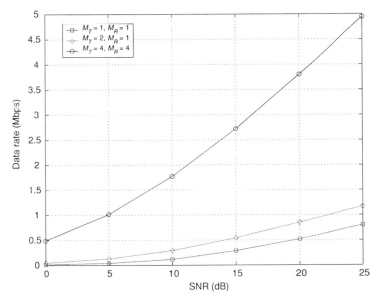

Figure 1.4: Data rate (at 95% reliability) vs SNR for different antenna configurations. Channel bandwidth is 200 KHz.

1994 proposed a technique for increasing the capacity of a wireless link using multiple antennas at both the transmitter and the receiver. These ideas along with the fundamental research done at Bell Labs [Telatar, 1995; Foschini, 1996; Foschini and Gans, 1998; Tarokh *et al.*, 1998] began a new revolution in information and communications theory in the mid 1990s. The goal is to approach performance limits and to explore efficient but pragmatic coding and modulation schemes for wireless links using multiple antennas. Clearly much more work has yet to be done and the field is attracting considerable research talent.

The leverage of ST wireless technology is significant. Figure 1.4 plots the maximum error-free data rate in a 200 KHz fading channel vs the signal to noise ratio (SNR)

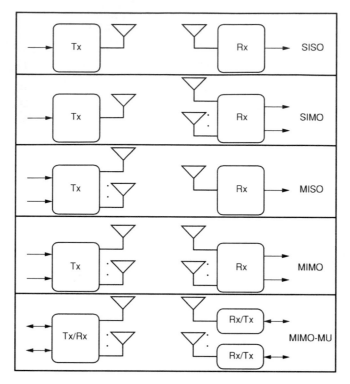

Figure 1.5: Antenna configurations in ST wireless systems (Tx: Transmitter, Rx: Receiver).

that is guaranteed at 95% reliability. Assuming a target receive SNR of 20 dB, current single antenna transmit and receive technology can offer a data rate of 0.5 Mbps. A two-transmit and one-receive antenna system would achieve 0.8 Mbps. A four-transmit and four-receive antenna system can reach 3.75 Mbps. It is worth noting that 3.75 Mbps is also achievable in a single antenna transmit and receive technology, but needs 10^5 times higher SNR or transmit power compared with a four-transmit and four-receive antenna configuration. The technology that can deliver such remarkable gains is the subject of this book.

1.2 Exploiting multiple antennas in wireless

Figure 1.5 illustrates different antenna configurations for ST wireless links. SISO (single input single output) is the familiar wireless configuration, SIMO (single input multiple output) has a single transmit antenna and multiple (M_R) receive antennas, MISO (multiple input single output) has multiple (M_T) transmit antennas and a single receive antenna and MIMO (multiple input multiple output) has multiple (M_T)

transmit antennas and multiple (M_R) receive antennas. The MIMO-MU (MIMO multiuser) configuration refers to the case where a base-station with multiple (M) antennas communicates with P users each with one or more antennas. Both transmit and receive configurations are shown. We sometimes abbreviate SIMO, MISO and MIMO configurations as XIXO.

1.2.1   Array gain

Array gain refers to the average increase in the SNR at the receiver that arises from the coherent combining effect of multiple antennas at the receiver or transmitter or both. Consider, as an example, a SIMO channel. Signals arriving at the receive antennas have different amplitudes and phases. The receiver can combine the signals coherently so that the resultant signal is enhanced. The average increase in signal power at the receiver is proportional to the number of receive antennas. In channels with multiple antennas at the transmitter (MISO or MIMO channels), array gain exploitation requires channel knowledge at the transmitter.

1.2.2   Diversity gain

Signal power in a wireless channel fluctuates (or fades). When the signal power drops significantly, the channel is said to be in a fade. Diversity is used in wireless channels to combat fading.

Receive antenna diversity can be used in SIMO channels [Jakes, 1974]. The receive antennas see independently faded versions of the same signal. The receiver combines these signals so that the resultant signal exhibits considerably reduced amplitude variability (fading) in comparison with the signal at any one antenna. Diversity is characterized by the number of independently fading branches, also known as the diversity order and is equal to the number of receive antennas in SIMO channels.

Transmit diversity is applicable to MISO channels and has become an active area for research [Wittneben, 1991; Seshadri and Winters, 1994; Kuo and Fitz, 1997; Olofsson *et al.*, 1997; Heath and Paulraj, 1999]. Extracting diversity in such channels is possible with or without channel knowledge at the transmitter. Suitable design of the transmitted signal is required to extract diversity. ST diversity coding [Seshadri and Winters, 1994; Guey *et al.*, 1996; Alamouti, 1998; Tarokh *et al.*, 1998, 1999b] is a transmit diversity technique that relies on coding across space (transmit antennas) to extract diversity in the absence of channel knowledge at the transmitter. If the channels of all transmit antennas to the receive antenna have independent fades, the diversity order of this channel is equal to the number of transmit antennas.

Utilization of diversity in MIMO channels requires a combination of the receive and transmit diversity described above. The diversity order is equal to the product of the number of transmit and receive antennas, if the channel between each transmit–receive antenna pair fades independently.

1.2.3   Spatial multiplexing (SM)

SM offers a linear (in the number of transmit–receive antenna pairs or $\min(M_R, M_T)$) increase in the transmission rate (or capacity) for the same bandwidth and with no additional power expenditure. SM is only possible in MIMO channels [Paulraj and Kailath, 1994; Foschini, 1996; Telatar, 1999a]. In the following we discuss the basic principles of SM for a system with two transmit and two receive antennas. The concept can be extended to more general MIMO channels.

The bit stream to be transmitted is demultiplexed into two half-rate sub-streams, modulated and transmitted simultaneously from each transmit antenna. Under favorable channel conditions, the spatial signatures of these signals induced at the receive antennas are well separated. The receiver, having knowledge of the channel, can differentiate between the two co-channel signals and extract both signals, after which demodulation yields the original sub-streams that can now be combined to yield the original bit stream. Thus SM increases transmission rate proportionally with the number of transmit–receive antenna pairs.

SM can also be applied in a multiuser format (MIMO-MU, also known as space division multiple access or SDMA). Consider two users transmitting their individual signals, which arrive at a base-station equipped with two antennas. The base-station can separate the two signals to support simultaneous use of the channel by both users. Likewise the base-station can transmit two signals with spatial filtering so that each user can decode its own signal adequately. This allows a capacity increase proportional to the number of antennas at the base-station and the number of users.

1.2.4   Interference reduction

Co-channel interference arises due to frequency reuse in wireless channels. When multiple antennas are used, the differentiation between the spatial signatures of the desired signal and co-channel signals can be exploited to reduce the interference. Interference reduction requires knowledge of the channel of the desired signal. However, exact knowledge of the interferer's channel may not be necessary.

Interference reduction (or avoidance) can also be implemented at the transmitter, where the goal is to minimize the interference energy sent towards the co-channel users while delivering the signal to the desired user. Interference reduction allows the use of aggressive reuse factors and improves network capacity.

We note that it may not be possible to exploit all the leverages simultaneously due to conflicting demands on the spatial degrees of freedom (or number of antennas). The degree to which these conflicts are resolved depends upon the signaling scheme and receiver design.

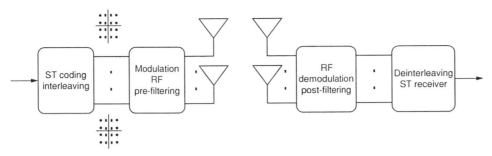

Figure 1.6: Schematic of a ST wireless communication system.

1.3   ST wireless communication systems

Figure 1.6 shows a typical ST wireless system with M_T transmit antennas and M_R receive antennas. The input data bits enter a ST coding block that adds parity bits for protection against noise and also captures diversity from the space and possibly frequency or time dimensions in a fading environment. After coding, the bits (or words) are interleaved across space, time and frequency and mapped to data symbols (such as quadrature amplitude modulation (QAM)) to generate M_T outputs. The M_T symbol streams may then be ST pre-filtered before being modulated with a pulse shaping function, translated to the passband via parallel RF chains and then radiated from M_T antennas. These signals pass through the radio channel where they are attenuated and undergo fading in multiple dimensions before they arrive at the M_R receive antennas. Additive thermal noise in the M_R parallel RF chains at the receiver corrupts the received signal. The mixture of signal plus noise is matched-filtered and sampled to produce M_R output streams. Some form of additional ST post-filtering may also be applied. These streams are then ST deinterleaved and ST decoded to produce the output data bits.

The difference between a ST communication system and a conventional system comes from the use of multiple antennas, ST encoding/interleaving, ST pre-filtering and post-filtering and ST decoding/deinterleaving.

We conclude this chapter with a brief overview of the areas discussed in the remainder of this book. Chapter 2 overviews ST propagation. We develop a channel representation as a vector valued ST random field and derive multiple representations and statistical descriptions of ST channels. We also describe real world channel measurements and models.

Chapter 3 introduces XIXO channels, derives channels from statistical ST channel descriptions, proposes general XIXO channel models and test channel models and ends with a discussion on XIXO channel estimation at the receiver and transmitter.

Chapter 4 studies channel capacity of XIXO channels under a variety of conditions: channel known and unknown to the transmitter, general channel models and frequency selective channels. We also discuss the ergodic and outage capacity of random XIXO channels.

Chapter 5 overviews the spatial diversity for XIXO channels, bit error rate performance with diversity and the influence of general channel conditions on diversity and ends with techniques that can transform spatial diversity at the transmitter into time or frequency diversity at the receiver.

Chapter 6 develops ST coding for diversity, SM and hybrid schemes for single carrier modulation where the channel is not known at the transmitter. We discuss performance criteria in frequency flat and frequency selective fading environments.

Chapter 7 describes ST receivers for XIXO channels and for single carrier modulation. We discuss maximum likelihood (ML), zero forcing (ZF), minimum mean square error (MMSE) and successive cancellation (SUC) receiver structures. Performance analysis is also provided.

Chapter 8 addresses exploiting channel knowledge by the transmitter through transmit pre-processing, both for the case where the channel is perfectly known and the case where only statistical or partial channel knowledge is available.

Chapter 9 overviews how XIXO techniques can be applied to orthogonal frequency division multiplexing (OFDM) and spread spectrum (SS) modulation schemes. It also discusses how ST coding for single carrier modulation can be extended to the space-frequency or space-code dimensions.

Chapter 10 addresses MIMO-MU where multiple users (each with one or more antennas) communicate with the base (with multiple antennas). A quick summary of capacity, signaling and receivers is provided.

Chapter 11 discusses how multiple antennas can be used to reduce co-channel interference for XIXO signal and interference models. A short review of interference diversity is also provided.

Chapter 12 overviews performance limits of ST channels with optimal and sub-optimal signaling and receivers.

2 ST propagation

2

2.1 Introduction

In this chapter we overview wireless channel behavior, the focus being on outdoor macrocellular environments. We describe the wireless channel, develop a scattering model for such channels and discuss real world channel measurements. We show how the wireless channel may be modeled as a vector valued ST random field and its statistical behavior captured through scattering functions. Finally, we briefly review ST degenerate wireless channels. Our presentation is greatly simplified, covering the key ideas, but avoids the rich (and complex) nuances in the field. Wireless propagation has been covered by a number of excellent texts [Jakes, 1974; Lee, 1982; Parsons, 1992; Rappaport, 1996; Bertoni, 1999]. Our goal here is to integrate the spatial dimension into these propagation models.

2.2 The wireless channel

A signal propagating through the wireless channel arrives at the destination along a number of different paths, collectively referred to as multipath. These paths arise from scattering, reflection and diffraction of the radiated energy by objects in the environment or refraction in the medium. The different propagation mechanisms influence path loss and fading models differently. However, for convenience we refer to all these distorting mechanisms as "scattering". Further, throughout the book, we assume a complex baseband representation for the signal and channel unless otherwise specified.

The signal power drops off due to three effects: mean propagation (path) loss, macroscopic fading and microscopic fading. The mean propagation loss in macrocellular environments comes from inverse square law power loss, absorption by water and foliage and the effect of ground reflection. Mean propagation loss is range dependent. Macroscopic fading results from a blocking effect by buildings and natural features and is also known as long term fading or shadowing. Microscopic fading results from the constructive and destructive combination of multipaths and is also known as short term fading or fast fading. Multipath propagation results in the spreading of the signal in different dimensions. These are delay spread, Doppler (or frequency) spread (this needs a time-varying multipath channel) and angle spread. These spreads have significant effects on the signal. Mean path loss, macroscopic fading, microscopic fading, delay spread, Doppler spread and angle spread are the main channel effects and are described below.

2.2.1   Path loss

In ideal free space propagation we have inverse square law power loss and the received signal power is given by [Jakes, 1974]

$$P_r = P_t \left(\frac{\lambda_c}{4\pi d} \right)^2 G_t G_r, \tag{2.1}$$

where P_t and P_r are the transmitted and received powers respectively, λ_c is the wavelength, G_t, G_r are the power gains of the transmit and receive antennas respectively and d is the range separation. Equation (2.1) is also known as the Friis equation [Feher, 1995]. In cellular environments, the main path is accompanied by a surface reflected path that destructively interferes with the primary path. The received power can now be approximated by

$$P_r = P_t \left(\frac{h_t h_r}{d^2} \right)^2 G_t G_r, \tag{2.2}$$

where h_t, h_r are the effective heights of the transmit and receive antennas respectively and we have made the assumption that $d^2 >> h_t h_r$. The effective path loss follows an inverse fourth power law (the path loss exponent is equal to 4) that results in a loss of 40 dB/decade. In real environments the path loss exponent varies from 2.5 to 6 and depends on the terrain and foliage. Several empirically based path loss models have been developed for macrocellular and microcellular environments such as the Okumura, Hata, COST-231 and Erceg models [Okumura *et al.*, 1968; Hata and Nagatsu, 1980; COST 231 TD(973) 119-REV 2 (WG2), 1991; Erceg *et al.*, 1999a].

2.2.2   Fading

In addition to path loss, the received signal exhibits fluctuations in signal level called fading. Fluctuation in signal level is typically composed of two multiplicative components – macroscopic and microscopic fading. Macroscopic fading represents the long term variation of the received signal power level, while microscopic fading represents short term variation. We now describe the different types of fading.

Macroscopic fading

Macroscopic fading is caused by shadowing effects of buildings or natural features and is determined by the local mean of a fast fading signal. The statistical distribution of the local mean has been studied experimentally. This distribution is influenced by antenna heights, the operating frequency and the specific type of environment. However, it has been observed [Jakes, 1974] that the received power averaged over microscopic fading approaches a normal distribution when plotted on a logarithmic scale (i.e., in decibels) and is called a log-normal distribution described by the probability density function (PDF):

$$f(x) = \frac{1}{\sqrt{2\pi}\sigma} e^{-\frac{(x-\mu)^2}{2\sigma^2}}. \tag{2.3}$$

The above equation is the probability density of the long-term signal power fluctuation (measured in decibels). μ and σ are respectively the mean and standard deviation of the signal power, expressed in decibels.

Microscopic fading

Microscopic fading refers to the rapid fluctuations of the received signal in space, time and frequency, and is caused by the signal scattering off objects between the transmitter and receiver. If we assume that fading is caused by the superposition of a large number of independent scattered components, then the in-phase and quadrature components of the received signal can be assumed to be independent zero mean Gaussian processes. The envelope of the received signal has a Rayleigh density function given by

$$f(x) = \frac{2x}{\Omega} e^{-\frac{x^2}{\Omega}} u(x), \tag{2.4}$$

where Ω is the average received power and $u(x)$ is the unit step function defined as

$$u(x) = \begin{cases} 1 & \text{if } x \geq 0, x \in \mathcal{R} \\ 0 & \text{if } x < 0, x \in \mathcal{R} \end{cases}. \tag{2.5}$$

If there is a direct (possibly a line-of-sight (LOS)) path present between transmitter and receiver, the signal envelope is no longer Rayleigh and the distribution of the signal amplitude is Ricean. The Ricean distribution is often defined in terms of the Ricean factor, K, which is the ratio of the power in the mean component of the channel to the power in the scattered (varying) component. The Ricean PDF of the envelope of the received signal is given by

$$f(x) = \frac{2x(K+1)}{\Omega} e^{\left(-K - \frac{(K+1)x^2}{\Omega}\right)} I_0\left(2x\sqrt{\frac{K(K+1)}{\Omega}}\right) u(x), \tag{2.6}$$

where Ω is the mean received power as defined earlier and $I_0(x)$ is the zero-order

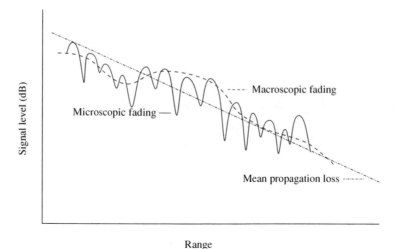

Figure 2.1: Signal power fluctuation vs range in wireless channels. Mean propagation loss increases monotonically with range. Local deviations may occur due to macroscopic and microscopic fading.

modified Bessel function of the first kind defined as

$$I_0(x) = \frac{1}{2\pi} \int_0^{2\pi} e^{-x \cos\theta} d\theta. \tag{2.7}$$

In the absence of a direct path ($K = 0$), the Ricean PDF in Eq. (2.6) reduces to the Rayleigh PDF in Eq. (2.4), since $I_0(0) = 1$. More sophisticated fading distributions such as the Nakagami distribution [Nakagami, 1960] (to characterize fading in high frequency channels) can be found in the literature, but we restrict our discussion to Rayleigh or Ricean fading in this book.

The above statistics are valid for microscopic fading in all three dimensions – space, time and frequency. Figure 2.1 shows the combined effects of path loss and macroscopic and microscopic fading on received power in a wireless channel.

Doppler spread – Time selective fading

Time-varying fading due to scatterer or transmitter/receiver motion results in a Doppler spread, i.e., a pure tone (frequency v_c in hertz) spreads over a finite spectral bandwidth ($v_c \pm v_{max}$). The Fourier transform of the time autocorrelation of the channel response to a continuous wave (CW) tone is defined as the Doppler power spectrum, $\psi_{Do}(v)$ with $v_c - v_{max} \leq v \leq v_c + v_{max}$ (see Fig. 2.2). $\psi_{Do}(v)$ is the average power of the channel output as a function of the Doppler frequency v.

If one assumes idealized, uniformly distributed scattering around a terminal with vertical **E**-field receive and transmit antennas, then the Doppler power spectrum has the classical U-shaped form and is approximated by the Jakes model [Jakes, 1974]. In reality, the Doppler spectrum can show considerable variation from this model. In fixed wireless applications the Doppler spectrum is approximately exponential [Baum *et al.*,

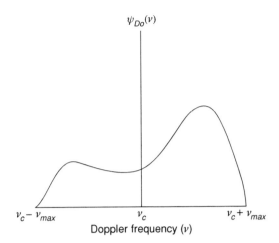

Figure 2.2: Typical Doppler (power) spectrum $\psi_{Do}(v)$ – average power as a function of Doppler frequency (v).

2000]. The root mean square (RMS) bandwidth of $\psi_{Do}(v)$ is called the Doppler spread, v_{RMS}, and is given by

$$v_{RMS} = \sqrt{\frac{\int_{\mathcal{F}} (v - \bar{v})^2 \psi_{Do}(v) dv}{\int_{\mathcal{F}} \psi_{Do}(v) dv}}, \tag{2.8}$$

where $\mathcal{F}$ represents the interval $v_c - v_{max} \leq v \leq v_c + v_{max}$ and $\bar{v}$ is the average frequency of the Doppler spectrum given by

$$\bar{v} = \frac{\int_{\mathcal{F}} v \psi_{Do}(v) dv}{\int_{\mathcal{F}} \psi_{Do}(v) dv}. \tag{2.9}$$

In the case of a direct path, the above spectrum is modified by an additional discrete frequency component corresponding to the relative velocity between the base-station and the terminal, and the AOA of the direct path. Time selective fading can be characterized by the coherence time, T_C, of the channel. Coherence time is typically defined as the time lag for which the signal autocorrelation coefficient reduces to 0.7. The coherence time is inversely proportional to the Doppler spread and can be approximated as

$$T_C \approx \frac{1}{v_{RMS}}. \tag{2.10}$$

The coherence time serves as a measure of how fast the channel changes in time – the larger the coherence time, the slower the channel fluctuation.

Delay spread – Frequency selective fading
In a multipath propagation environment, several delayed and scaled versions of the transmitted signal arrive at the receiver. An idealized classical model is a double negative

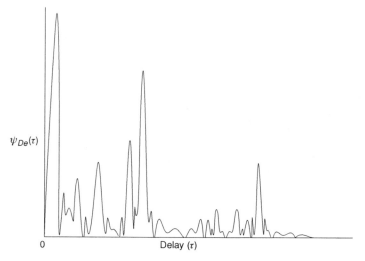

Figure 2.3: Typical delay (power) profile $\psi_{De}(\tau)$ – average power as a function of delay (τ).

exponential model: the delay separation between paths increases exponentially with path delay, and the path amplitudes also fall off exponentially with delay [Adachi *et al.*, 1986; Braun and Dersch, 1991]. The span of path delays is called the delay spread. In reality, the delay spread also shows considerable variability from the classical model. The RMS delay spread of the channel, τ_{RMS}, is defined as

$$\tau_{RMS} = \sqrt{\frac{\int_0^{\tau_{max}} (\tau - \overline{\tau})^2 \psi_{De}(\tau) d\tau}{\int_0^{\tau_{max}} \psi_{De}(\tau) d\tau}}, \tag{2.11}$$

where $\psi_{De}(\tau)$ is the multipath intensity profile or spectrum (the average power of the channel output as a function of delay τ, see Fig. 2.3), τ_{max} is the maximum path delay and $\overline{\tau}$ is the average delay spread given by

$$\overline{\tau} = \frac{\int_0^{\tau_{max}} \tau \psi_{De}(\tau) d\tau}{\int_0^{\tau_{max}} \psi_{De}(\tau) d\tau}. \tag{2.12}$$

Delay spread causes frequency selective fading as the channel acts like a tapped delay line filter. Frequency selective fading can be characterized in terms of coherence bandwidth, B_C, which is the frequency lag for which the channel's autocorrelation coefficient reduces to 0.7. The coherence bandwidth is inversely proportional to the RMS delay spread, and is a measure of the channel's frequency selectivity. Thus,

$$B_C \approx \frac{1}{\tau_{RMS}}. \tag{2.13}$$

When the coherence bandwidth is comparable to or less than the signal bandwidth, the channel is said to be frequency selective.

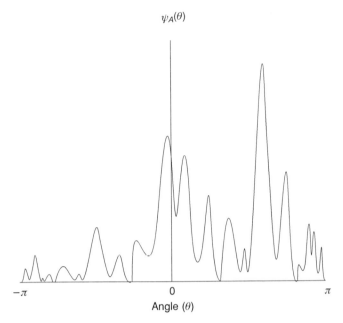

Figure 2.4: Typical angle (power) spectrum $\psi_A(\theta)$ – average power as a function of angle (θ).

Angle spread – Space selective fading

Angle spread at the receiver refers to the spread in AOAs of the multipath components at the receive antenna array. Similarly, angle spread at the transmitter refers to the spread in angles of departure (AODs) of the multipath that finally reach the receiver.

Denoting the AOA by θ and the average power as a function of AOA by $\psi_A(\theta)$, the angle spectrum (see Fig. 2.4), we can define the RMS angle spread, θ_{RMS}, as

$$\theta_{RMS} = \sqrt{\frac{\int_{-\pi}^{\pi}(\theta - \overline{\theta})^2 \psi_A(\theta)d\theta}{\int_{-\pi}^{\pi} \psi_A(\theta)d\theta}}, \tag{2.14}$$

where $\overline{\theta}$ is the mean AOA and is given by

$$\overline{\theta} = \frac{\int_{-\pi}^{\pi} \theta \psi_A(\theta)d\theta}{\int_{-\pi}^{\pi} \psi_A(\theta)d\theta}. \tag{2.15}$$

Angle spread causes space selective fading which means that signal amplitude depends on the spatial location of the antenna. Space selective fading is characterized by the coherence distance, D_C, which is the spatial separation for which the autocorrelation coefficient of the spatial fading drops to 0.7. The coherence distance is inversely proportional to the angle spread – the larger the angle spread, the shorter the coherence distance. Therefore,

$$D_C \propto \frac{1}{\theta_{RMS}}. \tag{2.16}$$

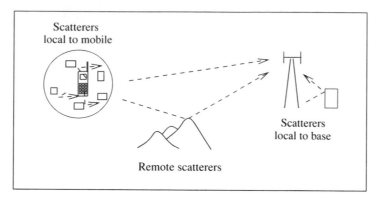

Figure 2.5: Classification of scatterers. Scattering is typically rich around the terminal and sparse at the base-station.

θ_{RMS} and D_C are intuitively well defined at a receive antenna for a signal launched from a reference transmit antenna. They can also be easily defined at a transmit antenna with reference to the target receive antenna.

2.3   Scattering model in macrocells

Multipath scattering underlies the three spreading effects described above (in addition, motion is also required to produce Doppler spread). In the following discussion we describe scattering effects for the reverse link channel (terminal to base-station, also known as the uplink), but the discussion applies equally to the forward link channel (base-station to terminal, also known as the downlink). Scatterers between the terminal and base-station can be categorized as follows.

Local to terminal scatterers
Local to terminal scattering is caused by buildings or other scatterers within the vicinity of the terminal (a few tens of meters) as shown in Fig. 2.5. Terminal motion and local scattering give rise to Doppler spread or equivalently time selective fading. While local scatterers contribute to Doppler spread, the delay spread they induce is usually insignificant because of the small scattering radius. Likewise, the angle spread induced at the base-station is small.

Remote scatterers
The emerging wavefront from the local scatterers may travel directly to the base-station or may be scattered towards the base-station by remote dominant scatterers, giving rise to a specular multipath. Remote scatterers can be either terrain features or high rise building complexes. Remote scattering can cause significant delay and angle spread. Remote scatterers at any fixed delay lie on an iso-delay ellipse (see Fig. 2.6)

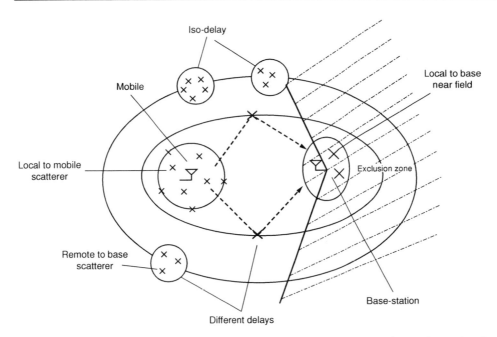

Figure 2.6: Scattering model for wireless channels. The terminal and base-station are located at the foci of the iso-delay ellipses.

corresponding to the excess path delay. Since remote scatterers such as hills or large building clusters are themselves composed of a number of smaller scatterers, these remote scatterers are best modeled as disks of smaller scatterers.

Local to base scatterers and exclusion zones

After undergoing scattering from local and remote scatterers (along with any direct components) the signal arrives at the base-station. Antennas at the base-station are typically elevated with narrow vertical beamwidths (typically 6°) and 120° horizontal beamwidth. Therefore, there is no contribution from scatterers from outside the vertical and horizontal antenna beamwidth, in contrast to the situation at the terminal where the near-omnidirectional antenna admits signals scattered from all directions. We model this effect with a scatterer exclusion zone around the base-station. However, there may be very close near-field scattering from antenna tower or roof corners (when base-station antennas are placed on roofs) and this can cause correlated scattering effects (see Section 2.8 for a more in-depth discussion).

   The overall scattering model therefore involves signals from the terminal, scattered initially by local to terminal scatterers. The emerging wavefront is further scattered by remote scatterers. After this secondary scattering, the signal arrives in the vicinity of the base-station. No remote scatterers are allowed within an exclusion zone around the base-station. The signal then finally arrives at the base-station antennas after any near-field scattering.

Observed channel behavior in macrocells

We briefly summarize observed wireless channel behavior.

K-factor: It is generally seen that K-factor varies from about 20 near the base-station to zero at large ranges. The K-factor typically shows an exponential fall off with range. Further details can be found in [Erceg *et al.*, 1992, 1999a, 1999b; Baum *et al.*, 2000].

Delay spread: Delay spread, τ_{RMS}, typically increases with distance from the terminal. This increase occurs due to the fact that at larger distances, multipaths with large delays have strengths comparable to the direct path, contributing to τ_{RMS}. In flat rural environments, τ_{RMS} is less than 0.05 μs, in urban areas τ_{RMS} is typically 0.2 μs, while in hilly terrains, τ_{RMS} of 2–3 μs has been observed. Therefore, the coherence bandwidth B_C varies from several megahertz to a few hundred kilohertz depending on the terrain.

Doppler spread: Doppler spread is usually invariant with range. However, if there is a significant LOS component (high K-factor), ν_{RMS} can fall, since the direct path has zero Doppler spread (the direct path may have non-zero Doppler shift). Doppler spread varies from a few hertz (static or pedestrian mobile) to about 200 Hz and depends on the carrier frequency, terminal speed and angle spread of the scatterers.

Angle spread: Angle spread depends strongly on the terrain and antenna height above the ground. At the base-station, θ_{RMS} may typically vary from a fraction of a degree in flat rural areas up to 20° in hilly and dense urban locations. Urban and hilly locations can exhibit specular arrivals with a cluster center spacing up to 120°. The coherence distance D_C (horizontal) varies between $3\lambda_c$ and $20\lambda_c$. At the terminal, angle spread is much larger. Scatterers at the terminal are distributed in all directions, though not necessarily uniformly. D_C at the terminal varies from $0.25\lambda_c$ to $5\lambda_c$. If an antenna array forms a beam in a particular direction (i.e., acts as a spatial filter), then the angle spread observed by the array cannot exceed its beamwidth. Further, D_C depends on the direction of the scatterer center with respect to the antenna baseline. Typically D_C is smaller along the broadside and larger along the endfire directions. These observations are critical to exploiting antenna diversity at the base-station.

An abstract model that captures all the scattering effects described above has been proposed in [Oestges and Paulraj, 2003]. This model has been successful in explaining a number of observed channel characteristics.

2.4   Channel as a ST random field

The wireless channel can be modeled as a linear time-varying system. Suppressing the time-varying nature of the channel for now, we denote the impulse response between the

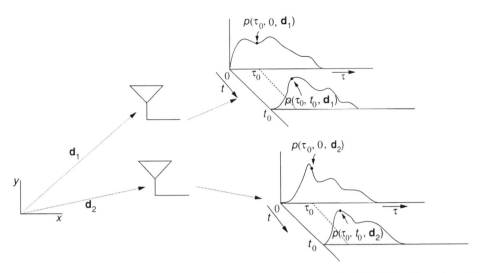

Figure 2.7: ST channel impulse response as a vector valued ST random field. Note that $p(\tau, t, \mathbf{d})$ is complex.

transmitter and receiver by $p(\tau)$, where $p(\tau)$ is itself often referred to as the "channel" and is the response at time τ to a unit impulse transmitted at time 0. Since wireless channels vary considerably with frequency, $p(\tau)$ is only meaningful if measured within a reasonably narrow passband that covers the frequencies of operation. In mobile communications, for example, $p(\tau)$ should be characterized over a 5–10% bandwidth passband channel, e.g., a 180 KHz bandwidth centered at 1.8 GHz. In the following, we assume that $p(\tau)$ is the complex envelope representation of the passband response. We also assume for now that the transmit and receive antennas are **E**-field, vertically polarized antennas. We can generalize $p(\tau)$ in a time- and space-varying environment to $p(\tau, t, \mathbf{d})$, which is defined as the response at a receiver whose antenna is located at position $\mathbf{d}$ at time t to an impulse launched at time $t - \tau$ from the transmit antenna placed at say the origin (see Fig. 2.7). In other words, $p(\tau)$ is indexed by $\mathbf{d}$ and $t -$ the space and time parameters. Further, $p(\tau, t, \mathbf{d})$ is assumed zero mean for simplicity. Note that $p(\tau)$ sampled at various delays (τ) can collectively be described by a vector (using any suitable basis) and consequently $p(\tau, t, \mathbf{d})$ may be thought of as a vector valued ST random field [Vanmarcke, 1983].

Clearly $p(\tau, t, \mathbf{d})$ depends on the transmit and receive antenna parameters such as the gain and phase patterns, but we do not explicitly model this dependence in the following discussion. The channel $p(\tau, t, \mathbf{d})$ can be represented in several alternative forms via Fourier transforms on the τ or t dimensions. Likewise, we can also define angle or wavenumber/wavevector transforms on the $\mathbf{d}$ dimension. See [Stüber, 1996; Durgin, 2000] for an extensive discussion. The behavior of $p(\tau, t, \mathbf{d})$ in general is very complicated. However, practical situations lend themselves to certain simplifying assumptions such as stationarity which we shall describe next. In the following, we assume the expectation operator, $\mathcal{E}$, to be in the ensemble sense. Great care is normally needed in treating the existence and convergence of such statistics, but it is beyond the scope of this brief overview.

2.4.1   Wide sense stationarity (WSS)

WSS implies that the second-order time statistics of the channel are stationary. This assumption is justified in mobile channels over short periods, T_u. We drop the space dimension (**d**) dependence temporarily for clarity. WSS implies that

$$\mathcal{E}\{p(\tau, t)p^*(\tau, t + \Delta t)\} = R_t(\tau, \Delta t), \tag{2.17}$$

i.e., the autocorrelation in time depends only on the lag Δt and not on t; $R_t(\tau, \Delta t)$ is called the lagged-time correlation function. We can define

$$U(\tau, \nu) = \int_{-\frac{T_u}{2}}^{\frac{T_u}{2}} p(\tau, t)e^{-j2\pi \nu t}dt, \tag{2.18}$$

which is the truncated Fourier transform (in the t dimension) of $p(\tau, t)$; $U(\tau, \nu)$ is the channel description in the delay(τ)-Doppler frequency(ν) domain. As $T_u \to \infty$, we get

$$\mathcal{E}\{U(\tau, \nu_1)U^*(\tau, \nu_2)\} = 0 \text{ if } \nu_1 \neq \nu_2, \tag{2.19}$$

which implies that the channel at different Doppler frequencies is uncorrelated [Bello, 1963].

2.4.2   Uncorrelated scattering (US)

This US model assumes that the scatterers contributing to the delay spread in the channel have independent fading, i.e.,

$$\mathcal{E}\{p(\tau_1, t)p^*(\tau_2, t)\} = 0 \text{ if } \tau_1 \neq \tau_2. \tag{2.20}$$

The US assumption implies stationarity in the transmission frequency domain, which is the frequency in the passband of the channel. Let

$$P(f, t) = \int_0^{\tau_{max}} p(\tau, t)e^{-j2\pi f\tau}d\tau, \tag{2.21}$$

where $P(f, t)$ is the channel description in the transmission frequency(f)–time(t) domain. From the US assumption we have

$$\mathcal{E}\{P(f, t)P^*(f + \Delta f, t)\} = R_f(\Delta f, t), \tag{2.22}$$

over the channel bandwidth. $R_f(\Delta f, t)$ is called the lagged-transmission frequency correlation function. The combination of the WSS and US assumptions leads to what is called the wide sense stationary uncorrelated scattering (WSSUS) model, which is

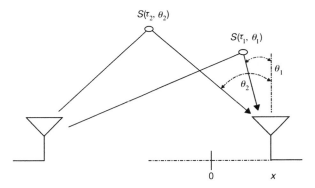

Figure 2.8: $p(\tau, x)$ can be modeled as the sum of responses from scatterers at (τ_i, θ_i) with amplitude $S(\tau_i, \theta_i)$.

stationary in time and transmission frequency domains and conversely has independent components in the Doppler frequency (ν) and delay (τ) dimensions.

2.4.3 Homogeneous channels (HO)

We now reintroduce the space dimension $\mathbf{d}$. A reasonable assumption in practice for spatial models is that the statistical behavior of $p(\tau, t, \mathbf{d})$ is locally stationary in space over several tens of the coherence distance D_C. This implies

$$\mathcal{E}\{p(\tau, t, \mathbf{d})p^*(\tau, t, \mathbf{d} + \Delta\mathbf{d})\} = R_\mathbf{d}(\tau, t, \Delta\mathbf{d}), \tag{2.23}$$

i.e., the autocorrelation of the channel response across space depends only on $\Delta\mathbf{d}$ and not on $\mathbf{d}$. Here $R_\mathbf{d}(\tau, t, \Delta\mathbf{d})$ is called the lagged-space correlation function.

Assuming that $\mathbf{d}$ lies along the x-axis and is parameterized by scalar x ($-D_u/2 \leq x \leq D_u/2$), where D_u is the span of stationarity, we can define the angle transform of the channel response as $S(\tau, t, \theta)$, where

$$p(\tau, t, x) = \int_{-\pi}^{\pi} S(\tau, t, \theta)e^{-j2\pi \sin(\theta)\frac{x}{\lambda_c}}d\theta. \tag{2.24}$$

$S(\tau, t, \theta)$ is the channel description in the delay(τ)–time(t)–angle(θ) domain. Assuming a discrete scattering model, with scatterers located at different path delays and AOAs, $S(\tau_i, \theta_i)$ (dropping t) can be interpreted as the scattering amplitude of the scatterer located at τ_i and θ_i (see Fig. 2.8). We shall use this interpretation to develop channel models in Chapter 3.

If $p(\tau, t, x)$ is homogeneous in x, we can show that for $D_u \to \infty$

$$\mathcal{E}\{S(\tau, t, \theta_1)S^*(\tau, t, \theta_2)\} = 0 \text{ if } \theta_1 \neq \theta_2. \tag{2.25}$$

This implies that signals arriving from scatterers at different angles are uncorrelated. The combination of the WSSUS channel with the HO assumption is termed the WSSUS-HO channel. Homogeneity requires scatterers to have statistically omnidirectional scattering and linear uniformity. If the scatterers are uniformly distributed in angle $[0, 2\pi]$ in a two-dimensional plane (cylindrically isotropic), it can be shown that the correlation between the channels at two antennas spaced Δx apart, $R_x(\Delta x)$ (we have suppressed the τ and t dimensions), satisfies [Stüber, 1996]

$$R_x(\Delta x) \propto J_0\left(\frac{2\pi|\Delta x|}{\lambda_c}\right), \tag{2.26}$$

where $J_0(x)$ is the zero-order Bessel function of the first kind given by

$$J_0(x) = \frac{1}{\pi} \int_0^\pi \cos(x \sin\theta)d\theta. \tag{2.27}$$

Further, $J_0(2\pi|\Delta x|/\lambda_c) \approx 0$ when $|\Delta x| \approx 0.4\lambda_c$. Note that $R_x(\Delta x)$ does not depend on the direction of Δx in the two-dimensional plane. Therefore, in practice, spatially uncorrelated fading can be obtained by spacing antennas $0.4\lambda_c$ apart in a rich scattering environment. To achieve a correlation coefficient close to 0.7, the appropriate antenna spacing is only $0.25\lambda_c$.

It is interesting to note that if the scatterers are uniformly distributed in three-dimensional space (spherically isotropic), one can show that [Abhayapala *et al.*, 1999]

$$R_x(\Delta x) \propto \frac{\sin(2\pi|\Delta x|/\lambda_c)}{2\pi|\Delta x|/\lambda_c}, \tag{2.28}$$

which implies that an antenna spacing of $\lambda_c/2$ is optimal for full decorrelation. Again, $R_x(\Delta x)$ is independent of direction in the three-dimensional volume.

2.5   Scattering functions

Earlier we studied the channel response in different domain representations and in terms of different one-dimensional lagged correlation functions. Scattering functions are useful two- or three-dimensional statistical characterizations of $p(\tau, t, \mathbf{d})$. In the following, we assume that the channel is WSSUS-HO.

Doppler-delay scattering function
We define this function as

$$\psi_{Do,De}(\nu, \tau) = \lim_{T_u \to \infty} \frac{\mathcal{E}\{|U(\tau, \nu)|^2\}}{T_u}. \tag{2.29}$$

We have suppressed the spatial dimension $\mathbf{d}$. The Doppler-delay scattering function $\psi_{Do,De}(\nu, \tau)$ represents the average power of the channel as a function of the Doppler frequency and delay. Under the WSS assumption, invoking the Wiener–Khintchine theorem, $\psi_{Do,De}(\nu, \tau)$ is the Fourier transform of $R_t(\tau, \Delta t)$ in the Δt dimension. Under

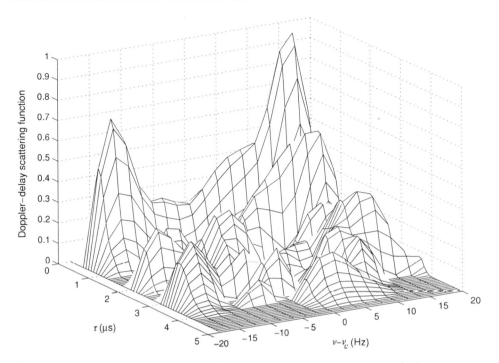

Figure 2.9: The Doppler-delay scattering function represents the average power in the Doppler-delay dimensions.

the WSSUS assumption, $\psi_{Do,De}(\nu, \tau)$ can be an arbitrary function of ν and τ. See Fig. 2.9 for a typical $\psi_{Do,De}(\nu, \tau)$.

Angle-delay scattering function

We define this function as

$$\psi_{A,De}(\theta, \tau) = \mathcal{E}\{|S(\tau, \theta)|^2\}, \tag{2.30}$$

where we have dropped the time dimension, t. The angle-delay scattering function $\psi_{A,De}(\theta, \tau)$ represents the average power of the channel as a function of the angle and delay dimensions. Note that homogeneity is needed for $\psi_{A,De}(\theta, \tau)$ to exist. Further, under the US assumption, $\psi_{A,De}(\theta, \tau)$ can be a complicated function of θ and τ. See Fig. 2.10 for a typical $\psi_{A,De}(\theta, \tau)$.

Doppler-angle-delay scattering function

We can now capture the complete channel statistics in a triple scattering function. Writing

$$Q(\tau, \nu, \theta) = \int_{-\frac{T_u}{2}}^{\frac{T_u}{2}} S(\tau, t, \theta) e^{-j2\pi t\nu} dt, \tag{2.31}$$

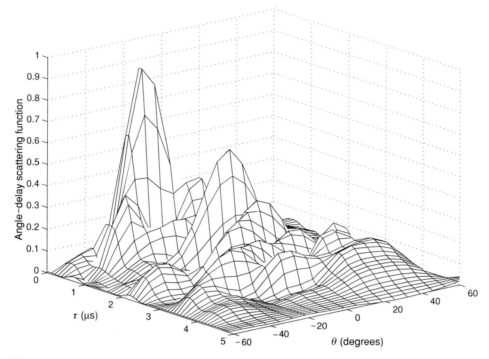

Figure 2.10: The angle-delay scattering function represents the average power in the angle-delay dimensions.

we can define

$$\psi_{Do,A,De}(\nu, \theta, \tau) = \lim_{T_u \to \infty} \frac{\mathcal{E}\{|Q(\tau, \nu, \theta)|^2\}}{T_u}. \tag{2.32}$$

$\psi_{Do,A,De}(\nu, \theta, \tau)$ represents the average power of the channel in the Doppler-frequency–angle-delay dimensions. Note that the WSS and HO assumptions are needed for $\psi_{Do,A,De}(\nu, \theta, \tau)$ to exist. Further, $\psi_{Do,A,De}(\nu, \theta, \tau)$ can be a complicated function of ν, θ and τ if the channel is WSSUS-HO. We note again that the scattering functions are defined at the receive antenna(s) and depend on the joint geometry of the transmit and receive antennas.

Marginal spectra

Note that the Doppler, delay and angle power spectra ($\psi_{Do}(\nu)$, $\psi_{De}(\tau)$ and $\psi_A(\theta)$, respectively) discussed in Section 2.2.2 are related to the scattering functions through

$$\psi_{Do}(\nu) = \int_0^{\tau_{max}} \psi_{Do,De}(\nu, \tau)d\tau, \tag{2.33}$$

$$\psi_{De}(\tau) = \int_{\mathcal{F}} \psi_{Do,De}(\nu, \tau)d\nu = \int_{-\pi}^{\pi} \psi_{A,De}(\theta, \tau)d\theta. \tag{2.34}$$

Similarly,

$$\psi_A(\theta) = \int_0^{\tau_{max}} \psi_{A,De}(\theta, \tau) d\tau. \tag{2.35}$$

The relation between the marginal power spectra and the composite scattering function $\psi_{Do,De,A}(\nu, \theta, \tau)$ follows likewise.

LOS component

The development of scattering functions assumes zero mean fading channels. However, real channels may have non-zero mean (LOS) components. These non-fading components by implication must have zero spread in Doppler frequency, delay and angle dimensions. A power representation for such components will be impulses at the appropriate Doppler, delay and angle values. The total scattering function is then the sum of the scattering function for the fading component and these LOS generated impulse components.

2.6   Polarization and field diverse channels

So far we have assumed that the transmitter and receiver employ unipolar **E**-field antennas. In this section we make brief comments on ST channels in which the transmitter and receiver use different polarization and field antennas.

Consider **E**-field antennas and a channel with no multipath. If we transmit on one polarization and receive on an orthogonal polarization (cross-pol as opposed to co-pol), the response will be zero in ideal conditions where both the transmit and receive antenna have perfect cross-polarization discrimination (XPD) (i.e., do not respond to the cross-polarization field) and the medium has zero cross-polarization coupling (XPC) (i.e., does not transfer energy into the cross-polarization direction). In practice neither condition is true.

For imperfect XPD alone, the responses at co- and cross-polarized receive antennas are fully envelope correlated. For imperfect XPC alone, the cross-pol channel is usually poorly correlated or uncorrelated with the co-pol antenna. Cross-polarization coupling increases with range in macrocells and XPC varies inversely with the K-factor. As the K-factor drops at large ranges, XPC increases, particularly in a rich scattering environment. Since both XPD and XPC describe cross-polarization mixing, they are often jointly referred to as XPD, but they have different fading correlation behavior. See [Turkmani, 1995; Andrews et al., 2001; Nabar et al., 2002a] for more details.

Again, assuming uni-polar antennas, if we use different field (**E** or **H**) antennas, a variety of new effects are observed. Without any scatterers, **E** and **H** channels show perfect envelope correlation. In the presence of scattering, **E**- and **H**-field antennas have different responses in element patterns and will show decorrelation as scattering

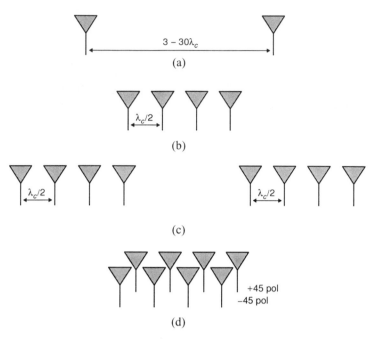

Figure 2.11: Some antenna array topologies at the base-station: (a) widely spaced antennas (good spatial diversity but excessive grating lobes); (b) a compact array (good beam pattern but poor spatial diversity); (c) a compromise solution that combines the benefits of (a) and (b); (d) a dual-polarized array.

increases. The scattering functions for different antennas will show significant variations, including the Doppler spectrum which has different characteristics [Lee, 1982].

2.7   Antenna array topology

Having discussed the role of **E**- and **H**-fields and polarization in antenna element patterns, we now consider the topology of the antenna array, which is another important factor influencing ST channels. There are significant differences between the base-station and the terminal. At the base-station we have noted that the coherence distance D_C (for uni-polar antennas) can be large ($3\lambda_c - 30\lambda_c$). Therefore, antennas spaced (see Fig. 2.11(a)) to capture diversity will be "spatially aliased" with conventional beamforming and no scattering showing excessive grating lobes. Aliased beams are not localized in angle and can receive or transmit interfering signals at even wide angles. These remarks also largely carry over to a scattering medium.

A "compact" array ($\lambda_c/2$ spaced) array (see Fig. 2.11(b)), on the other hand, will have good beam pattern characteristics, but poor diversity. The topology in Fig. 2.11(c) is a possible compromise that combines both advantages. Generally, capturing diversity is

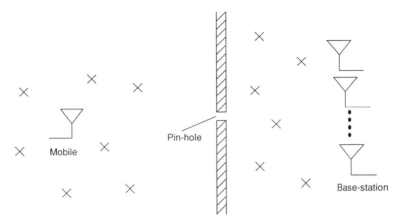

Figure 2.12: Pin-hole (or key-hole) model in ST channels. This leads to significant impact on ST channel capacity and diversity.

critical, and option (a) or (c) is typically used. If cross-polarized antennas are deployed (see Fig. 2.11(d)), we can have "compact" arrays for beamforming and still hope to pick up adequate diversity in the polarization domain (see [Turkmani, 1995]).

Antenna topology choices at the terminal are limited due to size constraints (e.g., a cell phone). However, the presence of rich scattering near the terminal allows spacing as small as $0.25\lambda_c$ to capture diversity.

2.8   Degenerate channels

In the discussion above, paths from the transmitter to the receiver propagate via scatterers that are distributed along and across the transmit–receive axis. However, in certain propagation conditions a pin-hole effect forces all paths to go through a single pin-hole, causing a degenerate channel condition. We explore such channels below.

Consider the simple model in Fig. 2.12. The signal emerging from the moving transmitter is scattered by scatterers around it (LHS or left-hand-side scatterers) and scatterers around the fixed receiver (RHS or right-hand-side scatterers). We place a shield with a pin-hole (or key-hole) between the transmit and receive scatterers. The signal emerging from the pin-hole [Gesbert *et al.*, 2000; Chizhik *et al.*, 2000, 2002; Andersen, 2001; Vaughan and Andersen, 2001; Molisch, 2002] is a point source for the RHS scatterers. We assume all scatterers are static, therefore Doppler spread is created by transmit motion alone.

Case 1: ST degeneracy
We assume that the terminal is in motion so that there is Doppler spread or time selective fading. Further, we assume no delay spread in LHS or RHS scatterers, i.e.,

$\tau_{RMS} = 0$. Therefore, the channel is frequency flat and we drop the τ dimension. The signal emerging from the pin-hole has time selective fading and is scattered by the RHS scatterers. The signal arrives at the base-station array with spatially "frozen" signatures since the RHS scatterers are static. Assuming an array of antennas at the base-station, the channel at the ith antenna is given by

$$p(t, i) = \widehat{p}(t)\widehat{p}(i), \tag{2.36}$$

where $\widehat{p}(t)$ is a Rayleigh time selective fading channel (same as at the pin-hole) and $\widehat{p}(i)$ is the frozen Rayleigh space selective fading channel. $p(t, i)$ is therefore a double Rayleigh fading random variable [Gesbert *et al.*, 2000]. The above model represents "multiplicative fading" in the ST domain.

Case 2: Time-frequency degeneracy

Once again we assume that the terminal is in motion, but the RHS scatterers induce delay spread due to significant differential travel times. The channel in the transmission frequency–time domain at any base-station antenna is given by

$$P(f, t) = \widehat{p}(t)\widehat{P}(f), \tag{2.37}$$

where $\widehat{P}(f)$ represents frequency selective Rayleigh fading. Thus the channel is multiplicative in the transmission frequency–time domain with a double Rayleigh fading distribution.

Case 3: Space-frequency degeneracy

If the scatterers to the LHS of the pin-hole have delay spread, the terminal is stationary and there is no delay spread from the RHS scatterers, we get space-frequency degeneracy, i.e., the channel in the transmission frequency domain at the ith receive antenna is given by

$$P(f, i) = \widehat{P}(f)\widehat{p}(i). \tag{2.38}$$

We now get a double Rayleigh fading channel in the space-frequency domain.

Pin-hole effects sometimes occur in cases where a distant transmit signal arrives as a zero angle spread wavefront. Signals penetrating into buildings through small windows have a similar degeneracy. Pin-holes are also known to occur in tunnels and street canyons.

Statistics of double Rayleigh fading

Consider $p(t, i) = \widehat{p}(t)\widehat{p}(i)$, where each component is Rayleigh fading. Let $\Omega_1 = \mathcal{E}\{\widehat{p}(t)^2\}$ and $\Omega_2 = \mathcal{E}\{\widehat{p}(i)^2\}$. It has been shown in [Erceg *et al.*, 1997; Simon, 2002] that the PDF of the envelope of $p(t, i)$ is

$$f(x) = \frac{2}{\pi\sqrt{\Omega_1\Omega_2}} K_0\left(\frac{x}{\sqrt{\Omega_1\Omega_2}}\right) u(x), \tag{2.39}$$

where $K_0(x)$ is the zero-order modified Bessel function of the second kind.

Scattering function separability

It is easy to see that the multiplicative fading models cause "separability" of the scattering functions. In general, if the channel has degeneracy in all three dimensions, we can show that

$$\psi_{Do,De,A}(\nu, \theta, \tau) = \psi_{Do}(\nu)\psi_{De}(\tau)\psi_A(\theta). \tag{2.40}$$

This means $\psi_{Do,De,A}(\nu, \tau, \theta)$ is no longer an arbitrary function of ν, τ and θ. Alternatively, separability may occur in only two dimensions, say θ and τ, in which case

$$\psi_{De,A}(\tau, \theta) = \psi_{De}(\tau)\psi_A(\theta). \tag{2.41}$$

Degenerate channels have interesting implications on the capacity and diversity of wireless channels that will be discussed later in Chapters 4 and 5.

2.9   Reciprocity and its implications

The reciprocity principle implies that the paths used by the signal from a transmit antenna (say the base-station) to a receive antenna (say the terminal) are identical if the signal is launched from the terminal and received at the base. Of course, the frequency, time and antenna locations have to be identical. Reciprocity implies that the channel (or impulse response) from the base-station to the terminal is the same as the channel from the terminal to the base-station. This immediately implies that the Doppler spectrum ($\psi_{Do}(\nu)$) and delay spectrum ($\psi_{De}(\tau)$) are also the same, whether measured at the base-station or the terminal. We need to take special care with regard to the angle spectrum, $\psi_A(\theta)$. We can state that $\psi_A(\theta)$ at the base-station on receive is the same as $\psi_A(\theta)$ at the base-station on transmit if we count only the paths that reach the terminal. However, $\psi_A(\theta)$ (and θ_{RMS}, D_C) will, in general, not be the same at the base and the terminal. Therefore, great care is needed in interpreting the reciprocity principle in the spatial domain.

3 ST channel and signal models

3.1 Introduction

In Chapter 2 we introduced ST propagation and characterized the bandpass physical channel as a vector valued random field, $p(\tau, t, \mathbf{d})$. We now consider typical ST wireless systems that use a specific narrow bandwidth of operation and one or more antennas at the transmitter and receiver at specific locations. We give an overview of SISO, SIMO, MISO and MIMO channel models, develop techniques for constructing channels from statistical channel descriptions, develop models for capturing macrocellular ST channel effects and discuss techniques for ST channel estimation.

3.2 Definitions

In this section we define SISO, SIMO, MISO and MIMO channels and their input–output models. We ignore additive noise at the receiver in the development below.

3.2.1 SISO channel

Let $h(\tau, t)$ be the time-varying channel impulse response from the input of the pulse-shaping filter $g(\tau)$ at the transmitter, through the propagation channel $p(\tau, t)$, to the output of the receiver matched-filter. We define $h(\tau, t)$ as the response at time t to an impulse at time $t - \tau$. The combination of the pulse-shaping filter and the matched-filter makes $h(\tau, t)$ a narrowband channel. For convenience we normally refer to $h(\tau, t)$ as the channel from the transmit antenna to the receive antenna, but it is strictly defined as above. The details can be found in several excellent textbooks [Wozencraft and Jacobs, 1965; Proakis, 1995; Cioffi, 2002].

Note that $h(\tau, t)$ is again the complex envelope of the bandpass impulse response function. Typically, wireless links in mobile radio use a bandwidth that varies between 0.01% and 0.1% of the center frequency, e.g., in the IS-95 system, the channel bandwidth is 1.25 MHz centered at 900 MHz or 1.8 GHz.

If a signal $s(t)$ is transmitted, the received signal $y(t)$ is given by

$$y(t) = \int_0^{\tau_{total}} h(\tau, t)s(t - \tau)d\tau = h(\tau, t) \star s(t),$$ (3.1)

where $\star$ denotes the convolution operator and where we have assumed a causal channel impulse response of duration τ_{total}. The signals $s(t)$ and $y(t)$ are also complex envelopes of a narrowband signal.

3.2.2 SIMO channel

Consider a SIMO channel with M_R receive antennas. The SIMO channel can be decomposed into M_R SISO channels. Denoting the impulse response between the transmit antenna and the ith ($i = 1, 2, \ldots, M_R$) receive antenna by $h_i(\tau, t)$, we see that the SIMO channel may be represented as an $M_R \times 1$ vector, $\mathbf{h}(\tau, t)$, given by

$$\mathbf{h}(\tau, t) = \left[h_1(\tau, t)\, h_2(\tau, t) \cdots h_{M_R}(\tau, t) \right]^T.$$ (3.2)

Further, when a signal $s(t)$ is launched from the transmit antenna, the signal received at the ith receive antenna, $y_i(t)$, is given by

$$y_i(t) = h_i(\tau, t) \star s(t), \quad i = 1, 2, \ldots, M_R.$$ (3.3)

Denoting the signals received at the M_R receive antennas by the $M_R \times 1$ vector $\mathbf{y}(t) = [y_1(t)\, y_2(t) \cdots y_{M_R}(t)]^T$, we see that the relation in Eq. (3.3) may be concisely expressed as

$$\mathbf{y}(t) = \mathbf{h}(\tau, t) \star s(t).$$ (3.4)

3.2.3 MISO channel

Consider a MISO system with M_T transmit antennas. Analogous to the SIMO channel discussed earlier, the MISO channel comprises M_T SISO links. Denoting the impulse response between the jth ($j = 1, 2, \ldots, M_T$) transmit antenna and the receive antenna by $h_j(\tau, t)$, the MISO channel may be represented by a $1 \times M_T$ vector $\mathbf{h}(\tau, t)$ given by

$$\mathbf{h}(\tau, t) = \left[h_1(\tau, t)\, h_2(\tau, t) \cdots h_{M_T}(\tau, t) \right].$$ (3.5)

Assuming $s_j(t)$ is the signal transmitted from the jth transmit antenna and $y(t)$ is the received signal, the input–output relation for the MISO channel is given by

$$y(t) = \sum_{j=1}^{M_T} h_j(\tau, t) \star s_j(t),$$ (3.6)

which may alternatively be expressed in vector notation as

$$y(t) = \mathbf{h}(\tau, t) \star \mathbf{s}(t),$$

(3.7)

where $\mathbf{s}(t) = [s_1(t)\ s_2(t)\ \cdots\ s_{M_T}(t)]^T$ is an $M_T \times 1$ vector.

3.2.4   MIMO channel

Consider a MIMO system with M_T transmit antennas and M_R receive antennas. Denoting the impulse response between the jth ($j = 1, 2, \ldots, M_T$) transmit antenna and the ith ($i = 1, 2, \ldots, M_R$) receive antenna by $h_{i,j}(\tau, t)$, the MIMO channel is given by the $M_R \times M_T$ matrix $\mathbf{H}(\tau, t)$ with

$$\mathbf{H}(\tau, t) = \begin{bmatrix} h_{1,1}(\tau, t) & h_{1,2}(\tau, t) & \cdots & h_{1,M_T}(\tau, t) \\ h_{2,1}(\tau, t) & h_{2,2}(\tau, t) & \cdots & h_{2,M_T}(\tau, t) \\ \vdots & \vdots & \ddots & \vdots \\ h_{M_R,1}(\tau, t) & h_{M_R,2}(\tau, t) & \cdots & h_{M_R,M_T}(\tau, t) \end{bmatrix}.$$

(3.8)

The vector $[h_{1,j}(\tau, t)\ h_{2,j}(\tau, t)\ \cdots\ h_{M_R,j}(\tau, t)]^T$ is the spatio-temporal signature or channel induced by the jth transmit antenna across the receive antenna array. Further, given that the signal $s_j(t)$ is launched from the jth transmit antenna, the signal received at the ith receive antenna, $y_i(t)$, is given by

$$y_i(t) = \sum_{j=1}^{M_T} h_{i,j}(\tau, t) \star s_j(t), \quad i = 1, 2, \ldots, M_R.$$

(3.9)

The input–output relation for the MIMO channel may be expressed in matrix notation as

$$\mathbf{y}(t) = \mathbf{H}(\tau, t) \star \mathbf{s}(t),$$

(3.10)

where $\mathbf{s}(t) = [s_1(t)\ s_2(t)\ \cdots\ s_{M_T}(t)]^T$ is an $M_T \times 1$ vector and $\mathbf{y}(t) = [y_1(t)\ y_2(t)\ \cdots\ y_{M_R}(t)]^T$ is a vector of dimension $M_R \times 1$.

3.3   Physical scattering model for ST channels

In this section we relate the multiple antenna wireless channels discussed in Section 3.2 to a physical scattering model and scattering functions discussed in Chapter 2. For convenience, we suppress the time-varying nature of the channel. We show how statistics such as the angle-delay scattering function, $\psi_{A,De}(\theta, \tau)$, developed in Chapter 2, can be used to generate SIMO, MISO and MIMO channel models under the "narrowband array" assumption described below.

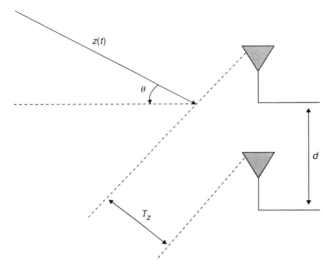

Figure 3.1: Schematic of a wavefront impinging across an antenna array. Under the narrowband assumption the antenna outputs are identical except for a complex scalar.

Narrowband array

Consider a signal wavefront, $z(t)$, impinging on an antenna array comprising two antennas spaced d apart at angle θ (see Fig. 3.1).

We assume that the impinging wavefront has a bandwidth B and is represented as

$$z(t) = \beta(t)e^{j2\pi \nu_c t}, \tag{3.11}$$

where $\beta(t)$ is the complex envelope representation of the signal (with bandwidth B) and ν_c is the carrier frequency. Under the narrowband assumption, we take the bandwidth B to be much smaller than the reciprocal of the transit time of the wavefront across the antenna array T_z, i.e., $B \ll 1/T_z$. Under this assumption, if the signal received at the first antenna, $y_1(t)$, is $z(t)$, then the signal received at the second antenna, $y_2(t)$, is given by

$$y_2(t) = z(t - T_z) = \beta(t - T_z)e^{j2\pi \nu_c(t - T_z)}, \tag{3.12}$$

where we have assumed identical element patterns. Under the narrowband assumption, $\beta(t - T_z) \approx \beta(t)$. Further,

$$e^{j2\pi \nu_c(t - T_z)} = e^{j2\pi \nu_c t}e^{-j2\pi \sin(\theta)\frac{d}{\lambda_c}}, \tag{3.13}$$

where λ_c is the wavelength of the signal wavefront. Hence, the output at the second antenna can be written as

$$\begin{aligned} y_2(t) &= \beta(t)e^{j2\pi \nu_c t}e^{-j2\pi \sin(\theta)\frac{d}{\lambda_c}} \\ &= y_1(t)e^{-j2\pi \sin(\theta)\frac{d}{\lambda_c}}. \end{aligned} \tag{3.14}$$

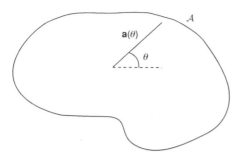

Figure 3.2: Schematic of an array manifold of an antenna array.

It is clear from Eq. (3.14) that the signals received at the two antennas are identical, except for a phase shift that depends on the array geometry and the AOA of the wavefront. The result can be extended to arrays with more than two antennas. We shall make use of this property in modeling SIMO, MISO and MIMO channels. Note that the narrowband assumption does not imply that the channel is frequency flat.

Array manifold

Consider an antenna array comprising M antennas in a free field environment (no scatterers and therefore no multipath either). A planar CW wavefront of frequency v_c arriving from an angle θ at the array will induce a spatial signature across the antenna array. This signature will be a function of array geometry, antenna element patterns and the AOA. We call this complex $M \times 1$ vector, $\mathbf{a}(\theta) = [a_1(\theta) \, a_2(\theta) \, \cdots \, a_M(\theta)]^T$, the array response vector. If we let $a_1(\theta) = 1$, $a_i(\theta)$ for $i > 1$ represent the phase (and if element patterns are not identical, then also magnitude) differences induced at the antennas by a wavefront at angle θ. The set of all array response vectors $\{\mathbf{a}(\theta): -\pi \leq \theta \leq \pi\}$ is called the array manifold $\mathcal{A}$. The array manifold is a function of the frequency v_c and is a well-defined smooth function of θ for an array in the absence of scatterers. See Fig. 3.2 for an example of the array manifold.

$\mathcal{A}$ can be measured (array calibration) by moving a CW source in azimuth around the array in a free field. However, in the presence of scatterers, which act as virtual antennas, the array manifold is a rapidly fluctuating function of θ. Moreover, the scatterer amplitude or position can vary with time, making array manifold calibration difficult, if not useless. Generally speaking, the array manifold is a useful conceptual construct for modeling ST channels. The rich theories of sub-space methods [Schmidt, 1981; Roy *et al.*, 1986; Paulraj *et al.*, 1986; Zoltowski and Stavrinides, 1989; Stoica and Nehorai, 1991] which leverage array manifold concepts do not apply to rich scattering environments due to the difficulties in getting an accurate estimate of $\mathcal{A}$. At the base-station in flat terrains, the angle spread may be small enough to allow array manifold and sub-space methods to be used. Even in this case, the array manifold can change due to precipitation (rain or snow) on the antenna support structures such as towers

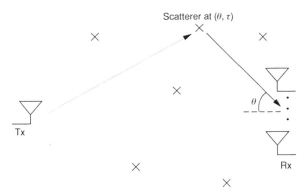

Figure 3.3: SIMO channel construction. The scatterer location induces path delay τ and AOA θ.

or roofs of buildings. Consequently it needs frequent calibration to remain accurate, thereby limiting its usefulness.

3.3.1 SIMO channel

Consider a SIMO channel with M_R receive antennas. We denote the combined effect of pulse-shaping at the transmitter and matched-filtering at the receiver by $g(\tau)$. The transmitted signal is scattered and arrives at the receiver at different angles and delays.

The average power of the scatterers is given by the angle-delay scattering function, $\psi_{A,De}(\theta, \tau) = \mathcal{E}\{|S(\theta, \tau)|^2\}$, defined in Chapter 2. We assume a simple single scattering model. $S(\theta_i, \tau_i)$ is the complex scatterer amplitude of the scatterer located at angle θ_i and delay τ_i. The scattering amplitude $S(\theta_i, \tau_i)$ is an independent identically distributed (IID) complex Gaussian random variable (US assumption). Each scattered wavefront impinging on the antenna array induces an $M_R \times 1$ signature given by the array response vector of the antenna array, $\mathbf{a}(\theta)$. Under the narrowband assumption discussed earlier, the array response of an impinging wavefront can be modeled as an $M_R \times 1$ vector that is solely a function of the AOA, θ, and the array geometry (see Fig. 3.3). The SIMO channel can then be expressed as

$$\mathbf{h}(\tau) = \int_{-\pi}^{\pi} \int_{0}^{\tau_{max}} S(\theta, \tau')\mathbf{a}(\theta)g(\tau - \tau')d\tau'd\theta, \tag{3.15}$$

where τ_{max} is the maximum delay spread of the physical channel. More channel constructions can be found in [Ertel *et al.*, 1998].

3.3.2 MISO channel

Consider a MISO system with M_T transmit antennas as shown in Fig. 3.4. To construct the physical MISO channel we invoke the reciprocity principle, which states that the forward channel from a transmit antenna to a scatterer is the same as the reverse channel

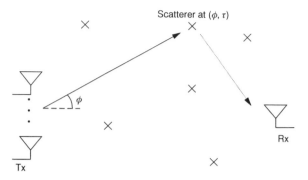

Figure 3.4: MISO channel construction.

from the scatterer to the same antenna. A scatterer located at angle ϕ and delay τ with respect to the antenna array has a complex amplitude $S(\phi, \tau)$. The relative positions of the antennas at the transmitter and receiver allow us to map $S(\theta, \tau)$ to $S(\phi, \tau)$. Hence, the $1 \times M_T$ MISO channel may be constructed as

$$\mathbf{h}(\tau) = \int_{-\pi}^{\pi} \int_0^{\tau_{max}} S(\phi, \tau')\mathbf{b}(\phi)^T g(\tau - \tau')d\tau' d\phi, \tag{3.16}$$

where $\mathbf{b}(\phi)$ is the $M_T \times 1$ array response vector induced by a wavefront at angle ϕ across the transmit antenna array.

3.3.3   MIMO channel

Consider a MIMO channel with M_T transmit antennas and M_R receive antennas. The physical MIMO channel is a combination of the SIMO and MISO channels discussed earlier. A scatterer located at angle θ and delay τ with respect to the receive array now appears to be located at angle ϕ with respect to the transmit antenna array. Thus, given the overall geometries of the transmit and receive arrays, any two of the variables ϕ, θ and τ can define the third. Likewise the angle-delay scattering function $\psi_{A,De}(\theta, \tau)$ observed at the receiver array will define the scattering function seen by the transmit array. Underlying scatterers being the same, the scattering amplitude is $S(\theta, \tau) = S(\phi, \tau)$. The $M_R \times M_T$ MIMO channel, $\mathbf{H}(\tau)$, can then be constructed as (ϕ is a function of θ and τ)

$$\mathbf{H}(\tau) = \int_{-\pi}^{\pi} \int_0^{\tau_{max}} S(\theta, \tau')\mathbf{a}(\theta)\mathbf{b}(\phi)^T g(\tau - \tau')d\tau' d\theta. \tag{3.17}$$

The single scattering model in Eq. (3.17) has a number of limitations and cannot adequately model all observed channel effects. A more general model is to assume multiple scattering, i.e., energy from the transmitter uses more than one scatterer to reach the receiver. If we use a double (or multiple) scattering model, the parameters θ, ϕ and τ in Eq. (3.17) become independent.

The scatterer location, antenna element patterns and geometry, and the scattering model together determine the correlation between elements of $\mathbf{H}$, the channel between the transmit and the receive antennas. With suitable choices of the above, including a double scattering model, we can show that the elements of $\mathbf{H}$ are independent zero mean circularly symmetric complex Gaussian (ZMCSCG) random variables. A complex Gaussian random variable $Z = X + jY$ is ZMCSCG if X and Y are independent real Gaussian random variables with zero mean and equal variance.

Classical IID channel model

We assume that the delay spread in the channel is negligible, i.e., $\tau_{RMS} \approx 0$. Equation (3.17) can be rewritten as

$$\mathbf{H}(\tau) = \left(\int_{-\pi}^{\pi} \int_{0}^{\tau_{max}} S(\theta, \tau')\mathbf{a}(\theta)\mathbf{b}(\phi)^T d\tau' d\theta \right) g(\tau)$$
$$= \mathbf{H}g(\tau). \tag{3.18}$$

We drop $g(\tau)$ and under assumptions discussed above the elements of $\mathbf{H}$ can be modeled to be ZMCSCG with unit variance. We then get $\mathbf{H} = \mathbf{H}_w$, the IID (spatially white) channel. Some properties of $\mathbf{H}_w$ are summarized below:

$$\mathcal{E}\{[\mathbf{H}_w]_{i,j}\} = 0, \tag{3.19}$$

$$\mathcal{E}\{|[\mathbf{H}_w]_{i,j}|^2\} = 1, \tag{3.20}$$

$$\mathcal{E}\{[\mathbf{H}_w]_{i,j}[\mathbf{H}_w]_{m,n}^*\} = 0 \text{ if } i \neq m \text{ or } j \neq n. \tag{3.21}$$

The MIMO channel $\mathbf{H}_w$ can be restricted to SIMO and MISO channels by dropping either columns or rows respectively. The resulting vector channels are $\mathbf{h}_w$ with dimension $M_R \times 1$ for SIMO and $1 \times M_T$ for MISO. The statistics of $\mathbf{h}_w$ follow from Eqs. (3.19)–(3.21). $\mathbf{H}_w$ and the extended models discussed later in the chapter are known as non-physical models since they do not incorporate the physical path structure.

Frequency flat vs frequency selective fading

If the bandwidth-delay spread product of the channel $B \times \tau_{RMS} \geq 0.1$, the channel is generally said to be frequency selective. Otherwise, the channel is frequency flat. The channel $\mathbf{H}(\tau)$ may be expressed in the frequency domain via a Fourier transform as

$$\widetilde{\mathbf{H}}(f) = \int_{0}^{\infty} \mathbf{H}(\tau)e^{-j2\pi f\tau}d\tau. \tag{3.22}$$

We have switched to $\widetilde{\mathbf{H}}(f)$ notation instead of the $P(f)$ used in Chapter 2. The variation of $\widetilde{\mathbf{H}}(f)$ with f will depend on the delay spread and hence on the coherence bandwidth B_C. If f_1 and f_2 are two frequencies such that $|f_1 - f_2| >> B_C$, we should expect that $\mathcal{E}\{\text{vec}(\widetilde{\mathbf{H}}(f_1))\text{vec}(\widetilde{\mathbf{H}}(f_2))^H\} = \mathbf{0}_{M_T M_R}$, i.e., that all cross-correlations are zero. The spatial statistics of $\widetilde{\mathbf{H}}(f)$ depend on the scattering environment and array geometry at both the transmitter and receiver. If we have appropriate scattering and antenna geometry, $\widetilde{\mathbf{H}}(f) = \widetilde{\mathbf{H}}_w(f)$. This implies that the channel is $\mathbf{H}_w$ at any given frequency and that it varies with frequency depending on coherence bandwidth B_C.

3.4   Extended channel models

In the real world, $\mathbf{H}$ can deviate significantly from $\mathbf{H}_w$ for a variety of reasons. These include inadequate antenna spacing or scattering leading to spatial correlation, the presence of a LOS component resulting in Ricean fading and gain imbalances between the channel elements through the use of polarized antennas. We study some of these effects below. For convenience we assume no delay spread, drop the τ dependence and refer to the MIMO channel as $\mathbf{H}$.

3.4.1   Spatial fading correlation

Correlated channels imply that elements of $\mathbf{H}$ are correlated and may be modeled by

$$\text{vec}(\mathbf{H}) = \mathbf{R}^{1/2}\text{vec}(\mathbf{H}_w), \tag{3.23}$$

where $\mathbf{H}_w$ is the spatially white $M_R \times M_T$ MIMO channel described earlier and $\mathbf{R}$ is the $M_T M_R \times M_T M_R$ covariance matrix defined as

$$\mathbf{R} = \mathcal{E}\{\text{vec}(\mathbf{H})\text{vec}(\mathbf{H})^H\}. \tag{3.24}$$

$\mathbf{R}$ is a positive semi-definite Hermitian matrix. If $\mathbf{R} = \mathbf{I}_{M_T M_R}$, then $\mathbf{H} = \mathbf{H}_w$. Although the model described above is capable of capturing any correlation effects between the elements of $\mathbf{H}$, a simpler and less generalized model is often adequate and is given by

$$\mathbf{H} = \mathbf{R}_r^{1/2}\mathbf{H}_w\mathbf{R}_t^{1/2}, \tag{3.25}$$

where $\mathbf{R}_t$ is the $M_T \times M_T$ transmit covariance matrix and $\mathbf{R}_r$ is the $M_R \times M_R$ receive covariance matrix. Both $\mathbf{R}_t$ and $\mathbf{R}_r$ are positive semi-definite Hermitian matrices. This model has fewer degrees of freedom than Eq. (3.23). Equation (3.25) implies that receive antenna correlation $\mathbf{R}_r$ is equal to the covariance of the $M_R \times 1$ receive vector channel when excited by any transmit antenna, and is therefore the same for all transmit antennas. The model can occur when the angle spectra of the scatterers at the receive array for signals arriving from any transmit antenna are identical. This condition arises if all the transmit antennas are closely located and have identical radiation patterns. These remarks also carry over to the transmit antenna correlation $\mathbf{R}_t$.

From Eqs. (3.23)–(3.25), we can show that $\mathbf{R}$, $\mathbf{R}_t$ and $\mathbf{R}_r$ are related by

$$\mathbf{R} = \mathbf{R}_t^T \otimes \mathbf{R}_r. \tag{3.26}$$

Note also that $\mathbf{H}_w$ is a full rank matrix with probability 1. In the presence of receive or transmit correlation, the rank of $\mathbf{H}$ is constrained by $\min(r(\mathbf{R}_r), r(\mathbf{R}_t))$, where $r(\mathbf{A})$ denotes the rank of $\mathbf{A}$.

3.4.2 LOS component

So far we have considered only Rayleigh fading in describing the MIMO channel. In the presence of a LOS component between the transmitter and receiver, the MIMO channel may be modeled approximately as the sum of a fixed component and a variable (or scattered) component, as follows [Rashid-Farrokhi et al., 2000]:

$$\mathbf{H} = \sqrt{\frac{K}{1+K}}\overline{\mathbf{H}} + \sqrt{\frac{1}{1+K}}\mathbf{H}_w, \tag{3.27}$$

where $\sqrt{K/(1+K)}\,\overline{\mathbf{H}} = \mathcal{E}\{\mathbf{H}\}$ is the LOS component of the channel and $\sqrt{1/(1+K)}\mathbf{H}_w$ is the fading component that assumes uncorrelated fading. The elements of $\overline{\mathbf{H}}$ are assumed to have unit power. K in Eq. (3.27) is the Ricean K-factor of the system and is essentially the ratio of the power in the LOS component of the channel to the power in the fading component. $K = 0$ corresponds to pure Rayleigh fading, while $K = \infty$ corresponds to a non-fading channel. Two alternative extreme prototypes of $\overline{\mathbf{H}}$ for a MIMO channel with $M_T = M_R = 2$ and uni-polarized antennas are

$$\overline{\mathbf{H}}_1 = \begin{bmatrix} e^{j\theta_1} & 0 \\ 0 & e^{j\theta_2} \end{bmatrix} \begin{bmatrix} 1 & 1 \\ 1 & 1 \end{bmatrix} \begin{bmatrix} e^{j\theta_3} & 0 \\ 0 & e^{j\theta_4} \end{bmatrix}, \tag{3.28}$$

$$\overline{\mathbf{H}}_2 = \begin{bmatrix} e^{j\theta_1} & 0 \\ 0 & e^{j\theta_2} \end{bmatrix} \begin{bmatrix} 1 & -1 \\ 1 & 1 \end{bmatrix} \begin{bmatrix} e^{j\theta_3} & 0 \\ 0 & e^{j\theta_4} \end{bmatrix}. \tag{3.29}$$

θ_i are arbitrary phase factors that model array geometry and orientation. Figure 3.5 shows the array geometry that can yield $\overline{\mathbf{H}}_1$ and $\overline{\mathbf{H}}_2$ channels. D is the transmitter–receiver separation (range) and d_t and d_r are the element separations at transmitter and receiver respectively. $\overline{\mathbf{H}}_1$ always results when $D >> d_t, d_r$ (see Fig. 3.5(a)) and $\overline{\mathbf{H}}_2$ can result when D is comparable to d_t or d_r (see Fig. 3.5(b)) or both. Perfect orthogonality of $\overline{\mathbf{H}}_2$ requires specific antenna locations and geometry. $\overline{\mathbf{H}}_2$ is likely only in multibase operations when transmit (or receive) antennas are located at different base-stations.

3.4.3 Cross-polarized antennas

The channel models discussed so far assume that the antennas at the base-station and terminal transmit and receive with identical polarizations. The use of antennas with differing polarizations at the transmitter and receiver leads to a gain (or power) and correlation imbalance between the elements of $\mathbf{H}$ [Nabar et al., 2002a]. As a consequence, elements of $\mathbf{H}$ show more complex behavior. For example, consider a 2×2 antenna system. The transmitter and receiver use antennas with $\pm 45°$ polarization as shown in Fig. 3.6.

The diagonal elements of $\mathbf{H}$ correspond to transmission and reception on the same polarization, while the off-diagonal elements correspond to transmission and reception

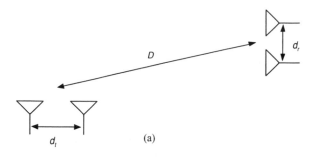

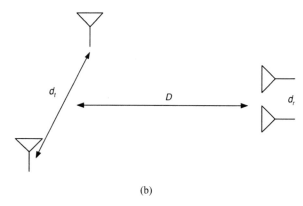

Figure 3.5: Channel dependence on the array geometry: (a) a poorly-conditioned channel; (b) a well-conditioned channel.

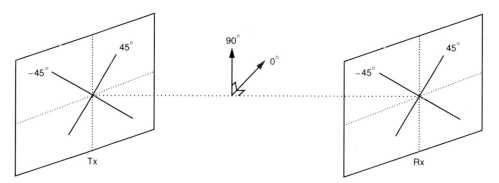

Figure 3.6: Dual-polarized antenna system. Signals are launched and received on orthogonal polarizations.

on orthogonal polarizations. The power in the individual channel elements is assumed to be

$$\mathcal{E}\{|h_{1,1}|^2\} = \mathcal{E}\{|h_{2,2}|^2\} = 1, \tag{3.30}$$
$$\mathcal{E}\{|h_{1,2}|^2\} = \mathcal{E}\{|h_{2,1}|^2\} = \alpha, \tag{3.31}$$

where α ($0 \leq \alpha \leq 1$) depends on the XPD of the antennas and the XPC of the propagation environment [Vaughan, 1990; Eggers *et al.*, 1993; Lempiäinen and Laiho-Steffens, 1999; Neubauer and Eggers, 1999], often collectively referred to as XPD. With no XPC, a good XPD results in $\alpha \approx 0$, while a poor XPD results in $\alpha \approx 1$. Assuming Rayleigh fading, the channel $\mathbf{H}$ with cross-polarized antennas may be modeled approximately as

$$\mathbf{H} = \mathbf{X} \odot \left(\mathbf{R}_r^{1/2} \mathbf{H}_w \mathbf{R}_t^{1/2} \right), \tag{3.32}$$

where

$$\mathbf{X} = \begin{bmatrix} 1 & \sqrt{\alpha} \\ \sqrt{\alpha} & 1 \end{bmatrix}, \tag{3.33}$$

and $\odot$ stands for the Hadamard product (if $\mathbf{A} = \mathbf{B} \odot \mathbf{C}$ then $[\mathbf{A}]_{i,j} = [\mathbf{B}]_{i,j}[\mathbf{C}]_{i,j}$). The covariance matrices $\mathbf{R}_r$ and $\mathbf{R}_t$ are defined in Eq. (3.25) and specify the correlation coefficients between the polarizations at the transmitter and receiver respectively. α, $\mathbf{R}_r$ and $\mathbf{R}_t$ depend on a variety of factors including XPD, XPC and antenna spacing. With no scattering and therefore no XPC, $\mathbf{H}$ will reduce to $\mathbf{X}$.

3.4.4 Degenerate channels

Consider a pin-hole channel that is a time and frequency flat channel, with M_R receive antennas and M_T transmit antennas (see Fig. 2.12). Since the pin-hole acts as a point source the MIMO channel becomes

$$\mathbf{H} = \mathbf{h}_r \mathbf{h}_t^T, \tag{3.34}$$

where $\mathbf{h}_r$ is the $M_R \times 1$ vector channel from the pin-hole to the receive array and $\mathbf{h}_t^T$ is the $1 \times M_T$ vector channel from the transmit antenna array to the pin-hole. The envelope of each element of $\mathbf{H}$ is double Rayleigh distributed (Eq. (2.39)). Further, since $\mathbf{H}$ is constructed from the product of two vectors, every realization of the channel $\mathbf{H}$ is rank-deficient with

$$r(\mathbf{H}) = 1. \tag{3.35}$$

3.5 Statistical properties of H

In this section we describe certain statistics of $\mathbf{H}$, in particular the statistics of the singular values and the squared Frobenius norm of $\mathbf{H}$.

3.5.1 Singular values of H

The $M_R \times M_T$ channel matrix $\mathbf{H}$, with rank r, has a singular value decomposition (SVD)

$$\mathbf{H} = \mathbf{U}\mathbf{\Sigma}\mathbf{V}^H, \tag{3.36}$$

where $\mathbf{U}$ and $\mathbf{V}$ are $M_R \times r$ and $M_T \times r$ matrices respectively, and satisfy $\mathbf{U}^H \mathbf{U} = \mathbf{V}^H \mathbf{V} = \mathbf{I}_r$, and $\mathbf{\Sigma} = \text{diag}\{\sigma_1, \sigma_2, \ldots, \sigma_r\}$ with $\sigma_i \geq 0$ and $\sigma_i \geq \sigma_{i+1}$, where σ_i is the ith singular value [Golub and Van Loan, 1989] of the channel. The columns of $\mathbf{V}$ and $\mathbf{U}$ are also known as the input and output singular vectors, respectively.

$\mathbf{H}\mathbf{H}^H$ is an $M_R \times M_R$ positive semi-definite Hermitian matrix. Let the eigen-decomposition of $\mathbf{H}\mathbf{H}^H$ be $\mathbf{Q}\mathbf{\Lambda}\mathbf{Q}^H$, where $\mathbf{Q}$ is an $M_R \times M_R$ matrix satisfying $\mathbf{Q}^H \mathbf{Q} = \mathbf{Q}\mathbf{Q}^H = \mathbf{I}_{M_R}$ and $\mathbf{\Lambda} = \text{diag}\{\lambda_1 \lambda_2 \cdots \lambda_{M_R}\}$ with $\lambda_i \geq 0$. We assume the eigenvalues λ_i are ordered so that $\lambda_i \geq \lambda_{i+1}$. Then,

$$\lambda_i = \begin{cases} \sigma_i^2, & \text{if } i = 1, 2, \ldots, r \\ 0, & \text{if } i = r+1, \; r+2, \ldots, M_R \end{cases}. \tag{3.37}$$

Since $\mathbf{H}$ is random, λ_i is also a random variable. If $\mathbf{H} = \mathbf{H}_w$, the elements of $\mathbf{H}_w \mathbf{H}_w^H$ are Wishart distributed [Muirhead, 1982]. The joint distribution of λ_i for the $\mathbf{H}_w$ channel model with $M_T = M_R = M$ is given by [Edelman, 1989]

$$f(x_1, x_2, \ldots, x_M) = \frac{(\pi/2)^{M^2-M}}{(\widetilde{\Gamma}_M(M))^2} e^{\left(-\sum_{i=1}^{M} x_i\right)} \prod_{i,j,i<j} (2x_i - 2x_j)^2 \prod_{i=1}^{M} u(x_i), \tag{3.38}$$

where $\widetilde{\Gamma}_M(M)$ is the complex multivariate gamma function defined as

$$\widetilde{\Gamma}_M(M) = \pi^{\frac{M^2-M}{2}} \prod_{i=1}^{M} (M - i + 1)!. \tag{3.39}$$

Further, we know that the smallest eigenvalue of $\mathbf{H}_w \mathbf{H}_w^H$, λ_M, is exponentially distributed [Edelman, 1989]

$$f(x_M) = M e^{-x_M M} u(x_M). \tag{3.40}$$

Thus, the squared minimum singular value of an $\mathbf{H}_w$ channel for $M_T = M_R$ has an exponential distribution, or equivalently the minimum singular value is Rayleigh distributed.

3.5.2   Squared Frobenius norm of H

The squared Frobenius norm of $\mathbf{H}$, $\|\mathbf{H}\|_F^2$, is defined as

$$\|\mathbf{H}\|_F^2 = \text{Tr}(\mathbf{H}\mathbf{H}^H) = \sum_{i=1}^{M_R} \sum_{j=1}^{M_T} |h_{i,j}|^2. \tag{3.41}$$

$\|\mathbf{H}\|_F^2$ may be interpreted as the total power gain of the channel and satisfies

$$\|\mathbf{H}\|_F^2 = \sum_{i=1}^{M_R} \lambda_i, \tag{3.42}$$

where λ_i $(i = 1, 2, \ldots, M_R)$ are the eigenvalues of $\mathbf{HH}^H$. Analogous to singular values of $\mathbf{H}$, $\|\mathbf{H}\|_F^2$ is also a random variable. The statistics of $\|\mathbf{H}\|_F^2$ determines diversity performance. The PDF of $\|\mathbf{H}\|_F^2$ when $\mathbf{H} = \mathbf{H}_w$ is given by

$$f(x) = \frac{x^{M_T M_R - 1}}{(M_T M_R - 1)!} e^{-x} u(x). \tag{3.43}$$

$\|\mathbf{H}_w\|_F^2$ is a Chi-squared random variable with $2M_T M_R$ degrees of freedom. The distribution of $\|\mathbf{H}\|_F^2$ under more complex channel conditions (Ricean fading, correlation, etc.) can be found in [Nabar *et al.*, 2002b]. The quantity of interest that we use in Chapter 5 to evaluate diversity performance is the moment generating function of the random variable $\|\mathbf{H}\|_F^2$ denoted by $\psi_{\|\mathbf{H}\|_F^2}(v)$. Assuming Rayleigh fading with $\mathbf{R} = \mathcal{E}\{\mathrm{vec}(\mathbf{H})\mathrm{vec}(\mathbf{H})^H\}$, $\psi_{\|\mathbf{H}\|_F^2}(v)$ is given by [Turin, 1960]

$$\begin{aligned}
\psi_{\|\mathbf{H}\|_F^2}(v) &= \mathcal{E}\left\{e^{-v\|\mathbf{H}\|_F^2}\right\} \\
&= \frac{1}{\det\left(\mathbf{I}_{M_T M_R} + v\mathbf{R}\right)} \\
&= \prod_{i=1}^{M_T M_R} \frac{1}{1 + v\lambda_i(\mathbf{R})},
\end{aligned} \tag{3.44}$$

where $\lambda_i(\mathbf{R})$ $(i = 1, 2, \ldots, M_T M_R)$ is the ith eigenvalue of $\mathbf{R}$ and the associated region of convergence (ROC) is given by $\Re\{v\} \geq \max_i -1/\lambda_i(\mathbf{R})$.

3.6   Channel measurements and test channels

A thorough understanding of typical ST channel characteristics is critical in order to design efficient communication systems. A number of measurements of MIMO channels have been carried out across the globe [Stridh *et al.*, 2000; Kermoal *et al.*, 2000; Swindlehurst *et al.*, 2001; Yu *et al.*, 2001a; Erceg *et al.*, 2002; Soma *et al.*, 2002; Kyritsi, 2002]. Figure 3.7 shows a measured $M_T = M_R = 2$ MIMO channel in the frequency–time domain for a fixed wireless application at 2.5 GHz. Parameters extracted from such measurements include path loss, the K-factor, fading signal correlation, delay spread and Doppler spread.

To facilitate the development of bench or laboratory testing of ST wireless systems we need hardware channel simulators. Test channel models specify such simulators and should span the range of channels likely to be encountered in practice by the system. Test models typically use a simplified version of a physical channel model. We describe a set of six channels known as the Stanford University Interim (SUI) models [Baum, 2001], reflective of the three terrains in continental USA, that have been developed for fixed broadband wireless applications. The terrains include a hilly terrain with moderate to heavy tree density (Terrain A), a flat terrain with light tree density (Terrain C)

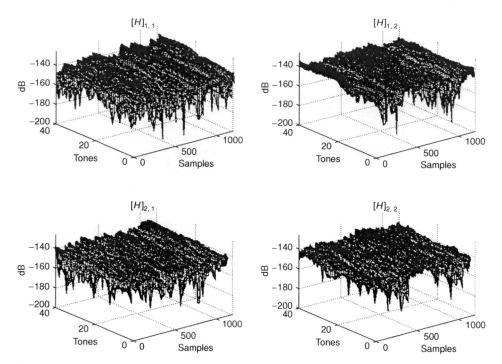

Figure 3.7: Measured time–frequency response of a $M_T = M_R = 2$ MIMO channel. $[H]_{i,j}$ is the channel response between the jth transmit and the ith receive antennas.

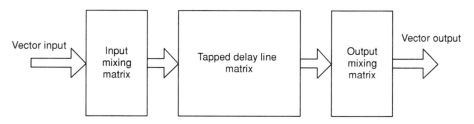

Figure 3.8: Schematic of a SUI channel.

and an intermediate terrain classified as Terrain B. The SUI channel models have been specifically designed for a cell size of 7 km, base-station antenna height of 30 m, receive antenna height of 6 m, base-station antenna beamwidth of 120° and optional receive antenna beamwidth of 360° (omni) or 30°. The generic structure of the SUI channel model is given in Fig. 3.8.

The input mixing matrix models the correlation between transmitted signals (transmit correlation) when multiple antennas are used. Multipath fading is modeled as a tapped delay line with three taps and non-uniform delays. The gain associated with each tap is characterized by a fading distribution (Ricean with $K > 0$ or Rayleigh with $K = 0$) and Doppler frequency. The output mixing matrix specifies the correlation between signals

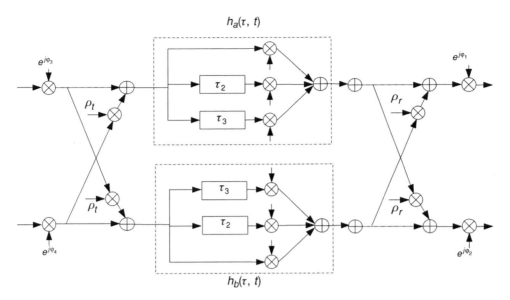

Figure 3.9: SUI channel for a $M_T = M_R = 2$.

at the output (receive correlation). Figure 3.9 shows the generation of a two-transmit two-receive antenna SUI channel.

The SUI channel, $\mathbf{H}(\tau, t)$ (neglecting pulse-shaping/matched-filtering) is given by

$$\mathbf{H}(\tau, t) = \begin{bmatrix} e^{j\phi_1} & 0 \\ 0 & e^{j\phi_2} \end{bmatrix} \begin{bmatrix} 1 & \rho_r \\ \rho_r & 1 \end{bmatrix} \begin{bmatrix} h_a(\tau, t) & 0 \\ 0 & h_b(\tau, t) \end{bmatrix} \begin{bmatrix} 1 & \rho_t \\ \rho_t & 1 \end{bmatrix} \begin{bmatrix} e^{j\phi_3} & 0 \\ 0 & e^{j\phi_4} \end{bmatrix}, \quad (3.45)$$

where

$$h_a(\tau, t) = h_a(0, t)\delta(\tau) + \sum_{i=1,2} h_a(i, t)\delta(\tau - \tau_i), \tag{3.46}$$

$$h_b(\tau, t) = h_b(0, t)\delta(\tau) + \sum_{i=1,2} h_b(i, t)\delta(\tau - \tau_i). \tag{3.47}$$

The tapped delay line weights $h_a(i, t)$ and $h_b(i, t)$ ($i = 0, 1, 2$) are independent complex Gaussian random variables with a K-factor (K_i), delay (τ_i), Doppler spread (ν_i) and power ($p_i = \mathcal{E}\{|h_a(i, t)|^2\} = \mathcal{E}\{|h_b(i, t)|^2\}$), all of which are specified for each of the six SUI models. The mixing weights ρ_t and ρ_r are correlation parameters that can be used to adjust the degree of fading signal correlation. The phases ϕ_i ($i = 1, 2, 3, 4$) can be chosen arbitrarily to reflect array geometry and AOA or AOD of the a and b path clusters. Table 3.1 summarizes values of the SUI parameters for the SUI-3 channel (Terrain B). For further details on SUI channel modeling and implementation the reader is referred to [Freedman *et al.*, 2001; Baum, 2001]. The SUI channel models can be restricted to SIMO and MISO cases by disconnecting an input or output respectively.

Table 3.1. *SUI-3 channel model parameters. The model is applicable to an intermediate terrain (between hilly and flat) with moderate tree density.*

Correlation = 0.25	Tap 1	Tap 2	Tap 3	
Delay	0	0.5	1	μs
Power	0	-5	-10	dB
K-factor	0.6	0	0	
Doppler	0.4	0.4	0.4	Hz

3.7   Sampled signal model

In this section we introduce the discrete time (or sampled) signal model for various antenna configurations (SISO, SIMO, MISO, MIMO) in conjunction with single carrier (SC) modulation. For the SISO channel, we derive the discrete time baseband input–output model from the continuous time relation. For all other antenna configurations, we directly develop the discrete time model.

3.7.1   Normalization

Proper normalization of signals, channels and noise is important for both the absolute and the comparative performance of communication systems. For SC modulation, we assume the channel bandwidth is 1 Hz and the symbol period is 1 second. We assume all signals and noise are modeled as the complex envelope of the underlying passband channel. Hopefully, areas where we do depart from the normalization below will be self-evident.

Channel

Recall that we assume that in frequency flat channels the average channel element (SISO link) energy is normalized i.e., $\mathcal{E}\{|h_{i,j}|^2\} = 1$. Further, we assume that delay spread channels have a multitap channel response and that the total average energy of all taps for a given channel element is normalized, i.e., the multipath effects do not change average channel power transfer efficiency. We also assume that the varying (fading) component of the channel is ZMCSCG. We depart from unit average energy normalization of the channel elements in Chapters 10 and 12, where path loss differences between different users are accounted for in the channel **H**.

Signal energy

Assume in the SISO and SIMO cases that the average transmit symbol energy is E_s. Since $T_s = 1$ second, E_s is therefore also the transmit power. Hence, we shall refer to

E_s in conjunction with SC modulation interchangeably as energy or power throughout this book. For the MISO or MIMO channels, we assume the average transmit energy per symbol period is constant and therefore the symbol energy per antenna is reduced by the number of antennas, i.e., the energy per symbol per antenna is E_s/M_T. We assume data symbols prior to coding are IID and are drawn from scalar constellations with zero mean and unit average energy. We depart from this unit average energy constraint in Chapters 4, 10 and 11, where E_s is absorbed into the data symbols resulting in variable average energy.

Noise

Since the transmission bandwidth is 1 Hz, the noise power in the band is the same as spectral power density N_o. Therefore, we interchangeably refer to N_o as noise power or noise spectral density. Noise is assumed to be ZMCSCG with variance N_o.

3.7.2   SISO sampled signal model

Let $h(\tau)$ denote the continuous time baseband channel response (we drop the t dependence for convenience). Recall that $h(\tau)$ includes the effects of pulse-shaping at the transmitter, matched-filtering at the receiver and the physical channel. Assuming that a sequence of data symbols $s[l]$ ($l = 0, 1, 2, \ldots$) is to be transmitted, the received signal $y(t)$ can be written as

$$
\begin{aligned}
y(t) &= h(\tau) \star \left(\sum_l \sqrt{E_s} s[l] \delta(t - lT_s) \right) + n(t) \\
&= \sum_l \sqrt{E_s} s[l] h(t - lT_s) + n(t),
\end{aligned} \tag{3.48}
$$

where T_s is the duration of a single symbol ($1/T_s \approx B$, the bandwidth of transmission) and $n(t)$ is additive noise. We have assumed scalar linear modulation such as pulse amplitude modulation (PAM) or quadrature amplitude modulation (QAM).

If this signal is sampled at instants $t = kT_s + \Delta$ ($k = 0, 1, 2, \ldots$), where Δ is the sampling delay, then the sampled signal response is

$$
y(kT_s + \Delta) = \sum_l \sqrt{E_s} s[l] h((k - l)T_s + \Delta) + n(kT_s + \Delta), \tag{3.49}
$$

and may be rewritten as

$$
y[k] = \sum_l \sqrt{E_s} s[l] h[k - l] + n[k], \quad k = 0, 1, 2, \ldots . \tag{3.50}
$$

$h[l]$ ($l = 0, 1, 2, \ldots, L - 1$) is the T_s spaced sampled channel. L is the channel length measured in sampling periods. $h[l]$ depends on $h(\tau)$ and the sampling delay Δ. In turn, $h(\tau)$ depends on the underlying physical channel $p(\tau)$ and the pulse-shaping/matched-filter channel $g(\tau)$. The components $h[l]$ will in general be correlated, even if the underlying multipath scattering components are uncorrelated.

In a frequency selective channel, matched-filtering to the transmit pulse followed by sampling at the symbol rate can result in a sub-optimal receiver. The received signal $y(t)$ must be oversampled, so we take multiple samples per symbol period T_s. Typically an oversampling factor of 2 is adequate. Oversampling offers a number of other advantages, including improved jitter tolerance, reduced noise enhancement and diversity performance of linear equalizers (see [Lee and Messerschmitt, 1993; Proakis, 1995; Cioffi, 2002] for details).

We assume that the noise samples $n[k]$ are temporally white ZMCSCG with variance N_o. The noise samples are not independent if we use non-shift-orthogonal matched-filtering at the receiver. Nevertheless, we shall make this assumption for convenience, i.e., $\mathcal{E}\{n[k]n[l]^*\} = N_o\delta[k-l]$. Clearly, noise samples spaced $T_s/2$ apart that occur if we oversample by a factor of 2 are not independent.

Frequency flat channel

For a frequency flat channel $h[k] = 0$ for $k \neq 0$ (assuming no sample delay) and denoting $h[0]$ by h, the input–output relation for the channel simplifies to

$$y[k] = \sqrt{E_s}hs[k] + n[k]. \tag{3.51}$$

Frequency selective channel

For the frequency selective case, the channel is given by $h[l]$ ($l = 0, 1, \ldots, L-1$), where L is the channel length. The received signal sample at time index k is

$$y[k] = \sqrt{E_s}[h[L-1]\cdots h[1]\,h[0]]\begin{bmatrix} s[k-L+1] \\ \vdots \\ s[k-1] \\ s[k] \end{bmatrix} + n[k]. \tag{3.52}$$

T successive received signal samples are therefore

$$[y[k]\cdots y[k+T-1]] = \sqrt{E_s}[h[L-1]\cdots h[0]]\mathcal{S} + [n[k]\cdots n[k+T-1]], \tag{3.53}$$

where $\mathcal{S}$ is a Hankel matrix of dimension $L \times T$ given by

$$\mathcal{S} = \begin{bmatrix} s[k-L+1] & s[k-L+2] & \cdots & s[k-L+T] \\ \vdots & \vdots & \cdots & \vdots \\ s[k-1] & s[k] & \cdots & s[k+T-2] \\ s[k] & s[k+1] & \cdots & s[k+T-1] \end{bmatrix}. \tag{3.54}$$

The input–output relation for frequency selective fading in Eq. (3.53) can alternatively be expressed as

$$
\begin{bmatrix} y[k] \\ \vdots \\ y[k+T-1] \end{bmatrix} = \sqrt{E_s}\,\mathcal{H} \begin{bmatrix} s[k-L+1] \\ \vdots \\ s[k+T-1] \end{bmatrix} + \begin{bmatrix} n[k] \\ \vdots \\ n[k+T-1] \end{bmatrix}, \tag{3.55}
$$

where $\mathcal{H}$ is a Toeplitz matrix of dimension $T \times (T+L-1)$ given by

$$
\mathcal{H} = \begin{bmatrix} h[L-1] & \cdots & h[0] & 0 & \cdots & 0 \\ 0 & h[L-1] & \cdots & h[0] & \cdots & 0 \\ \vdots & \ddots & \ddots & \ddots & \ddots & \vdots \\ 0 & \cdots & 0 & h[L-1] & \cdots & h[0] \end{bmatrix}. \tag{3.56}
$$

We note that $\mathcal{H}$ is always a fat matrix, independent of the number of symbols, T. With oversampling of 2, we can reformulate $h[i]$ to be a 2×1 vector

$$
\overline{\mathbf{h}}[i] = \begin{bmatrix} h[i]|_{\Delta=0} \\ h[i]|_{\Delta=T_s/2} \end{bmatrix}. \tag{3.57}
$$

Equation (3.56) can now be rewritten with $\mathcal{H}$ as a $2T \times (T+L-1)$ matrix. $\mathcal{H}$ can be made tall, with the appropriate choice of T. As long as $\{h[i]|_{\Delta=0}\}$ and $\{h[i]|_{\Delta=T_s/2}\}$ do not share common zeros [Paulraj *et al.*, 1998], we receive useful leverages from oversampling. In what follows, we do not specifically address oversampling, but we assume it is used wherever appropriate.

3.7.3 SIMO sampled signal model

Consider a SIMO channel with M_R receive antennas. As in the SISO channel, we consider the cases of frequency flat and frequency selective fading separately.

Frequency flat channel
The channel is modeled by a $M_R \times 1$ vector $\mathbf{h} = [h_1\ h_2\ \cdots\ h_{M_R}]^T$, where h_i is the channel transfer function (SISO) between the transmit antenna and the ith receive antenna. The signal model is

$$
\mathbf{y}[k] = \sqrt{E_s}\,\mathbf{h}s[k] + \mathbf{n}[k], \tag{3.58}
$$

where $\mathbf{y}[k]$ is the received signal vector of dimension $M_R \times 1$, $s[k]$ is the transmitted signal and $\mathbf{n}[k]$ is the ZMCSCG noise vector of dimension $M_R \times 1$ that is spatially and temporally white, i.e., $\mathcal{E}\{\mathbf{n}[k]\mathbf{n}[l]^H\} = N_o\mathbf{I}_{M_R}\delta[k-l]$.

Frequency selective channel
We denote the channel by the $M_R \times 1$ vector, $\mathbf{h}[l] = [h_1[l]\ h_2[l]\ \cdots\ h_{M_R}[l]]^T$, where $h_i[l]$ ($l = 0, 1, 2, \ldots, L-1$) is the channel between the transmit antenna and the ith

receive antenna. L is the maximum channel length of all component M_R SISO links. Let $\mathbf{y}[k] = [y_1[k]\ y_2[k]\ \cdots\ y_{M_R}[k]]^T$ be the received signal vector of dimension $M_R \times 1$ at time index k. The input–output relation for the channel is given by

$$\mathbf{y}[k] = \sqrt{E_s}[\mathbf{h}[L-1]\cdots\mathbf{h}[1]\ \mathbf{h}[0]] \begin{bmatrix} s[k-L+1] \\ \vdots \\ s[k] \end{bmatrix} + \begin{bmatrix} n_1[k] \\ \vdots \\ n_{M_R}[k] \end{bmatrix}. \tag{3.59}$$

T contiguous received vector samples can be stacked to yield

$$[\mathbf{y}[k]\cdots\mathbf{y}[k+T-1]] = \sqrt{E_s}\mathbf{H}\mathcal{S} + \mathcal{N}, \tag{3.60}$$

where $\mathcal{S}$ is defined in Eq. (3.54) and $\mathbf{H}$ and $\mathcal{N}$ are $M_R \times L$ and $M_R \times T$ matrices given by

$$\mathbf{H} = \begin{bmatrix} h_1[L-1] & \cdots & h_1[1] & h_1[0] \\ h_2[L-1] & \cdots & h_2[1] & h_2[0] \\ \vdots & \cdots & \vdots & \vdots \\ h_{M_R}[L-1] & \cdots & h_{M_R}[1] & h_{M_R}[0] \end{bmatrix}, \quad \mathcal{N} = \begin{bmatrix} n_1[k] & n_1[k+1] & \cdots & n_1[k+T-1] \\ n_2[k] & n_2[k+1] & \cdots & n_2[k+T-1] \\ \vdots & \vdots & \cdots & \vdots \\ n_{M_R}[k] & n_{M_R}[k+1] & \cdots & n_{M_R}[k+T-1] \end{bmatrix}. \tag{3.61}$$

3.7.4   MISO sampled signal model

Frequency flat channel

The channel is modeled by a $1 \times M_T$ row vector $\mathbf{h} = [h_1\ h_2\ \cdots\ h_{M_T}]$. The signal model is

$$y[k] = \sqrt{\frac{E_s}{M_T}}\mathbf{h}\mathbf{s}[k] + n[k], \tag{3.62}$$

where $y[k]$ is the received signal at time index k, $\mathbf{s}[k] = [s_1[k]\ s_2[k]\ \cdots\ s_{M_T}[k]]^T$ is the transmitted data signal vector of dimension $M_T \times 1$ at time instant k and $n[k]$ is ZMCSCG noise with variance N_o.

Frequency selective channel

The frequency selective MISO channel is expressed by the $1 \times M_T$ vector, $\mathbf{h}[l] = [h_1[l]\ h_2[l]\ \cdots\ h_{M_T}[l]]$, where $h_j[l]$ $(l = 0, 1, \ldots, L-1)$ is the channel between the jth transmit antenna and the receive antenna. L is the maximum channel length of all the component SISO links. The received signal (scalar), $y[k]$, at time index k, can be written as

$$y[k] = \sqrt{\frac{E_s}{M_T}}[\mathbf{h}_1\ \mathbf{h}_2\ \cdots\ \mathbf{h}_{M_T}] \begin{bmatrix} \mathbf{s}_1[k] \\ \mathbf{s}_2[k] \\ \vdots \\ \mathbf{s}_{M_T}[k] \end{bmatrix} + n[k], \tag{3.63}$$

where

$$
\mathbf{h}_j = [\, h_j[L-1] \cdots h_j[1]\, h_j[0]\,], \quad \mathbf{s}_j[k] =
\begin{bmatrix}
s_j[k-L+1] \\
\vdots \\
s_j[k-1] \\
s_j[k]
\end{bmatrix}.
\tag{3.64}
$$

T contiguous samples of the received signal can be stacked and written as

$$
[\, y[k] \cdots y[k+T-1]\,] = \sqrt{\frac{E_s}{M_T}} [\mathbf{h}_1 \cdots \mathbf{h}_{M_T}]
\begin{bmatrix}
\mathcal{S}_1 \\
\vdots \\
\mathcal{S}_{M_T}
\end{bmatrix}
+ [\, n[k] \cdots n[k+T-1]\,],
\tag{3.65}
$$

where

$$
\mathcal{S}_j =
\begin{bmatrix}
s_j[k-L+1] & s_j[k-L+2] & \cdots & s_j[k-L+T] \\
\vdots & \vdots & \cdots & \vdots \\
s_j[k-1] & s_j[k] & \cdots & s_j[k+T-2] \\
s_j[k] & s_j[k+1] & \cdots & s_j[k+T-1]
\end{bmatrix}.
\tag{3.66}
$$

3.7.5   MIMO sampled signal model

Frequency flat channel

The channel is modeled by a matrix $\mathbf{H}$ of dimension $M_R \times M_T$. The signal model is

$$
\mathbf{y}[k] = \sqrt{\frac{E_s}{M_T}} \mathbf{H}\mathbf{s}[k] + \mathbf{n}[k],
\tag{3.67}
$$

where $\mathbf{y}[k]$ is the received signal vector with dimension $M_R \times 1$, $\mathbf{s}[k]$ is the transmit signal vector with dimension $M_T \times 1$ and $\mathbf{n}[k]$ is the $M_R \times 1$ spatio-temporally white ZMCSCG noise vector with variance N_o in each dimension. Since the output at any instant of time is independent of inputs at previous times, we can drop the time index k for clarity and express the input–output relation simply as

$$
\mathbf{y} = \sqrt{\frac{E_s}{M_T}} \mathbf{H}\mathbf{s} + \mathbf{n}.
\tag{3.68}
$$

The same is true for flat fading SISO, SIMO and MISO channels discussed earlier.

Frequency selective channel

We represent the channel by the $M_R \times M_T$ matrix $\mathbf{H}[l]$ $(l = 0, 1, 2, \ldots, L-1)$, where L is the maximum channel length of all component $M_R M_T$ SISO links. The channel between the ith receive and jth transmit antenna is given by $h_{i,j}[l]$ $(i = 1, 2, \ldots, M_R,$

$j = 1, 2, \ldots, M_T$). The received signal vector at time k, $\mathbf{y}[k]$, of dimension $M_R \times 1$ may be expressed as

$$\mathbf{y}[k] = \sqrt{\frac{E_s}{M_T}} \begin{bmatrix} \mathbf{h}_{1,1} & \cdots & \mathbf{h}_{1,M_T} \\ \vdots & \vdots & \vdots \\ \mathbf{h}_{M_R,1} & \cdots & \mathbf{h}_{M_R,M_T} \end{bmatrix} \begin{bmatrix} \mathbf{s}_1[k] \\ \vdots \\ \mathbf{s}_{M_T}[k] \end{bmatrix} + \mathbf{n}[k], \tag{3.69}$$

where

$$\mathbf{h}_{i,j} = [\, h_{i,j}[L-1] \cdots h_{i,j}[0]\,], \ \mathbf{s}_j[k] = \begin{bmatrix} s_j[k-L+1] \\ \vdots \\ s_j[k] \end{bmatrix}. \tag{3.70}$$

T continuous received vector samples may be stacked and written as

$$[\,\mathbf{y}[k] \cdots \mathbf{y}[k+T-1]\,] = \sqrt{\frac{E_s}{M_T}} \begin{bmatrix} \mathbf{h}_{1,1} & \cdots & \mathbf{h}_{1,M_T} \\ \vdots & \vdots & \vdots \\ \mathbf{h}_{M_R,1} & \cdots & \mathbf{h}_{M_R,M_T} \end{bmatrix} \begin{bmatrix} \mathcal{S}_1 \\ \vdots \\ \mathcal{S}_{M_T} \end{bmatrix} + \mathcal{N}, \tag{3.71}$$

where $\mathcal{S}_j$ and $\mathcal{N}$ are as defined in Eqs. (3.66) and (3.61) respectively.

3.8   ST multiuser and ST interference channels

So far we have only considered ST channels for a single user model (one transmitter and one receiver) with multiple antennas at either or both ends. These models can be extended to ST multiuser and interference channels. We assume frequency flat channels and SC modulation. Extensions to frequency selective channels are readily possible.

3.8.1   ST multiuser channel

In the multiple access channel, P single antenna users transmit signals that arrive at a multiantenna receiving base. This is a summed SIMO model. Since the users are geographically dispersed, joint encoding of the user information is not possible. Joint decoding at the base is possible. The signal model is given by

$$\mathbf{y} = \sum_{i=1}^{P} \sqrt{E_{s,i}} \mathbf{h}_i s_i + \mathbf{n}, \tag{3.72}$$

where $E_{s,i}$ is the energy of the ith user, s_i is the signal transmitted by user i and $\mathbf{h}_i$ is the SIMO channel from user i to the base-station. A peak power constraint on $E_{s,i}$ is typical. The elements of $\mathbf{h}_i$ are normalized to different values to model path loss variations from the users to the base. The model can be generalized to a summed MIMO channel for multiantenna user terminals.

In the broadcast channel, a multiantenna base transmits signals that arrive at P single antenna receiving terminals. This is a summed MISO model. Since the users are geographically dispersed, joint decoding of the received signal is not possible. Joint encoding at the base is necessary. The signal model is given by

$$y_i = \sum_{j=1}^{P} \sqrt{E_{s,j}} \mathbf{h}_i \mathbf{s}_j + n_i, \tag{3.73}$$

where $E_{s,j}$ is the energy of the jth user, $\mathbf{s}_j$ is the vector signal transmitted to the jth user and y_i and n_i are respectively the received signal and noise at user i. $\mathbf{h}_i$ is the channel from the base to the ith user. A sum power constraint on $E_{s,j}$ is typical. Further, $\mathbf{h}_i$ are normalized to different values to model different path losses from the base to the users. The model can be generalized to a summed MIMO case for multiantenna user terminals. See Chapter 10 for explanatory figures and more discussion.

3.8.2   ST interference channel

The interference ST channel model consists of single user SIMO, MISO and MIMO channels discussed earlier in the chapter, with interference from or to other users located in co-channel cells.

SIMO interference channel

This refers to the reverse link channel where we have one multiantenna receiving base and one single antenna transmitting terminal. However, there are N additional co-channel users in other cells. The model is again a summed SIMO model. No joint encoding of users is possible. The composite signal arrives at the multiantenna base and the desired signal is decoded along with spatial processing. The signal model is given by

$$\mathbf{y} = \sqrt{E_{s,0}} \mathbf{h}_0 s_0 + \sum_{i=1}^{N} \sqrt{E_{s,i}} \mathbf{h}_i s_i + \mathbf{n}, \tag{3.74}$$

where $E_{s,i}$, $\mathbf{h}_i$ $(M_R \times 1)$ and s_i represent respectively the energy, channel and signal of the ith user. The index 0 refers to the desired user in the above equation.

MISO interference channel

This refers to the forward link with one multiantenna transmitting base and one intended single antenna receiving terminal. However, there are N co-channel interfering bases in other cells. The model is a summed MISO model. Joint encoding across bases is not assumed (though possible in principle). However, co-operative spatial processing (beamforming) at each base can be used to reduce interference to other cells. The signal model is given by (0 is the index of the desired user)

$$y = \sqrt{E_{s,0}}\mathbf{h}_0\mathbf{s}_0 + \sum_{i=1}^{N}\sqrt{E_{s,i}}\mathbf{h}_i\mathbf{s}_i + n, \qquad (3.75)$$

where $E_{s,0}$, $\mathbf{h}_0$ $(1 \times M_T)$ and $\mathbf{s}_0$ are the energy, channel and signal of the desired user. $E_{s,i}$, $\mathbf{s}_i$ and $\mathbf{h}_i$ refer to the ith interfering base. Note, $\mathbf{h}_i$ is the channel from the ith interfering base to user 0. A peak power constraint on $E_{s,i}$ (note this is a vector signal) is typical. Further, $\mathbf{h}_i$ are normalized to model path loss from base i to user 0. The model can be generalized to a summed MIMO model for multiantenna user terminals.

See Chapter 11 for explanatory figures and more discussion.

3.9   ST channel estimation

In this section we very briefly review ST channel estimation at the receiver and transmitter. We begin by considering channel estimation at the receiver, which is commonly needed in communication systems.

3.9.1   Estimating the ST channel at the receiver

In SISO systems, the channel is estimated by the receiver using training signals emitted by the transmitter. A number of training techniques have been developed and are specific to each modulation scheme. The receiver knows the training signal sequence $F[k]$ $(k = 0, 1, 2, \ldots, J - 1)$, of length J in advance. Denoting the channel by $\mathbf{h} = [h[L - 1] \cdots h[1]\, h[0]]$, the received signal is

$$[y[k] \ldots y[k + T - 1]] = \mathbf{h}\mathcal{F} + [n[k] \ldots n[k + T - 1]], \qquad (3.76)$$

following Eq. (3.53), where we have replaced $\mathcal{S}$ by $\mathcal{F}$ ($\mathcal{F}$ is appropriately constructed from $F[k]$ $(k = 0, 1, \ldots, J - 1)$. The channel estimate, $\widehat{\mathbf{h}}$, is obtained using a least squares approach

$$\widehat{\mathbf{h}} = [y[k] \ldots y[k + T - 1]]\mathcal{F}^{\dagger}. \qquad (3.77)$$

$F[k]$ is typically chosen to have good autocorrelation properties. Depending on the SNR at the receiver and the desired channel estimation accuracy, the duration (or energy) of the training signal has to be selected. See [Meyr et al., 1997] for details. The desired channel estimation accuracy depends on the modulation order used, a useful rule of thumb being that the channel estimation error should be 10 dB below the additive noise power. If the channel has delay spread, more channel parameters have to be estimated and additional training signal energy has to be expended to estimate the channel. Further, the frequency of channel estimation depends on the Doppler spread, i.e., the more rapidly the channel changes, the more frequently we need to estimate. In SIMO systems the training procedure is the same. No additional training energy is needed as each receive antenna picks up its own signal for channel estimation.

Channel estimation often uses interpolation techniques, where the channel is estimated at discrete points in time or frequency (spaced well below T_C or B_C respectively) and the channel at the other points is interpolated through some suitable scheme [Lo *et al.*, 1991].

In blind techniques for channel estimation no explicit training signals are used, instead the receiver estimates the channels from the signals received during normal data (information) transmission. Substantial work has been done on blind estimation of ST channels and uses a number of leverages including cyclo-stationarity, finite alphabet, constant modulus, etc., to estimate the channel. The use of blind methods has generally not been popular in practical systems. Semi-blind methods that mix both training and blind based techniques show more promise. More details are beyond the scope of this book but the interested reader is referred to [Tong *et al.*, 1994; Ding and Li, 1994; van der Veen *et al.*, 1995; Paulraj *et al.*, 1998; Larsson, 2001; Larsson *et al.*, 2001; Bölcskei *et al.*, 2002b].

Training with multiple transmit antennas

All the above comments on training apply. In addition, the multiple transmit antennas will need additional training effort, since more parameters (proportional to the number of transmit antennas) have to be estimated [Marzetta, 1999; Hassibi and Hochwald, 2000; Tong, 2001; Bölcskei *et al.*, 2002b]. We try to ensure that the training signals from the multiple antennas are mutually orthogonal in some dimension – for example, time (different time slots) or frequency (different tones in orthogonal frequency division multiplexing (OFDM)) or code (different orthogonal codes). Though orthogonality is not strictly required, orthogonal signals provide the best estimation accuracy for a given transmit power under most circumstances. The same channel estimation technique expressed in Eq. (3.76) applies, except that $\mathbf{h}$ and $\mathcal{F}$ are suitably structured to reflect multiple antenna training structure as per Eq. (3.71). Typically the training sequences should have good auto- and cross-correlation properties.

The number of samples collected during training (per receive antenna) must now be $T \geq M_T \times L$. If we assume a block fading model (i.e., the channel is constant during a coherence period T_C and changes abruptly to a new value at T_C intervals), the maximum number of symbols per coherence period is $T_C \times B$. Thus T can be at most $T_C \times B$. Therefore if $M_T \times L > T_C \times B$, we can never get perfect channel estimation at the receiver even in the absence of noise.

Channel estimation for frequency and time selective ST channels has been studied. Usually the receiver estimates the channel at adequately (Nyquist) spaced frequencies or times (sampling). The full channel is then determined through interpolation.

3.9.2   Estimating the ST channel at the transmitter

In SISO wireless links, knowledge of the channel at the transmitter is typically used for adapting the modulation rate or for power control. This only needs the magnitude (or gain) of the forward channel. In MISO and MIMO channels, knowledge of the channel (**h** or **H**) can be leveraged in additional ways, such as beamforming or pre-filtering, to provide significant value. Therefore, there is significant motivation for channel knowledge at the transmitter in ST channels. In MIMO-MU (SDMA) channel knowledge is necessary to steer signals selectively at users. Depending on the application, differing levels of accuracy in channel information are needed.

We assume a two-way (duplex) communication link. For convenience, we assume that we are interested in estimating the forward channel at the base-station. Forward channel estimation at the base is not directly possible as the signal travels through the channel only after leaving the base transmitter. Two general techniques are used in channel estimation at the transmitter. In the first approach the forward channel is estimated at the terminal receiver after the signal has traveled through the channel and then sent back (feedback) to the base-station on the reverse link. In the second approach, we leverage the reciprocity principle in duplex transmission. The base-station first estimates the reverse link channel, and uses this estimate for the forward link channel. The two techniques are described below.

Channel estimation at transmitter using feedback

In this approach the forward link ST channel is estimated at the terminal (the channel of interest to the base-station transmitter) and is sent to the base-station on the reverse link. This feedback will involve some delay (or lag), δ_{lag}. Since wireless channels are time-varying, we need

$$\delta_{lag} \ll T_C, \tag{3.78}$$

where T_C is the coherence time. Therefore, δ_{lag}/T_C determines channel accuracy at the transmitter. In a fast changing channel, we need more frequent estimation and feedback. The resulting overhead on the reverse channel can be prohibitive [Gerlach, 1995].

One approach to reducing the feedback overhead is to send a slow changing statistic of the channel such as the correlation matrices $\mathbf{R}_t$ or $\mathbf{R}_r$. Another option can be to feedback only partial channel information such as channel condition number.

Channel estimation at transmitter using reciprocity

Let us first discuss a SISO case. Let $h_f(t_f, f_f, i_f)$ be the forward SISO channel from the base-station transmitter antenna to the terminal and $h_r(t_r, f_r, i_r)$ be the reverse SISO channel (see Fig. 3.10). t_f, f_f and i_f refer to the time, frequency and antenna index used on the forward link. t_r, f_r and i_r are similarly defined for the reverse link. The antenna index specifies the antenna used at the base and the terminal.

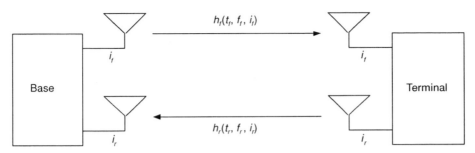

Figure 3.10: Duplexing in ST channels. If the time, frequency of operation and antennas of the forward and reverse links are the same, the channels are identical.

The reciprocity principle states that if the time, frequency and antennas for channel use are the same ($t_f = t_r$, $f_f = f_r$, $i_f = i_r$), then the channels in the forward and reverse links are identical, i.e.,

$$h_r(t_r, f_r, i_r) = h_f(t_f, f_f, i_f). \tag{3.79}$$

However, duplexing schemes support simultaneous two-way links and need to isolate these links to prevent interference. Therefore we need to force some difference in time, or frequency and/or spatial parameters. In turn this causes errors in estimating the base-station transmit (forward) channel from the base-station receive (reverse) channel. A few approaches are discussed below.

In time division duplexing (TDD), the forward and reverse channels use the same frequency and antennas for the duplex links, but use different time slots (ping-pong) to communicate. Let $\delta_t = t_f - t_r$ be the duplexing time delay. It follows that the forward and reverse channels can be equated only if

$$\delta_t << T_C, \tag{3.80}$$

where T_C is the coherence time of the channel. Clearly, the more stringent the requirements of accuracy in channel estimates, the smaller δ_t / T_C will need to be.

In frequency division duplexing (FDD), the forward and reverse channels use the same time and antennas to communicate, but use different frequencies on the links. Let $\delta_f = f_f - f_r$ be the duplexing frequency difference. It follows that the forward and reverse channels can be equated if

$$\delta_f << B_C, \tag{3.81}$$

where B_C is the coherence bandwidth of the channel. There will still be a phase difference given by $2\pi \delta_f T_{f-r}$, where T_{f-r} is the total travel time. In practice, due to physical limits on duplexing filters that isolate the reverse and forward links, δ_f is about 5% of the operating frequency v_c. This usually means that $\delta_f >> B_C$. Therefore, the reciprocity principle in general cannot be exploited in FDD for transmit estimation.

In antenna division duplexing (ADD), the forward and reverse channels use the same frequency and time, but use different antennas (or beams) on each link to communicate. Let δ_d be the separation (also called duplexing location difference) between the antennas indexed by i_f and i_r. It follows that the forward and reverse channels can be equated if

$$\delta_d \ll D_C, \tag{3.82}$$

where D_C is the coherence distance of the channel. This may be impossible to meet physically when D_C is itself as small as $\lambda_c/2$. When ADD can be applied, a correction will still be needed to cover array antenna element and geometry differences. Another disadvantage is that ADD does not provide sufficient isolation between the two links and is almost never used directly as a duplexing scheme.

Many communication systems use a combination of time/frequency/antenna separation in the duplex links, making reciprocity infeasible. Only pure TDD offers a realistic opportunity for exploiting reciprocity for channel estimation at the transmitter. However, even here there are a number of real world complications, including lack of reciprocity in the transmit and receive electronics and so great care must be exercised to arrive at reliable transmit channel estimates.

If reciprocity is truly applicable, the reverse SIMO channel will be the same as the forward MISO channel. Likewise, the reverse MIMO channel will be the same as the MIMO forward channel.

Exploiting channel invariances

Above, we saw how reciprocity is in general a very poor leverage for estimating the transmit channel. We now discuss how, if we focus on the spatial dimension of the channel alone, we may be able to exploit certain invariances to improve channel estimates. We assume a frequency flat channel in the following.

Application to TDD   We saw that in the presence of the pin-hole effect in Chapter 2, channels become degenerate and from Eq. (2.36) we can write (we use composite channel **h** instead of **p** and a vector notation)

$$\mathbf{h}(t) = \hat{h}(t)\mathbf{h}, \tag{3.83}$$

where **h** is a spatial vector channel and is time invariant. In a TDD system with a degenerate channel, the spatial components **h** of forward and reverse channels are identical even if $\delta_t \gg T_C$, i.e.,

$$\mathbf{h}_f(t) = \widehat{h}(t_f)\mathbf{h}, \tag{3.84}$$
$$\mathbf{h}_r(t) = \widehat{h}(t_r)\mathbf{h}. \tag{3.85}$$

Therefore, the transmit channel can be estimated within a complex scalar ambiguity caused by $\widehat{h}(t_f) \neq \widehat{h}(t_r)$.

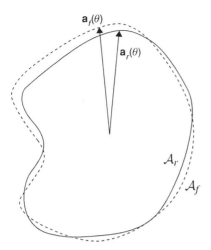

Figure 3.11: Compact aperture, the array manifolds of the forward and reverse links in FDD are closely aligned.

Application to FDD   If we restrict the channel to have a single planar (negligible angle spread, i.e., no scatterers) wavefront arriving at the base-station antenna array, we can extend the invariance principle to a FDD system. From the principle of reciprocity the AOA on the reverse link path is the AOD on the forward link, i.e., if $\mathbf{h}_f$ and $\mathbf{h}_r$ are the spatial channels on the forward and reverse links then

$$\mathbf{h}_f = \alpha_f \mathbf{a}_f(\theta), \tag{3.86}$$

$$\mathbf{h}_r = \alpha_r \mathbf{a}_r(\theta), \tag{3.87}$$

where θ is the AOA/AOD, α_f and α_r are uncorrelated complex scalars and $\mathbf{a}_f(\theta) \in \mathcal{A}_f$ is the array response vector corresponding to θ on the array manifold $\mathcal{A}_f$ for the forward channel with frequency f_f. Likewise, $\mathbf{a}_r(\theta) \in \mathcal{A}_r$ is the array response vector corresponding to θ on the array manifold $\mathcal{A}_r$ for the reverse channel with frequency f_r. Once θ is estimated from the reverse link data, we can determine $\mathbf{a}_f(\theta)$ and $\mathbf{a}_r(\theta)$, assuming $\mathcal{A}_f$ and $\mathcal{A}_r$ are known. For small, compact arrays (aperture $\approx 2\lambda_c$), and when $\delta_f < 0.05 v_c$, we can assume $\mathcal{A}_f$ and $\mathcal{A}_r$ are very similar (see Fig. 3.11). In that case $\mathbf{a}_r(\theta) \approx \mathbf{a}_f(\theta)$ and hence

$$\mathbf{h}_f \approx \mathbf{h}_r, \tag{3.88}$$

within a complex scalar (ambiguity is caused by $\alpha_f \neq \alpha_r$). This technique can also be extended to a multiple wavefront case. In this situation

$$\mathbf{h}_f(t) = \sum_i \alpha_{f,i}(t) \mathbf{a}_f(\theta_i),$$

$$\mathbf{h}_r(t) = \sum_i \alpha_{r,i}(t) \mathbf{a}_r(\theta_i), \tag{3.89}$$

where $\alpha_{f,i}(t)$ and $\alpha_{r,i}(t)$ are the uncorrelated path gains.

The reverse and forward array response vectors for any given path are approximately the same. If the number of paths is less than the number of antennas, as $\mathbf{h}_f(t)$ and $\mathbf{h}_r(t)$ evolve in time, we get

$$\text{span}\{\mathbf{h}_f(t)\} \approx \text{span}\{\mathbf{h}_r(t)\}, \tag{3.90}$$

and can infer that the sub-spaces spanned by the forward and reverse channels are approximately the same. In other words, the reverse and forward channel responses lie in the same approximate sub-space. The potential for exploiting such invariance is, however, limited since if the number of paths exceeds the number of antennas, the sub-space structure is no longer valid.

Structured ST channel estimation

The invariance in channel structure or statistics can be exploited to improve channel estimation at the receiver. We discuss one example. A reasonable assumption in many cases is a specular multipath environment, where the paths (say N_P) arrive at the receiver at discrete time delays and where the arrival time of these paths is the same at each of the M_R receive antennas. Consider a SIMO channel $\mathbf{H}$ defined in Eq. (3.61). If $N_P \leq M_R$, it can be shown that the row span of $\mathbf{H}$ is of dimension N_P (i.e., $\mathbf{H}$ loses row rank) and, moreover, the row sub-space remains invariant over several coherence intervals. This invariance can be extended to MISO and MIMO channels.

Likewise, assuming planar wavefront arrivals (likely at the base stations) and $N_P \leq L$, the column span of $\mathbf{H}$ is of dimension N_P (i.e., $\mathbf{H}$ loses column rank) and, moreover, the column sub-space remains invariant over several coherence intervals. This invariance can also be extended to MISO and MIMO channels.

The above invariances can be exploited as follows. The row and column sub-space of the channel estimated in the recent past remains unchanged at the current time and can be used to improve channel estimation by forcing appropriate sub-space constraints [Ng, 1998]. Also, these structures can be used to generalize transmit–receive channel relationships in Eq. (3.90) to delay spread channels.

4 Capacity of ST channels

4.1 Introduction

In this chapter we study the fundamental limit on the spectral efficiency that can be supported reliably in ST wireless channels. The maximum error-free data rate that a channel can support is called the channel capacity. The channel capacity for additive white Gaussian noise (AWGN) channels was first derived by Claude Shannon in 1948 in his celebrated paper "A mathematical theory of communication" [Shannon, 1948]. In contrast to scalar AWGN channels, ST channels exhibit fading and encompass a spatial dimension. The capacity results for ST channels have been developed only in the past few years. We summarize the key concepts in this chapter.

We discuss the capacity of ST channels for several different cases: channel known and channel unknown to the transmitter (perfect channel knowledge at the receiver is always assumed), deterministic and random fading channels, frequency flat and frequency selective channels, IID Gaussian and extended channels. A few important asymptotic results are also discussed.

4.2 Capacity of the frequency flat deterministic MIMO channel

We study the capacity of a MIMO channel and note that SIMO and MISO channels are sub-sets of the MIMO case. Consider a MIMO channel with M_T transmit antennas and M_R receive antennas. We assume that the channel has a bandwidth of 1 Hz and is frequency flat over this band (we extend our discussion of MIMO capacity to the case of frequency selectivity in Section 4.7). Denoting the $M_R \times M_T$ channel transfer matrix by $\mathbf{H}$, the input–output relation for the MIMO channel as derived in Chapter 3 is given by

$$\mathbf{y} = \sqrt{\frac{E_s}{M_T}} \mathbf{H}\mathbf{s} + \mathbf{n}, \tag{4.1}$$

where $\mathbf{y}$ is the $M_R \times 1$ received signal vector, $\mathbf{s}$ is the $M_T \times 1$ transmitted signal vector, $\mathbf{n}$ is the ZMCSCG noise with covariance matrix $\mathcal{E}\{\mathbf{nn}^H\} = N_o \mathbf{I}_{M_R}$ and E_s is the total average energy available at the transmitter over a symbol period (this equals the total average transmit power since the symbol period is 1 s). The covariance matrix of $\mathbf{s}$, $\mathbf{R}_{ss} = \mathcal{E}\{\mathbf{ss}^H\}$, ($\mathbf{s}$ is assumed to have zero mean) must satisfy[1] $\mathrm{Tr}(\mathbf{R}_{ss}) = M_T$ in order to constrain the total average energy transmitted over a symbol period.

In the following, we assume that the channel $\mathbf{H}$ is known to the receiver. Channel knowledge at the receiver is maintained via training and tracking. Although the channel $\mathbf{H}$ is random or stochastic, we first study the capacity of a sample realization of the channel, i.e., $\mathbf{H}$ is deterministic. The capacity of the MIMO channel is defined as [Foschini, 1996; Telatar, 1999a]

$$C = \max_{f(\mathbf{s})} I(\mathbf{s}; \mathbf{y}), \tag{4.2}$$

where $f(\mathbf{s})$ is the probability distribution of the vector $\mathbf{s}$, and $I(\mathbf{s}; \mathbf{y})$ is the mutual information between vectors $\mathbf{s}$ and $\mathbf{y}$. Note that

$$I(\mathbf{s}; \mathbf{y}) = H(\mathbf{y}) - H(\mathbf{y}|\mathbf{s}), \tag{4.3}$$

where $H(\mathbf{y})$ is the differential entropy of the vector $\mathbf{y}$, while $H(\mathbf{y}|\mathbf{s})$ is the conditional differential entropy of the vector $\mathbf{y}$, given knowledge of the vector $\mathbf{s}$. Since the vectors $\mathbf{s}$ and $\mathbf{n}$ are independent, $H(\mathbf{y}|\mathbf{s}) = H(\mathbf{n})$, Eq. (4.3) simplifies to

$$I(\mathbf{s}; \mathbf{y}) = H(\mathbf{y}) - H(\mathbf{n}). \tag{4.4}$$

Maximizing the mutual information $I(\mathbf{s}; \mathbf{y})$ reduces to maximizing $H(\mathbf{y})$. The covariance matrix of $\mathbf{y}$, $\mathbf{R}_{yy} = \mathcal{E}\{\mathbf{yy}^H\}$, satisfies

$$\mathbf{R}_{yy} = \frac{E_s}{M_T} \mathbf{H} \mathbf{R}_{ss} \mathbf{H}^H + N_o \mathbf{I}_{M_R}, \tag{4.5}$$

where $\mathbf{R}_{ss} = \mathcal{E}\{\mathbf{ss}^H\}$ is the covariance matrix of $\mathbf{s}$. We know that amongst all vectors $\mathbf{y}$ with a given covariance matrix $\mathbf{R}_{yy}$, the differential entropy $H(\mathbf{y})$ is maximized when $\mathbf{y}$ is ZMCSCG [Neeser and Massey, 1993]. This in turn implies that $\mathbf{s}$ must be a ZMCSCG vector, the distribution of which is completely characterized by $\mathbf{R}_{ss}$. The differential entropies of the vectors $\mathbf{y}$ and $\mathbf{n}$ are given by

$$H(\mathbf{y}) = \log_2(\det(\pi e \mathbf{R}_{yy})) \text{ bps/Hz}, \tag{4.6}$$

$$H(\mathbf{n}) = \log_2(\det(\pi e N_o \mathbf{I}_{M_R})) \text{ bps/Hz}. \tag{4.7}$$

Therefore, $I(\mathbf{s}; \mathbf{y})$ in Eq. (4.4) reduces to [Telatar, 1999a]

$$I(\mathbf{s}; \mathbf{y}) = \log_2 \det\left(\mathbf{I}_{M_R} + \frac{E_s}{M_T N_o} \mathbf{H} \mathbf{R}_{ss} \mathbf{H}^H\right) \text{ bps/Hz}, \tag{4.8}$$

---

[1] We have a different normalization on $\mathbf{s}$ compared with the signal model in Section 3.7.5.

and it follows from Eq. (4.2) that the capacity of the MIMO channel is given by

$$C = \max_{\text{Tr}(\mathbf{R}_{ss})=M_T} \log_2 \det \left(\mathbf{I}_{M_R} + \frac{E_s}{M_T N_o} \mathbf{H} \mathbf{R}_{ss} \mathbf{H}^H \right) \quad \text{bps/Hz}. \tag{4.9}$$

The capacity C in Eq. (4.9) is also often referred to as the error-free spectral efficiency, or the data rate per unit bandwidth that can be sustained reliably over the MIMO link. Thus given a bandwidth of W Hz, the maximum achievable data rate over this bandwidth using the MIMO channel is simply WC bps.

4.3   Channel unknown to the transmitter

If the channel has no preferred direction and is completely unknown to the transmitter, the vector $\mathbf{s}$ may be chosen to be statistically non-preferential, i.e., $\mathbf{R}_{ss} = \mathbf{I}_{M_T}$. This implies that the signals are independent and equi-powered at the transmit antennas. The capacity of the MIMO channel in the absence of channel knowledge at the transmitter is given by

$$C = \log_2 \det \left(\mathbf{I}_{M_R} + \frac{E_s}{M_T N_o} \mathbf{H} \mathbf{H}^H \right). \tag{4.10}$$

This is not the Shannon capacity in the true sense, since a genie with channel knowledge can choose a signal covariance matrix that outperforms $\mathbf{R}_{ss} = \mathbf{I}_{M_T}$. Nevertheless, we shall refer to the expression in Eq. (4.10) as the capacity. Given that $\mathbf{H}\mathbf{H}^H = \mathbf{Q}\mathbf{\Lambda}\mathbf{Q}^H$ (see Chapter 3), the capacity of the MIMO channel can be expressed as

$$C = \log_2 \det \left(\mathbf{I}_{M_R} + \frac{E_s}{M_T N_o} \mathbf{Q}\mathbf{\Lambda}\mathbf{Q}^H \right). \tag{4.11}$$

Using the identity $\det(\mathbf{I}_m + \mathbf{AB}) = \det(\mathbf{I}_n + \mathbf{BA})$ for matrices $\mathbf{A}$ $(m \times n)$ and $\mathbf{B}$ $(n \times m)$ and $\mathbf{Q}^H \mathbf{Q} = \mathbf{I}_{M_R}$, Eq. (4.11) simplifies to

$$C = \log_2 \det \left(\mathbf{I}_{M_R} + \frac{E_s}{M_T N_o} \mathbf{\Lambda} \right), \tag{4.12}$$

or equivalently

$$C = \sum_{i=1}^{r} \log_2 \left(1 + \frac{E_s}{M_T N_o} \lambda_i \right), \tag{4.13}$$

where r is the rank of the channel and λ_i $(i = 1, 2, \ldots, r)$ are the positive eigenvalues of $\mathbf{H}\mathbf{H}^H$. Equation (4.13) expresses the capacity of the MIMO channel as the sum of the capacities of r SISO channels, each having power gain λ_i $(i = 1, \ldots, r)$ and transmit power E_s/M_T.

Hence, the use of multiple antennas at the transmitter and receiver in a wireless link opens multiple scalar spatial data pipes (also known as modes) between transmitter and receiver. We note that in the absence of channel knowledge the individual channel modes are not accessible and that equal transmit energy is allocated to each spatial data pipe. Later, when we consider the case of channel knowledge at the transmitter, we shall see how the individual spatial data pipes may be accessed.

Orthogonal channels maximize capacity

Given a fixed total channel power transfer, i.e., $\|\mathbf{H}\|_F^2 = \sum_{i=1}^r \lambda_i = \zeta$, what is the nature of the channel $\mathbf{H}$ that maximizes capacity?

Consider a full-rank MIMO channel with $M_T = M_R = M$, so that $r = M$. The capacity C in Eq. (4.13) is concave in the variables $\lambda_i (i = 1, \ldots, M)$ and is maximized subject to the constraint $\sum_{i=1}^M \lambda_i = \zeta$, when $\lambda_i = \lambda_j = \zeta/M$ $(i, j = 1, 2, \ldots, M)$. Therefore, for maximum capacity, $\mathbf{H}$ must be an orthogonal matrix, i.e., $\mathbf{H}\mathbf{H}^H = \mathbf{H}^H\mathbf{H} = (\zeta/M)\mathbf{I}_M$ and the resulting capacity is

$$C = M \log_2 \left(1 + \frac{\zeta E_s}{N_o M^2} \right). \tag{4.14}$$

Further, if the elements of $\mathbf{H}$ satisfy $\|\mathbf{H}_{i,j}\|^2 = 1$, then $\|\mathbf{H}\|_F^2 = M^2$ and

$$C = M \log_2 \left(1 + \frac{E_s}{N_o} \right). \tag{4.15}$$

The capacity of an orthogonal MIMO channel is therefore M times the scalar channel capacity.

4.4   Channel known to the transmitter

So far we have studied the capacity of MIMO channels when the channel is known perfectly to the receiver and is unknown to the transmitter. As we have seen, equal power allocation across the transmit antenna array is logical under this scenario. We now ask if we can increase channel capacity if the channel is also known to the transmitter.

Channel knowledge at the transmitter can be maintained via feedback from the receiver or through the reciprocity principle in a duplex system. See Section 3.9 for a discussion.

From Eq. (4.13) we conclude that in absence of channel knowledge at the transmitter, the capacity of the $M_R \times M_T$ MIMO channel is equivalent to the capacity of r parallel spatial sub-channels, with equal power allocated to each sub-channel. When the channel is known at both the transmitter and receiver, the individual channel modes may be accessed through linear processing at the transmitter and receiver (see [Foschini, 1996; Telatar, 1999a; Lozano and Papadias, 2002] for more details).

Consider a ZMCSCG signal vector $\tilde{\mathbf{s}}$ of dimension $r \times 1$ where r is the rank of the channel $\mathbf{H}$ to be transmitted. The vector is multiplied (see Fig. 4.1) by the matrix $\mathbf{V}$ prior

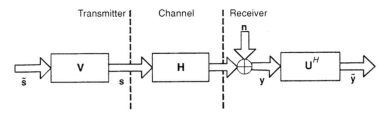

Figure 4.1: Schematic of modal decomposition of **H** when the channel is known to the transmitter and receiver.

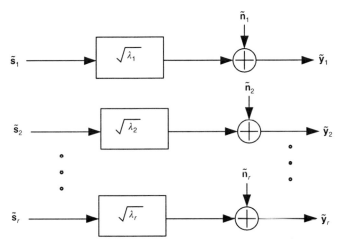

Figure 4.2: Schematic of modal decomposition of **H** when the channel is known to the transmitter and receiver.

to transmission (recall from Chapter 3 that $\mathbf{H} = \mathbf{U}\boldsymbol{\Sigma}\mathbf{V}^H$). At the receiver, the received signal vector $\mathbf{y}$ is multiplied by the matrix $\mathbf{U}^H$. The effective input–output relation for this system is given by

$$\widetilde{\mathbf{y}} = \sqrt{\frac{E_s}{M_T}}\mathbf{U}^H\mathbf{H}\mathbf{V}\widetilde{\mathbf{s}} + \mathbf{U}^H\mathbf{n}$$

$$= \sqrt{\frac{E_s}{M_T}}\boldsymbol{\Sigma}\widetilde{\mathbf{s}} + \widetilde{\mathbf{n}}, \tag{4.16}$$

where $\widetilde{\mathbf{y}}$ is the transformed received signal vector of dimension $r \times 1$ and $\widetilde{\mathbf{n}}$ is the ZMCSCG $r \times 1$ transformed noise vector with covariance matrix $\mathcal{E}\{\widetilde{\mathbf{n}}\widetilde{\mathbf{n}}^H\} = N_o\mathbf{I}_r$. The vector $\widetilde{\mathbf{s}}$ must satisfy $\mathcal{E}\{\widetilde{\mathbf{s}}\widetilde{\mathbf{s}}^H\} = M_T$ to constrain the total transmit energy. Equation (4.16) shows that with channel knowledge at the transmitter, **H** can be explicitly decomposed (see Fig. 4.2) into r parallel SISO channels satisfying

$$\widetilde{y}_i = \sqrt{\frac{E_s}{M_T}}\sqrt{\lambda_i}\widetilde{s}_i + \widetilde{n}_i, \quad i = 1, 2, \ldots, r. \tag{4.17}$$

The capacity of the MIMO channel is the sum of the individual parallel SISO channel capacities and is given by

$$C = \sum_{i=1}^{r} \log_2 \left(1 + \frac{E_s \gamma_i}{M_T N_o} \lambda_i \right), \tag{4.18}$$

where $\gamma_i = \mathcal{E}\{|\tilde{s}_i|^2\}$ $(i = 1, 2, \ldots, r)$ reflects the transmit energy in the ith sub-channel and satisfies $\sum_{i=1}^{r} \gamma_i = M_T$.

Since the transmitter can access the spatial sub-channels, it can allocate variable energy across the sub-channels to maximize the mutual information. The mutual information maximization problem now becomes

$$C = \max_{\sum_{i=1}^{r} \gamma_i = M_T} \sum_{i=1}^{r} \log_2 \left(1 + \frac{E_s \gamma_i}{M_T N_o} \lambda_i \right). \tag{4.19}$$

The objective for the maximization is concave in the variables γ_i $(i = 1, \ldots, r)$ and can be maximized using Lagrangian methods. The optimal energy allocation policy, γ_i^{opt}, satisfies

$$\gamma_i^{opt} = \left(\mu - \frac{M_T N_o}{E_s \lambda_i} \right)_+, \quad i = 1, \ldots, r, \tag{4.20}$$

$$\sum_{i=1}^{r} \gamma_i^{opt} = M_T, \tag{4.21}$$

where μ is a constant and $(x)_+$ implies

$$(x)_+ = \begin{cases} x & \text{if } x \geq 0 \\ 0 & \text{if } x < 0 \end{cases}. \tag{4.22}$$

The optimal energy allocation is found iteratively through the "waterpouring algorithm" [Cover and Thomas, 1991; Chuah $et\ al.$, 1998, 2002; Telatar, 1999a] which is described briefly below.

Waterpouring algorithm

Setting the iteration count p to 1, we first calculate the constant μ in Eq. (4.20):

$$\mu = \frac{M_T}{(r - p + 1)} \left[1 + \frac{N_o}{E_s} \sum_{i=1}^{r-p+1} \frac{1}{\lambda_i} \right]. \tag{4.23}$$

Using the value of μ found above, the power allocated to the ith sub-channel can be calculated using

$$\gamma_i = \left(\mu - \frac{M_T N_o}{E_s \lambda_i} \right), \quad i = 1, 2, \ldots, r - p + 1. \tag{4.24}$$

If the energy allocated to the channel with the lowest gain is negative, i.e., $\gamma_{r-p+1} < 0$, we discard this channel by setting $\gamma_{r-p+1}^{opt} = 0$ and rerun the algorithm with the iteration

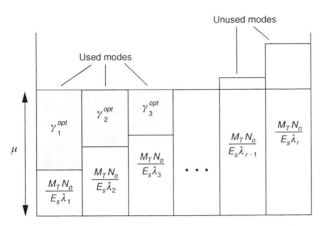

Figure 4.3: Schematic of the waterpouring algorithm. γ_i^{opt} is the optimal energy allocated to the ith spatial sub-channel and $\gamma_i^{opt} = (\mu - M_T N_o / E_s \lambda_i)_+$.

count p incremented by 1. The optimal waterpouring power allocation strategy is found when power allocated to each spatial sub-channel is non-negative. Figure 4.3 summarizes pictorially the outcome of the waterpouring algorithm.

The capacity of the MIMO channel when the channel is known to the transmitter is necessarily greater than (or equal to) the capacity when the channel is unknown to the transmitter.

Channel capacity when the channel is unknown to both the transmitter and receiver is an area of ongoing research [Marzetta and Hochwald, 1999; Zheng and Tse, 2002; Hassibi and Marzetta, 2002] and is not discussed in this book.

Optimal $\mathbf{R}_{ss}$

Once the optimal power allocation across the spatial sub-channels is determined, we can determine the optimal $\mathbf{R}_{ss}$ sought in Eq. (4.9). Noting from Fig. 4.1 that

$$\mathbf{s} = \mathbf{V}\widetilde{\mathbf{s}}, \tag{4.25}$$

the optimal covariance matrix $\mathbf{R}_{ss}^{opt}$ is given by

$$\mathbf{R}_{ss}^{opt} = \mathbf{V}\mathbf{R}_{\widetilde{ss}}^{opt}\mathbf{V}^H, \tag{4.26}$$

where $\mathbf{R}_{\widetilde{ss}}^{opt}$ is an $r \times r$ diagonal matrix (since the elements of $\widetilde{\mathbf{s}}$ are independent) given by

$$\mathbf{R}_{\widetilde{ss}}^{opt} = \text{diag}\{\gamma_1^{opt}, \gamma_2^{opt}, \ldots, \gamma_r^{opt}\}. \tag{4.27}$$

Referring back to Fig. 4.2, we can maximize capacity when we allocate power in each mode according to $\mathbf{R}_{\widetilde{ss}}^{opt}$.

4.4.1   Capacities of SIMO and MISO channels

The capacities of SIMO and MISO channels are special cases of MIMO channel capacity and can be evaluated using Eq. (4.13) or Eq. (4.19) depending on whether channel knowledge is available at the transmitter.

SIMO channel capacity

Consider a SIMO channel $\mathbf{h}$, with M_R receive antennas ($M_T = 1$). For this channel, $r = 1$ and $\lambda_1 = \|\mathbf{h}\|_F^2$. Thus, the capacity of the SIMO channel, C_{SIMO}, when the channel is unknown to the transmitter, is given by

$$C_{SIMO} = \log_2 \left(1 + \frac{E_s}{N_o} \|\mathbf{h}\|_F^2 \right). \tag{4.28}$$

The SIMO channel comprises only one spatial data pipe. If we assume $\mathbf{h}$ satisfies $|h_i|^2 = 1$ ($i = 1, 2, \ldots, M_R$), then $\|\mathbf{h}\|_F^2 = M_R$ and $C_{SIMO} = \log_2(1 + (E_s/N_o)M_R)$. Thus, the addition of receive antennas yields only a logarithmic increase in capacity in SIMO channels. Furthermore, we note that knowledge of the channel at the transmitter for SIMO channels provides no capacity benefit.

MISO channel capacity

Consider a MISO channel with M_T transmit antennas ($M_R = 1$). Representing the $1 \times M_T$ channel vector by $\mathbf{h}$, we note that $r = 1$ and $\lambda_1 = \|\mathbf{h}\|_F^2$. The capacity of the MISO channel in the absence of channel knowledge at the transmitter, C_{MISO}, is given by

$$C_{MISO} = \log_2 \left(1 + \frac{E_s}{M_T N_o} \|\mathbf{h}\|_F^2 \right). \tag{4.29}$$

If $|h_i|^2 = 1$ ($i = 1, 2, \ldots, M_T$), then $C = \log_2(1 + E_s/N_o)$ and there is no improvement in capacity over a SISO channel. Like in Section 4.4, since the channel is unknown to the transmitter, we assume $\mathbf{R}_{ss} = \mathbf{I}_{M_T}$. However, as we shall see in the following section, the capacity of MISO channels in a fading environment ($\mathbf{h} = \mathbf{h}_w$) will be superior to the capacity of a SISO channel. Comparing Eqs. (4.28) and (4.29) it is clear that $C_{MISO} < C_{SIMO}$ when the channel is unknown to the transmitter for the same $\|\mathbf{h}\|_F^2$. This is because of the inability of the transmitter in MISO channels to exploit transmit array gain when the channel is not known to it. When the channel is known to the transmitter, all power can be directed into the single spatial mode. From Eq. (4.19) it follows that the capacity of the MISO channel when the channel is known to the transmitter is given by

$$C_{MISO} = \log_2 \left(1 + \frac{E_s}{N_o} \|\mathbf{h}\|_F^2 \right). \tag{4.30}$$

Therefore, the capacity of a MISO channel equals the capacity of a SIMO channel when the channel is known to the transmitter for the same $\|\mathbf{h}\|_F^2$. As in SIMO channels, MISO channels offer only a logarithmic increase in capacity with the number of antennas.

4.5   Capacity of random MIMO channels

So far we have restricted our discussion of MIMO capacity to the case of a sample deterministic channel realization. In this section we consider the capacity of random MIMO channels. For now, we assume the elements of $\mathbf{H}$, $h_{i,j}$ ($i = 1, \ldots, M_R$, $j = 1, \ldots, M_T$) are normalized so that $\mathcal{E}\{|h_{i,j}|\}^2 = 1$. Each realization of the fading channel has a maximum information rate associated with it as per Eqs. (4.13) and (4.19), depending on whether the channel is known or unknown to the transmitter.

The term E_s/N_o in Eqs. (4.13) and (4.19) which we shall refer to henceforth as ρ may be interpreted as the average SNR at each of the receive antennas, as demonstrated by the following. The signal received at the ith receive antenna, y_i, is given by

$$y_i = \sqrt{\frac{E_s}{M_T}}\mathbf{h}_i \mathbf{s} + n_i, \qquad (4.31)$$

where $\mathbf{h}_i$, a vector of dimension $1 \times M_T$, represents the ith row of $\mathbf{H}$ and n_i is the ith element of $\mathbf{n}$. Since $\mathcal{E}\{|h_{i,j}|^2\} = 1$ and $\text{Tr}(\mathbf{R_{ss}}) = M_T$, it follows that $\mathcal{E}\{|y_i|^2\} = E_s + N_o$ and hence the average received SNR at the ith receive antenna is $\rho = E_s/N_o$. It follows that $E_s/M_T N_o$ in Eq. (4.9) may be interpreted as the average SNR at any receive antenna contributed by a single transmit antenna (assuming equal power allocation). Clearly, if $M_T = M_R = 1$, ρ is the average SNR at the receiver.

4.5.1   Capacity of H_w channels for large M

Consider a MIMO channel with $M_T = M_R = M$ and $\mathbf{H} = \mathbf{H}_w$. Using the strong law of large numbers [Papoulis, 1984; Leon-Garcia, 1994], we see that

$$\frac{1}{M}\mathbf{H}_w\mathbf{H}_w^H \to \mathbf{I}_M \quad \text{as} \quad M \to \infty. \qquad (4.32)$$

Hence, the capacity of this channel in the absence of channel knowledge at the transmitter approaches

$$C \to M \log_2(1 + \rho). \qquad (4.33)$$

Thus, asymptotically (in M) the capacity of the spatially white MIMO channel becomes deterministic and increases linearly with M for a fixed SNR. Also, for every 3 dB increase in SNR we get M bps/Hz increase in capacity for a MIMO channel (compared with 1 bps/Hz for a SISO channel).

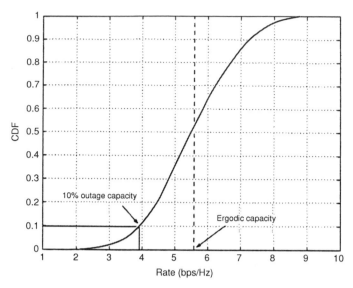

Figure 4.4: CDF of information rate for the $\mathbf{H}_w$ MIMO channel with $M_T = M_R = 2$ and a SNR of 10 dB.

4.5.2   Statistical characterization of the information rate

Since the channel $\mathbf{H}$ is random, the information rate associated with the MIMO channel is a random variable. Figure 4.4 shows the cumulative distribution function (CDF) of the information rate of a flat fading MIMO channel with $M_T = M_R = 2$, obtained through Monte Carlo methods for $\rho = 10$ dB when the channel is unknown to the transmitter and the sample channels are drawn from $\mathbf{H}_w$ (we therefore have $\mathbf{R}_{ss} = \mathbf{I}_2$). In analyzing the capacity of fading channels two commonly used statistics are the ergodic capacity and the outage capacity [Biglieri *et al.*, 1998; Telatar, 1999a]. We shall discuss the physical significance of these quantities in the ensuing discussion.

Ergodic capacity

The ergodic capacity $\overline{C}$ of a MIMO channel is the ensemble average of the information rate over the distribution of the elements of the channel matrix $\mathbf{H}$. The ergodic capacity gains practical significance when for every channel use, the channel is drawn from an independent realization (ergodic channel). The significance of the ergodic capacity is that in an ergodic channel, we can signal at the rate defined by ergodic capacity with vanishing error assuming we use asymptotically optimal codebooks. In this sense ergodic capacity is the Shannon capacity of the channel. When the channel is unknown to the transmitter, the ergodic capacity, $\overline{C}$, is given by

$$\overline{C} = \mathcal{E} \left\{ \sum_{i=1}^{r} \log_2 \left(1 + \frac{\rho}{M_T} \lambda_i \right) \right\}. \tag{4.34}$$

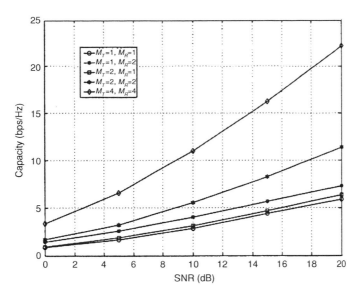

Figure 4.5: Ergodic capacity for different antenna configurations. Note that the SIMO channel has a higher ergodic capacity than the MISO channel.

The ergodic capacity of the channel as depicted in Fig. 4.4 is 5.7 bps/Hz. Note that the ergodic capacity is the mean information rate and is not necessarily equal to the median information rate (50 percentile). The median information rate in Fig. 4.4 is approximately 5.5 bps/Hz. Figure 4.5 shows the ergodic capacity of several MIMO configurations as a function of ρ. As expected, the ergodic capacity increases with increasing ρ and also with M_T and M_R. We note that the ergodic capacity of a SIMO ($M \times 1$) channel will be greater than the ergodic capacity of a MISO ($1 \times M$) channel when the channel is unknown to the transmitter. This follows from Eqs. (4.28) and (4.29) for the capacity of deterministic SIMO and MISO channels respectively, where SIMO outperforms MISO.

Ergodic capacity when the channel is known to the transmitter is the ensemble average of the capacity achieved when the waterpouring optimization is performed for each realization of **H**, and is given by

$$\overline{C} = \mathcal{E} \left\{ \sum_{i=1}^{r} \log_2 \left(1 + \frac{\rho}{M_T} \gamma_i^{opt} \lambda_i \right) \right\}. \tag{4.35}$$

Figure 4.6 compares the ergodic capacity of a MIMO channel with $M_R = M_T = 4$ and $\mathbf{H} = \mathbf{H}_w$, for the cases where the channel is known and unknown to the transmitter. As expected, the ergodic capacity when the channel is known to the transmitter is always higher than the ergodic capacity when the channel is unknown. This advantage reduces at higher SNR and is true for $\mathbf{H}_w$ MIMO channels with $M_T = M_R = M$. This can be explained as follows. At high SNR, from Eq. (4.9), the capacity of the MIMO

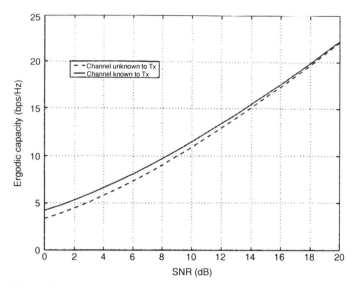

Figure 4.6: Ergodic capacity of a $M_T = M_R = 4$ MIMO channel with and without channel knowledge at the transmitter. The difference in ergodic capacity decreases with SNR.

channel may be approximated as

$$C \approx \max_{\mathrm{Tr}(\mathbf{R}_{ss})=M} \log_2 \det\left(\mathbf{R}_{ss}\right) + \log_2 \det\left(\frac{\rho}{M}\mathbf{H}_w\mathbf{H}_w^H\right). \tag{4.36}$$

With the channel known, $\det\left(\mathbf{R}_{ss}\right)$ is maximized, subject to the constraint $\mathrm{Tr}\left(\mathbf{R}_{ss}\right) = M$ when $\mathbf{R}_{ss} = \mathbf{I}_M$, which is also the optimal covariance matrix when the channel is unknown at the transmitter. Hence, the waterpouring solution will approach the uniform power allocation strategy at high SNR for the $\mathbf{H}_w$ MIMO channel with $M_T = M_R = M$. The capacity gap between channel known and unknown increases at higher M, but will reduce at high enough SNR.

Using multivariate statistics, it is possible to derive a lower bound on the ergodic capacity of $\mathbf{H}_w$ MIMO channels. We omit the proof of this bound. The interested reader is referred to [Oyman et al., 2002b] for further details. The lower bound on the ergodic capacity of MIMO channels is given by

$$\overline{C} \geq M \log_2\left(1 + \frac{\rho}{M_T}\exp\left(\frac{1}{M}\sum_{j=1}^{M}\sum_{p=1}^{N-j}\frac{1}{p} - \gamma\right)\right), \tag{4.37}$$

where $\gamma \approx 0.577\,215\,66$ is Euler's constant, $M = \min(M_T, M_R)$ and $N = \max(M_T, M_R)$. Therefore, at high SNR, ergodic capacity increases by M bps/Hz for every 3 dB increase in SNR. Figure 4.7 compares the ergodic capacity of a MIMO channel with $M_T = M_R = 2$ with the lower bound in Eq. (4.37), and demonstrates that the lower bound is tight at low SNR and tightens further at high SNR. These observations also hold true for higher antenna configurations. Alternative analytic expressions for the ergodic

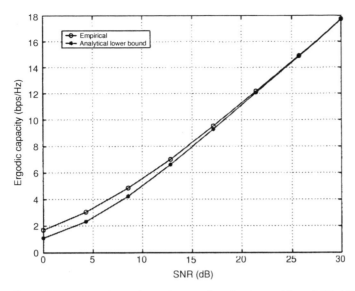

Figure 4.7: Comparison of ergodic capacity of a $M_T = M_R = 2$ **H**$_w$ MIMO channel with the lower bound.

capacity of MIMO channels can be found in [Telatar, 1999b; Gauthier *et al.*, 2000; Grant, 2002; Scaglione, 2002; Martin and Ottersten, 2002].

Outage capacity

Outage analysis quantifies the level of performance (in this case capacity) that is guaranteed with a certain level of reliability. We define the $q\%$ outage capacity $C_{out,q}$ as the information rate that is guaranteed for $(100 - q)\%$ of the channel realizations, i.e., $P(C \leq C_{out,q}) = q\%$ [Ozarow *et al.*, 1994; Biglieri *et al.*, 1998]. The 10% outage capacity for the **H**$_w$ MIMO channel with $M_T = M_R = 2$ is given in Fig. 4.4. Figure 4.8 shows the 10% outage capacity for several MIMO configurations when the channel is unknown to the transmitter. As in the case of ergodic capacity, we see that the outage capacity increases with SNR and is higher for larger antenna configurations. The latter effect arises from the observation in Section 4.5.1 that, as the number of antennas grows for an **H**$_w$ channel, the capacity shows less variability. Therefore, the 10% outage capacity approaches the median information rate. However, for smaller number of antennas, M, the 10% outage capacity is much smaller than the median information rate. Outage capacity is a useful characterization when the channel is unknown to the transmitter and **H** is random but held constant for each use of the channel. In this case, for any rate there is a non-zero probability that the given realization does not support it, resulting in packet error and giving rise to an interesting tradeoff between rate and outage probability. We shall explore this aspect in greater detail in Chapter 11, where performance limits of MIMO channels are discussed.

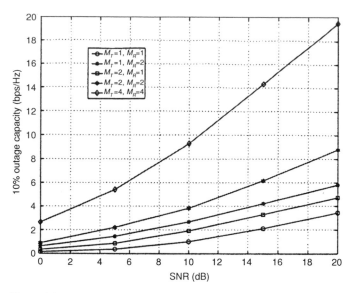

Figure 4.8: 10% outage capacity for different antenna configurations. Outage capacity improves with larger antenna configurations.

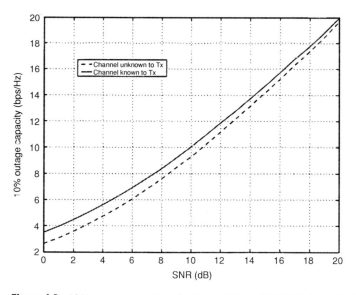

Figure 4.9: 10% outage capacity of a $M_T = M_R = 4$ MIMO channel with and without channel knowledge at the transmitter.

Outage capacity may also be defined when the channel is known to the transmitter. Figure 4.9 compares the 10% outage capacity for an $\mathbf{H}_w$ MIMO channel with $M_T = M_R = 4$, with and without channel knowledge at the transmitter, and shows that channel knowledge at the transmitter improves outage capacity. The gap is expected to increase at lower outage rates or with higher antenna configurations.

4.6 Influence of Ricean fading, fading correlation, XPD and degeneracy on MIMO capacity

The $\mathbf{H}_w$ channel results in a rich scattering environment with sufficient antenna spacing at transmitter and receiver. In practice, however, the $\mathbf{H}_w$ assumption may not be true for several reasons: insufficient scattering or spacing between antennas causing the fading to be correlated, the use of polarized antennas which leads to gain imbalances between the elements of $\mathbf{H}$, or the presence of a LOS component that causes Ricean fading. These effects have been modeled in Chapter 3. We now study their influence on the capacity of the MIMO channel with illustrative examples. In the following, we assume that the channel is known perfectly to the receiver and is unknown to the transmitter.

4.6.1 Influence of the spatial fading correlation

As seen in Chapter 3 the effects of spatial fading correlation for a Rayleigh flat fading channel can be reasonably captured by modeling the MIMO channel $\mathbf{H}$ as

$$\mathbf{H} = \mathbf{R}_r^{1/2}\mathbf{H}_w\mathbf{R}_t^{1/2}, \tag{4.38}$$

where the matrices $\mathbf{R}_r$ and $\mathbf{R}_t$ are positive definite Hermitian matrices that specify the receive and transmit correlations respectively. Furthermore, $\mathbf{R}_r$ and $\mathbf{R}_t$ are normalized so that $[\mathbf{R}_r]_{i,i} = 1$ $(i = 1, 2, \ldots, M_R)$ and $[\mathbf{R}_t]_{j,j} = 1$ $(j = 1, 2, \ldots, M_T)$, resulting in $\mathcal{E}\{|h_{i,j}|^2\} = 1$. The capacity of the MIMO channel in the presence of the spatial fading correlation without channel knowledge at the transmitter follows from simple substitution:

$$C = \log_2 \det\left(\mathbf{I}_{M_R} + \frac{\rho}{M_T}\mathbf{R}_r^{1/2}\mathbf{H}_w\mathbf{R}_t\mathbf{H}_w^H\mathbf{R}_r^{H/2}\right). \tag{4.39}$$

Assume that $M_R = M_T = M$ and the receive and transmit correlation matrices $\mathbf{R}_r$ and $\mathbf{R}_t$ are full rank. At high SNR the capacity of the MIMO channel can be written as

$$C \approx \log_2 \det\left(\frac{\rho}{M}\mathbf{H}_w\mathbf{H}_w^H\right) + \log_2 \det(\mathbf{R}_r) + \log_2 \det(\mathbf{R}_t). \tag{4.40}$$

From Eq. (4.40) it is clear that $\mathbf{R}_r$ and $\mathbf{R}_t$ have the same impact on the capacity of the MIMO channel. We now examine the conditions on $\mathbf{R}_r$ ($\mathbf{R}_t$ will be similar) that maximize capacity. The eigenvalues of $\mathbf{R}_r$, $\lambda_i(\mathbf{R}_r)$ $(i = 1, 2, \ldots, M)$ are constrained such that $\sum_{i=1}^{M} \lambda_i(\mathbf{R}_r) = M$. The arithmetic mean–geometric mean inequality [Courant and Robbins, 1996] implies

$$\prod_{i=1}^{M} \lambda_i(\mathbf{R}_r) \leq 1. \tag{4.41}$$

However, $\det(\mathbf{R}_r) = \prod_{i=1}^{M} \lambda_i(\mathbf{R}_r)$. This implies that $\log_2 \det(\mathbf{R}_r) \leq 0$, and is zero only if all eigenvalues of $\mathbf{R}_r$ are equal, i.e., $\mathbf{R}_r = \mathbf{I}_M$. Hence, we can conclude that fading

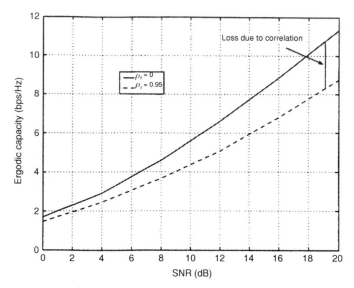

Figure 4.10: Ergodic capacity with low and high receive correlation. The loss in ergodic capacity is about 3.3 bps/Hz when $\rho_r = 0.95$.

signal correlation is detrimental to MIMO capacity and that the loss in ergodic or outage capacity at high SNR is given by $(\log_2 \det(\mathbf{R}_r) + \log_2 \det(\mathbf{R}_t))$ bps/Hz.

Figure 4.10 shows the ergodic capacity of a MIMO channel with $M_T = M_R = 2$ and variable receive correlation. We assume $\mathbf{R}_t = \mathbf{I}_2$. The receive correlation matrix is chosen according to

$$\mathbf{R}_r = \begin{bmatrix} 1 & \rho_r \\ \rho_r^* & 1 \end{bmatrix}. \tag{4.42}$$

As predicted by $\log_2 \det(\mathbf{R}_r)$, we observe a loss of 3.3 bps/Hz at high SNR for the correlated channel ($\rho_r = 0.95$), compared with the $\mathbf{H}_w$ channel ($\rho_r = 0$). Further details/results may be found in [Chuah et al., 2002]. Note that under extreme correlation when either or both of $\mathbf{R}_r$ and $\mathbf{R}_t$ become rank 1, the channel $\mathbf{H}$ also becomes rank 1. The channel now has a single spatial mode (scalar) channel and the capacity increase is logarithmic in $M_T M_R$ as against linear in $\min(M_T, M_R)$ for the $\mathbf{H}_w$ channel.

4.6.2   Influence of the LOS component

From Chapter 3 we know that the MIMO channel in the presence of Ricean fading can be modeled as the sum of a fixed (LOS) matrix and a fading matrix as follows:

$$\mathbf{H} = \sqrt{\frac{K}{1+K}}\overline{\mathbf{H}} + \sqrt{\frac{1}{1+K}}\mathbf{H}_w, \tag{4.43}$$

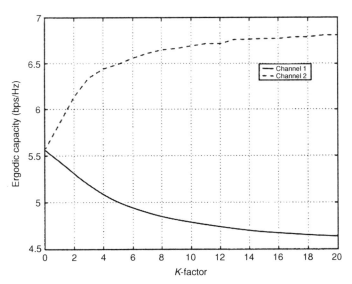

Figure 4.11: Ergodic capacity vs K-factor for a MIMO channel with $\overline{\mathbf{H}}_1$ and $\overline{\mathbf{H}}_2$ LOS components. The channel geometry has a significant impact on capacity at a high K-factor.

where $\sqrt{K/(1 + K)}\,\overline{\mathbf{H}} = \mathcal{E}\{\mathbf{H}\}$ is the fixed component of the channel and $\sqrt{1/(1 + K)}\mathbf{H}_w$ is the fading component of the channel. K is the Ricean factor of the channel and is the ratio of the total power in the fixed component of the channel to the power in the fading component. As explained in Chapter 3, $K = 0$ in the presence of pure Rayleigh fading, and the channel approaches a non-fading link as $K \to \infty$. Clearly $\mathbf{H}_w$ dominates channel behavior for low values of K, while $\overline{\mathbf{H}}$ dominates system behavior with an increasing degree of Ricean fading.

To study the influence of the K-factor on the capacity of MIMO channels, we examine two channels with different fixed components, $\overline{\mathbf{H}}_1$ and $\overline{\mathbf{H}}_2$ respectively, ignoring phase factors. Let

$$\overline{\mathbf{H}}_1 = \begin{bmatrix} 1 & 1 \\ 1 & 1 \end{bmatrix}, \tag{4.44}$$

$$\overline{\mathbf{H}}_2 = \begin{bmatrix} 1 & -1 \\ 1 & 1 \end{bmatrix}. \tag{4.45}$$

The physical context in which these two channels arise has been discussed in Chapter 3. Figure 4.11 shows the ergodic capacity for the two channels as a function of the K-factor for an SNR of 10 dB.

Clearly the $\overline{\mathbf{H}}_2$ channel outperforms the $\overline{\mathbf{H}}_1$ channel with an increasing K-factor. This follows since the $\overline{\mathbf{H}}_2$ channel is orthogonal while the $\overline{\mathbf{H}}_1$ channel is rank-deficient. Hence, the geometry of the fixed component of the channel matrix plays a critical role in channel capacity at a high K-factor. Further details may be found in [Godavarti *et al.*, 2001a,b].

4.6.3   Influence of XPD in a non-fading channel

Next, we shall briefly review the effect of XPD on MIMO capacity. Assume a very high K-factor and an $\overline{\mathbf{H}}_1$ type channel with $M_T = M_R = 2$ and cross-polarized antennas at the transmitter and receiver. This channel is no longer random and may be modeled as

$$\mathbf{H} = \begin{bmatrix} 1 & \sqrt{\alpha} \\ \sqrt{\alpha} & 1 \end{bmatrix}, \tag{4.46}$$

where $0 \le \alpha \le 1$ reflects the level of XPD. Poor XPD results in a high value of α (close to 1) and vice versa. Denoting the channel capacity in the presence of perfect XPD by $C(\alpha = 0)$ and in the absence of XPD by $C(\alpha = 1)$ we see that

$$C(\alpha = 0) = 2 \log_2 \left(1 + \frac{\rho}{2} \right), \tag{4.47}$$

$$C(\alpha = 1) = \log_2 \left(1 + 2\rho \right). \tag{4.48}$$

At very low SNR ($\rho << 1$), using $\log_2(1 + x) \approx x \log_2 e$ for $x \ll 1$,

$$C(\alpha = 0) \approx \rho \log_2 e, \tag{4.49}$$

and

$$C(\alpha = 1) \approx 2\rho \log_2 e. \tag{4.50}$$

Hence, high XPD is detrimental to capacity at low SNR. On the other hand, in the high SNR regime ($\rho >> 1$) we observe that

$$C(\alpha = 0) \approx 2 \log_2 \left(\frac{\rho}{2} \right), \tag{4.51}$$

$$C(\alpha = 1) \approx 1 + \log_2(\rho). \tag{4.52}$$

Clearly, high XPD enhances MIMO channel capacity at high SNR. Figure 4.12 shows the capacity as a function of SNR when $\alpha = 1$ (no XPD) and $\alpha = 0$ (perfect XPD). High XPD is better than low XPD at high SNR, and vice versa at low SNR.

Real channels may experience all three influences – spatial correlation, Ricean fading and XPD. The capacity of such composite channels is studied in [Erceg *et al.*, 2002; Soma *et al.*, 2002].

4.6.4   Influence of degeneracy

Consider the pin-hole channel described in Section 2.8. The effective channel $\mathbf{H}$ is given by (see Eq. (3.34))

$$\mathbf{H} = \mathbf{h}_r \mathbf{h}_t^T, \tag{4.53}$$

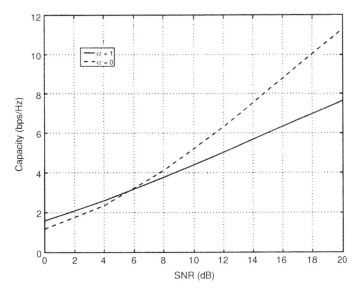

Figure 4.12: Capacity of a MIMO channel with perfect XPD ($\alpha = 0$) and no XPD ($\alpha = 1$). Good XPD restores MIMO capacity at high SNR.

where $\mathbf{h}_r$ is the $M_R \times 1$ vector channel from the pin-hole to the receive antenna array, and $\mathbf{h}_t^T$ is the $1 \times M_T$ vector channel from the transmit antenna array to the pin-hole. Since $r(\mathbf{H}) = 1$, there is only one spatial sub-channel between the transmitter and receiver. The corresponding channel capacity is given by

$$
\begin{aligned}
C &= \log_2 \det \left(1 + \frac{\rho}{M_T} \|\mathbf{H}\|_F^2 \right) \\
&= \log_2 \left(1 + \frac{\rho}{M_T} \|\mathbf{h}_r\|_F^2 \|\mathbf{h}_t\|_F^2 \right).
\end{aligned}
\tag{4.54}
$$

Hence, as with SIMO and MISO channels we can expect only a logarithmic increase in capacity with increasing SNR, although the underlying channel is a MIMO channel. Figure 4.13 compares the ergodic capacity as a function of SNR for the pin-hole channel with the regular $\mathbf{H}_w$ channel for several antenna configurations. The degradation in capacity is evident.

4.7   Capacity of frequency selective MIMO channels

So far we have assumed the channel is frequency flat. We now consider frequency selective channels. The capacity of a frequency selective fading MIMO channel can be calculated by dividing the frequency band of interest (say 1 Hz) into N narrower sub-channels, each having bandwidth $1/N$ Hz, such that each sub-channel is frequency flat (see Figure 4.14).

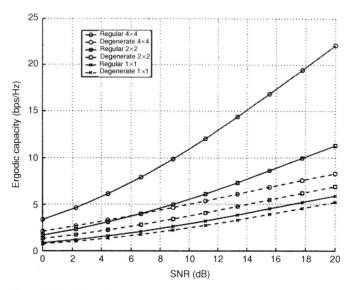

Figure 4.13: Channel degeneracy significantly degrades MIMO capacity.

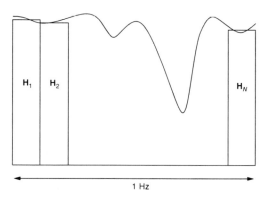

Figure 4.14: The capacity of a frequency selective MIMO channel is the sum of the capacities of frequency flat sub-channels.

Let the ith sub-channel be $\mathbf{H}_i$ ($i = 1, 2, \ldots, N$). The input–output relation for this sub-channel is given by

$$\mathbf{y}_i = \sqrt{\frac{E_s}{M_T}} \mathbf{H}_i \mathbf{s}_i + \mathbf{n}_i, \tag{4.55}$$

where $\mathbf{y}_i$ is the $M_R \times 1$ received signal vector, $\mathbf{s}_i$ is the $M_T \times 1$ transmitted signal vector and $\mathbf{n}_i$ is the $M_R \times 1$ noise vector for the ith sub-channel. The total input–output relation becomes

$$\mathcal{Y} = \sqrt{\frac{E_s}{M_T}} \mathcal{H}\mathcal{S} + \mathcal{N}, \qquad (4.56)$$

where $\mathcal{Y} = [\mathbf{y}_1^T \ \mathbf{y}_2^T \ \cdots \ \mathbf{y}_N^T]^T$ is $M_R N \times 1$, $\mathcal{S} = [\mathbf{s}_1^T \ \mathbf{s}_2^T \ \cdots \ \mathbf{s}_N^T]^T$ is $M_T N \times 1$, $\mathcal{N} = [\mathbf{n}_1^T \ \mathbf{n}_2^T \ \cdots \ \mathbf{n}_N^T]^T$ is $M_R N \times 1$ and $\mathcal{H}$ is an $M_R N \times M_T N$ block diagonal matrix with $\mathbf{H}_i$ as the block diagonal elements. The covariance matrix of $\mathcal{S}$ with $\mathbf{R}_{ss} = \mathcal{E}\{\mathcal{S}\mathcal{S}^H\}$ is constrained so that $\mathrm{Tr}(\mathbf{R}_{ss}) = N M_T$. This constrains the total average transmit power to E_s. It follows from Eq. (4.9) that the capacity of the frequency selective MIMO channel equals

$$C_{FS} = \frac{1}{N} \max_{\mathrm{Tr}(\mathbf{R}_{ss}) = N M_T} \log_2 \det\left(\mathbf{I}_{M_R N} + \frac{E_s}{M_T N_o}\mathcal{H}\mathbf{R}_{ss}\mathcal{H}^H\right) \ \text{bps/Hz.} \qquad (4.57)$$

Channel unknown to the transmitter

If the channel is unknown to the transmitter we should choose $\mathbf{R}_{ss} = \mathbf{I}_{M_T N}$, which implies that transmit power is allocated evenly across space (transmit antennas) and frequency. The capacity of a deterministic channel then reduces to

$$C_{FS} \approx \frac{1}{N} \sum_{i=1}^{N} \log_2 \det\left(\mathbf{I}_{M_R} + \frac{E_s}{M_T N_o}\mathbf{H}_i\mathbf{H}_i^H\right) \ \text{bps/Hz.} \qquad (4.58)$$

If the frequency response of the channel is flat, i.e., $\mathbf{H}_i = \mathbf{H}$ $(i = 1, 2, \ldots, N)$, then

$$C_{FS} = \log_2 \det\left(\mathbf{I}_{M_R} + \frac{\rho}{M_T}\mathbf{H}\mathbf{H}^H\right), \qquad (4.59)$$

which is the capacity of the frequency flat MIMO channel. Further, if all $\mathbf{H}_i$ have independent and identical distributions (i.e., the coherence bandwidth is $1/N$) then by the strong law of large numbers

$$C_{FS} \to C_{FS}^{\infty} \ \text{as} \ N \to \infty, \qquad (4.60)$$

i.e., the capacity of a sample realization of the frequency selective channel approaches a fixed quantity.

If the channel is random, we define ergodic and outage capacity to characterize the statistics of the information rate. The ergodic capacity for the frequency selective MIMO channel $\overline{C}_{FS}$ is

$$\overline{C}_{FS} = \mathcal{E}\left\{\frac{1}{N} \sum_{i=1}^{N} \log_2 \det\left(\mathbf{I}_{M_R} + \frac{\rho}{M_T}\mathbf{H}_i\mathbf{H}_i^H\right)\right\}. \qquad (4.61)$$

The outage capacity for frequency selective MIMO channels is similarly defined. The outage capacity of a frequency selective channel will be higher than the outage capacity of a frequency flat channel (at low outage rates). This is attributed to the

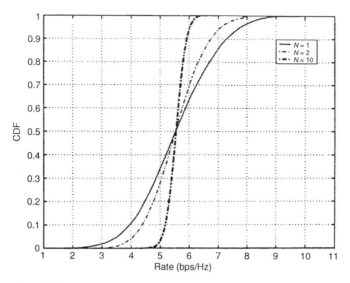

Figure 4.15: CDF of the information rate of an increasingly frequency selective MIMO channel. Outage performance improves with frequency selectivity.

increased tightening of the CDF of capacity due to frequency diversity. Under sufficient regularity conditions

$$C_{FS}^{\infty} \to \overline{C}_{FS}. \tag{4.62}$$

Therefore, asymptotically (in N) the capacity of a sample realization of the frequency selective MIMO channel equals its ergodic capacity. Figure 4.15 depicts this effect through the CDFs of the information rate of a frequency selective MIMO channel ($M_T = M_R = 2$), with increasing N at an SNR of 10 dB. The CDF of the information rate tightens with increasing N, improving outage capacity at low outage. The influence of multiple physical parameters such as delay spread, cluster angle spread and total angle spread on the ergodic and outage capacity of frequency selective MIMO channels has been studied in [Bölcskei *et al.*, 2002a].

Channel known to transmitter

In the case of a frequency flat channel, spectral efficiency can be improved if the channel is known to the transmitter. For the frequency flat channel, we saw that the waterpouring solution distributes the transmit energy optimally across the spatial sub-channels to maximize spectral efficiency. In the case of the frequency selective MIMO channel, energy must be distributed across space and frequency so as to maximize spectral efficiency. This form of waterpouring is known as space-frequency waterpouring [Raleigh and Cioffi, 1998]. Note that waterpouring is properly defined only for orthogonal channels. This is met if we use OFDM techniques (Chapter 9) to orthogonalize delay spread (intersymbol interference, ISI) channels and modal decomposition to orthogonalize

MIMO channels.

If the composite channel $\mathcal{H}$ is known to the transmitter the channel may be decomposed into $r(\mathcal{H})$ space-frequency modes. The capacity of the channel is then given by

$$C_{FS} = \frac{1}{N} \max_{\sum_{i=1}^{r(\mathcal{H})} \gamma_i = N M_T} \sum_{i=1}^{r(\mathcal{H})} \log_2 \left(1 + \frac{E_s \gamma_i}{M_T N_o} \lambda_i (\mathcal{H}\mathcal{H}^H) \right), \qquad (4.63)$$

where $\lambda_i(\mathcal{H}\mathcal{H}^H)$ $(i = 1, 2, \ldots, r(\mathcal{H}))$ represent the positive eigenvalues of $\mathcal{H}\mathcal{H}^H$ and γ_i is the energy allocated to ith space-frequency mode. The waterpouring algorithm described in Section 4.4 provides us with the optimal energy allocation policy from which the optimal mutual information maximizing space-frequency covariance matrix $\mathbf{R}_{ss}^{opt}$ can be derived. We can define an ergodic and outage capacity for frequency selective channels when the channel is known to the transmitter, like in frequency flat fading in Section 4.5.

5 Spatial diversity

5.1 Introduction

In Chapter 4 we studied how multiple antennas can enhance channel capacity, and in particular how MIMO systems can provide a linear increase in capacity, making MIMO very attractive in practical systems. In this chapter we discuss how antennas can also offer diversity. First, the role of diversity in improving the slope of the symbol error rate curve in fading channels is discussed. Next, the diversity performance of SIMO, MISO and MIMO channels is analyzed. The last two cases are analyzed with and without channel knowledge at the transmitter. Finally, diversity performance with antenna correlation, gain imbalance between channel elements and Ricean fading is discussed. We then extend these results to delay spread channels, where path or frequency diversity, in addition to space diversity, is available. Finally, we show how spatial diversity at the transmitter can be transformed into time or frequency diversity.

5.2 Diversity gain

Wireless links are impaired by the random fluctuations in signal level across space, time and frequency known as fading. Diversity provides the receiver with multiple (ideally independent) looks at the same transmitted signal. Each look constitutes a diversity branch. With an increase in the number of independent diversity branches, the probability that all branches are in a fade at the same time reduces sharply. Thus diversity techniques stabilize the wireless link leading to an improvement in link reliability or error rate. In this chapter we shall use the average symbol error rate (SER) as a measure of performance.

Assume that a symbol s, drawn from a scalar constellation with unit average energy, is to be transmitted. Furthermore, assume that there are M identical independent Rayleigh fading links between the transmitter and receiver (we shall explore more general fading channels in Section 5.6). These could be multiple coherence bandwidths if frequency diversity is to be exploited, or multiple coherence time intervals if time diversity is exploited. To leverage diversity, the transmitter must transmit the same symbol s across all links. With frequency flat fading across all diversity branches, the receiver sees multiple independently faded versions of the transmitted signal s, which are given by

$$y_i = \sqrt{\frac{E_s}{M}} h_i s + n_i, \quad i = 1, \ldots, M, \tag{5.1}$$

where E_s/M is the symbol energy available to the transmitter for each of the M diversity branches, y_i is the received signal corresponding to the ith diversity branch, h_i is the channel transfer function corresponding to the ith diversity branch, and n_i is additive ZMCSCG noise with variance N_o. Furthermore, we assume $\mathcal{E}\{n_i n_j^*\} = 0$, which ensures that the additive noise is uncorrelated across the diversity branches.

Given multiple faded versions of the transmitted signal s, the post-processing SNR η at the receiver can be maximized through a technique known as maximal ratio combining (MRC) [Proakis, 1995; Poor and Wornell, 1998]. Assuming perfect channel knowledge at the receiver, the M received signals are combined according to

$$z = \sum_{i=1}^{M} h_i^* y_i, \tag{5.2}$$

and thus η is given by

$$\eta = \frac{1}{M} \sum_{i=1}^{M} |h_i|^2 \rho, \tag{5.3}$$

where $\rho = E_s/N_o$ may be interpreted as the average SNR at the receive antenna in a SISO fading link. Assuming ML detection at the receiver, the corresponding probability of symbol error is given by [Proakis, 1995; Cioffi, 2002]

$$P_e \approx \overline{N_e} Q\left(\sqrt{\frac{\eta d_{min}^2}{2}}\right), \tag{5.4}$$

where $\overline{N_e}$ and d_{min} are the number of nearest neighbors and minimum distance of separation of the underlying scalar constellation respectively. Applying the Chernoff bound, $Q(x) \leq e^{-\frac{x^2}{2}}$, and combining Eqs. (5.3) and (5.4), P_e may be upper-bounded as

$$P_e \leq \overline{N_e} \; e^{-\left(\sum_{i=1}^{M} |h_i|^2\right)\frac{\rho d_{min}^2}{4M}}. \tag{5.5}$$

The probability of symbol error derived above corresponds to a particular realization of the channel. The M links making up the channel fade with time and therefore the value of diversity may be studied by considering the average uncoded SER $\overline{P_e} = \mathcal{E}\{P_e\}$. Assuming h_i $(i = 1, 2, \ldots, M)$ are independent ZMCSCG random variables with unit variance, the average probability of symbol error $\overline{P_e}$ is upper-bounded by

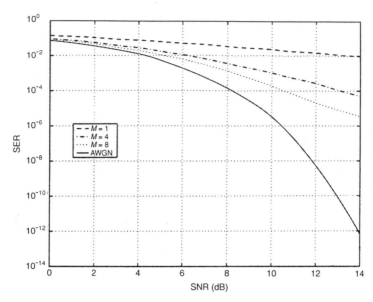

Figure 5.1: Effect of diversity on the SER performance in fading channels. The slope of the SER vs SNR curve increases with increasing M, the number of diversity branches.

(see Eq. (3.44))

$$\overline{P}_e \le \overline{N}_e \prod_{i=1}^{M} \frac{1}{1 + \rho d_{min}^2/4M} \cdot \tag{5.6}$$

In the high SNR regime ($\rho \gg 1$) Eq. (5.6) may be simplified as

$$\overline{P}_e \le \overline{N}_e \left(\frac{\rho d_{min}^2}{4M} \right)^{-M} . \tag{5.7}$$

From Eq. (5.7) it follows that in the absence of diversity ($M = 1$), $\overline{P}_e \le \overline{N}_e (\rho d_{min}^2/4)^{-1}$. Diversity affects the slope of the SER vs SNR (ρ) curve on a log–log scale. The magnitude of the slope equals the diversity order M. This is demonstrated in Fig. 5.1, where we plot SER as a function of SNR for a fading link, with binary phase shift keying (BPSK) modulation for $M = 1$, $M = 4$ and $M = 8$ (the SER is equivalent to the bit error rate (BER) for BPSK modulation). For comparison we also plot the SER as a function of SNR for an AWGN SISO channel with channel transfer function 1 (no fading).

With increasing diversity order the SER approaches the SER of an AWGN link. In fact, asymptotically as $M \to \infty$, Eq. (5.6) becomes

$$\overline{P}_e \le \overline{N}_e \, e^{-\frac{\rho d_{min}^2}{4}}, \tag{5.8}$$

where we have used the identity $\lim_{n\to\infty}(1 + x/n)^n \to e^x$. A more rigorous proof of this result based on exact error probability can be found in [Ventura-Travest *et al.*, 1997].

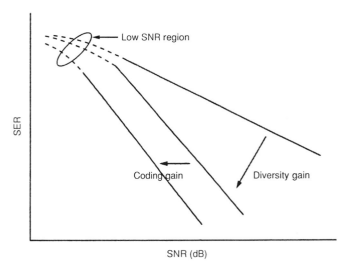

Figure 5.2: Schematic highlighting the difference between coding gain and diversity gain. The SNR advantage due to diversity gain increases with SNR but remains constant with coding gain.

Equation (5.8) gives precisely the Chernoff upper bound on the probability of symbol error for the AWGN channel, which confirms that fading is completely mitigated in the presence of infinite diversity.

In the example above, we have assumed repetition coding to transmit the symbols across various diversity branches, which leads to a loss in spectral efficiency. Multiple antennas may be used to extract diversity gain without sacrificing spectral efficiency. We shall discuss these schemes in greater detail in Sections 5.3 and 5.4. It is also important to note that since redundancy introduced through coding delivers coding gain in an AWGN channel, in the presence of fading both diversity and coding gain may be available. We shall briefly discuss the nature of these two gains.

5.2.1   Coding gain vs diversity gain

While both diversity and coding improve system performance (decrease error rate), the nature of these gains is very different. While diversity gain manifests itself in increasing the magnitude of the slope of the SER curve, coding gain shifts the error rate curve to the left. The SER for a system employing both coding and diversity techniques at high SNR can be approximated by

$$\overline{P}_e \approx \frac{c}{(\gamma_c \ \rho)^M}, \tag{5.9}$$

where c is a scaling constant specific to the modulation employed and the nature of the channel, while γ_c ($\gamma_c \geq 1$) denotes coding gain and M is the diversity order of the system. Figure 5.2 highlights the differences between coding gain and diversity gain.

The SNR advantage due to diversity gain increases with increasing diversity order and lower target error rate. On the other hand, the coding gain is typically constant at a high enough SNR.

5.2.2 Spatial diversity vs time/frequency diversity

From the preceding discussion, it is clear that diversity is a powerful technique for mitigating fading in wireless systems. The utilization of time/frequency diversity incurs an expense – time in the case of time diversity and bandwidth in the case of frequency diversity to introduce redundancy. Further, it follows from Eq. (5.3) that the average receive SNR using such techniques is the same as that for an AWGN channel. Spatial diversity is an attractive alternative that does not sacrifice time or bandwidth, while also providing array gain or increased average received SNR. The exact nature of the scheme that extracts spatial diversity depends on the antenna configuration (SIMO, MISO or MIMO). While receive diversity techniques may be employed in SIMO systems, the use of multiple antennas at the transmitter requires the use of more sophisticated diversity techniques. We shall study these below.

5.3 Receive antenna diversity

Consider a system with a single antenna at the transmitter and multiple antennas at the receiver (SIMO channel). Assuming flat fading conditions, the channel vector $\mathbf{h}$ for such a system is given by

$$\mathbf{h} = \begin{bmatrix} h_1 & h_2 & \cdots & h_{M_R} \end{bmatrix}^T, \qquad (5.10)$$

where M_R is the number of receive antennas. Once again, assuming that the symbol s to be transmitted is drawn from a scalar constellation with unit average energy, the input–output relation for the channel may be expressed as

$$\mathbf{y} = \sqrt{E_s}\mathbf{h}s + \mathbf{n}, \qquad (5.11)$$

where $\mathbf{y}$ is the $M_R \times 1$ received signal vector and $\mathbf{n}$ is ZMCSCG noise with $\mathcal{E}\{\mathbf{n}\mathbf{n}^H\} = N_o\mathbf{I}_{M_R}$. To maximize the received SNR, the receiver performs MRC, i.e.,

$$
\begin{aligned}
z &= \sqrt{E_s}\mathbf{h}^H\mathbf{h}s + \mathbf{h}^H\mathbf{n} \\
&= \sqrt{E_s}\|\mathbf{h}\|_F^2 s + \mathbf{h}^H\mathbf{n},
\end{aligned}
\qquad (5.12)
$$

where z is the receiver output. We assume perfect channel knowledge at the receiver. Thus, the effective channel is a scalar channel and standard scalar detection techniques may be applied to detect the transmitted signal. Since the noise vector $\mathbf{n}$ is spatially

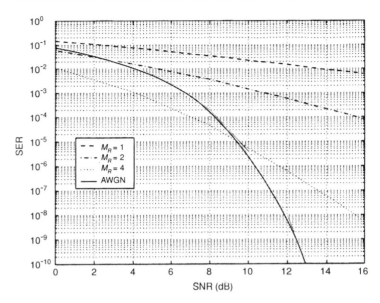

Figure 5.3: Performance of receive diversity with an increasing number of receive antennas. Array gain is also present.

white, the SNR at the receiver η is given by

$$\eta = \|\mathbf{h}\|_F^2 \rho. \tag{5.13}$$

If the separation between the antennas at the receiver is greater than the coherence distance (D_C) and if we assume a rich scattering environment, then $\mathbf{h} = \mathbf{h}_w$. Following the derivation in Section 5.2 (Eqs. (5.4)–(5.6)), the average probability of symbol error for such a channel is given by

$$\overline{P}_e \le \overline{N}_e \prod_{i=1}^{M_R} \frac{1}{1 + \rho d_{min}^2/4}. \tag{5.14}$$

In the high SNR regime Eq. (5.14) may be simplified as

$$\overline{P}_e \le \overline{N}_e \left(\frac{\rho d_{min}^2}{4} \right)^{-M_R}. \tag{5.15}$$

Thus the diversity order of the system is equal to the number of antennas at the receiver M_R. Furthermore, since $\mathcal{E}\{\|\mathbf{h}\|_F^2\} = M_R$ for $\mathbf{h} = \mathbf{h}_w$, the average SNR at the receiver, $\overline{\eta} = \mathcal{E}\{\eta\}$, is given by

$$\overline{\eta} = M_R \rho. \tag{5.16}$$

Hence, in addition to diversity gain, the average SNR at the receiver is enhanced by a factor of M_R over a standard SISO link due to array gain which is expressed as $10 \log_{10} M_R$ (in decibels). Figure 5.3 shows the SER performance of a receive diversity scheme assuming BPSK transmission for varying antenna configurations. The symbol error rate improves with the number of receive antennas in the system. Also, with four receive antennas, the system outperforms an AWGN link for a target BER greater than 10^{-5}. This can be explained by the role of array gain (which is 6 dB for four antennas) which provides the initial advantage. However, at lower target BER, the penalty due to fading overwhelms the array gain advantage. The effect of array gain is similar to that of coding gain (i.e., causes a left parallel shift in the curves) and is strictly independent of SNR.

We have seen that receive diversity techniques are capable of extracting full diversity gain and array gain. The performance improvement is proportional to the number of receive antennas used. However, deploying multiple antennas at the terminal receiver is often not feasible due to cost or space limitations. Instead, the use of multiple antennas at the transmitter in combination with transmit antenna diversity techniques is becoming increasingly popular. We shall explore transmit antenna diversity in further detail in the ensuing section.

5.4   Transmit antenna diversity

Exploiting spatial diversity in systems with multiple antennas at the transmitter requires that the signal be pre-processed or pre-coded prior to transmission. There has been increased interest in these techniques since the 1990s [Seshadri and Winters, 1994; Guey et al., 1996; Tarokh et al., 1998, 1999b; Alamouti, 1998; Papadias, 1999; Hochwald et al., 2001].

The need for pre-processing the transmit signal is explained below. Consider a symbol s that is to be transmitted over a system with two transmit antennas and a single receive antenna. A simplistic attempt to exploit diversity would be to transmit the signal from both transmit antennas at the same time. Assuming a flat fading environment, where the channel signatures corresponding to the transmit antennas are given by h_1 and h_2, the received signal y may be expressed as

$$y = \sqrt{\frac{E_s}{2}}(h_1 + h_2)s + n, \tag{5.17}$$

where E_s is the average energy available at the transmitter over a symbol period evenly divided between the transmit antennas and n is the ZMCSCG noise at the receiver. Noting that the sum of two complex Gaussian random variables is also complex Gaussian [Papoulis, 1984], $\frac{1}{\sqrt{2}}(h_1 + h_2)$ is ZMCSCG with unit variance. As a result, Eq. (5.17) may be expressed as

$$y = \sqrt{E_s}hs + n, \tag{5.18}$$

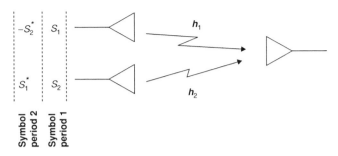

Figure 5.4: A schematic of the transmission strategy in the Alamouti scheme. The transmission strategy orthogonalizes the channel irrespective of the channel realization.

where h is ZMCSCG with $\mathcal{E}\{|h|^2\} = 1$. Therefore, this naive technique does not provide diversity.

We now show below how transmit diversity may be leveraged in both the presence and absence of channel knowledge at the transmitter for MISO and MIMO systems.

5.4.1   Channel unknown to the transmitter: MISO

We assume two antennas at the transmitter and a single receive antenna. We consider a simple but ingenious transmit diversity technique – the Alamouti scheme [Alamouti, 1998]. In this technique, two different symbols s_1 and s_2 are transmitted simultaneously from antennas 1 and 2 respectively during the first symbol period, followed by signals $-s_2^*$ and s_1^* from antennas 1 and 2 respectively during the next symbol period (see Fig. 5.4).

We assume that the channel remains constant over the two symbol periods, and is frequency flat. Therefore, $\mathbf{h} = [h_1, h_2]$ and the signals y_1 and y_2 received over the two symbol periods are given by

$$y_1 = \sqrt{\frac{E_s}{2}} h_1 s_1 + \sqrt{\frac{E_s}{2}} h_2 s_2 + n_1, \tag{5.19}$$

$$y_2 = -\sqrt{\frac{E_s}{2}} h_1 s_2^* + \sqrt{\frac{E_s}{2}} h_2 s_1^* + n_2, \tag{5.20}$$

where n_1 and n_2 are ZMCSCG noise with $\mathcal{E}\{|n_1|^2\} = \mathcal{E}\{|n_2|^2\} = N_o$ and $E_s/2$ is the average transmit energy per symbol period per antenna. The receiver forms a rearranged signal vector $\mathbf{y}$ as follows:

$$\mathbf{y} = \begin{bmatrix} y_1 \\ y_2^* \end{bmatrix}. \tag{5.21}$$

The vector $\mathbf{y}$ can be expressed as

$$
\mathbf{y} = \sqrt{\frac{E_s}{2}} \begin{bmatrix} h_1 & h_2 \\ h_2^* & -h_1^* \end{bmatrix} \begin{bmatrix} s_1 \\ s_2 \end{bmatrix} + \begin{bmatrix} n_1 \\ n_2^* \end{bmatrix}
$$

$$
= \sqrt{\frac{E_s}{2}} \mathbf{H}_{eff} \mathbf{s} + \mathbf{n}, \tag{5.22}
$$

where $\mathbf{s} = [s_1 \ s_2]^T$ and $\mathbf{n} = [n_1 \ n_2^*]^T$. The effective channel matrix $\mathbf{H}_{eff}$ is orthogonal (i.e., $\mathbf{H}_{eff}^H \mathbf{H}_{eff} = \|\mathbf{h}\|_F^2 \mathbf{I}_2$). If $\mathbf{z} = \mathbf{H}_{eff}^H \mathbf{y}$, we get

$$
\mathbf{z} = \sqrt{\frac{E_s}{2}} \|\mathbf{h}\|_F^2 \mathbf{I}_2 \mathbf{s} + \tilde{\mathbf{n}}, \tag{5.23}
$$

where $\mathcal{E}\{\tilde{\mathbf{n}}\} = \mathbf{0}_{2,1}$ and $\mathcal{E}\{\tilde{\mathbf{n}}\tilde{\mathbf{n}}^H\} = \|\mathbf{h}\|_F^2 N_o \mathbf{I}_2$. Hence, the effective channel for symbols s_i $(i = 1, 2)$ becomes

$$
z_i = \sqrt{\frac{E_s}{2}} \|\mathbf{h}\|_F^2 s_i + \tilde{n}_i, \quad i = 1, 2, \tag{5.24}
$$

and the received SNR, η, per symbol is given by

$$
\eta = \frac{\|\mathbf{h}\|_F^2 \rho}{2}. \tag{5.25}
$$

Assuming $\mathbf{h} = \mathbf{h}_w$, and following the derivation in Eqs. (5.4)–(5.6), $\overline{P}_e$ in the high SNR regime may be upper-bounded according to

$$
\overline{P}_e \leq \overline{N}_e \left(\frac{\rho d_{min}^2}{8} \right)^{-2}. \tag{5.26}
$$

The Alamouti scheme therefore extracts a diversity order of 2 (full M_T diversity), even in the absence of channel knowledge at the transmitter. A somewhat similar scheme (referred to as ST spreading) has been proposed [Papadias, 1999; Hochwald et al., 2001] for wideband CDMA systems and is now part of the CDMA 2000 standard.

Since $\mathcal{E}\{\|\mathbf{h}\|_F^2\} = 2$ for $\mathbf{h} = \mathbf{h}_w$, the average SNR at the receiver $\overline{\eta} = \rho$. Therefore, the absence of channel knowledge at the transmitter does not allow array gain. Figure 5.5 compares the performance of the Alamouti scheme ($M_T = 2$, $M_R = 1$) with a receive diversity scheme ($M_T = 1$, $M_R = 2$) for BPSK modulation. While both schemes outperform a SISO fading link and extract the same diversity gain (the SER curves have the same slope), receive diversity outperforms the Alamouti scheme because of array gain.

Transmit diversity techniques in the absence of channel knowledge at the transmitter may be designed to extract spatial diversity in systems with more than two transmit antennas. We study these in Chapter 6.

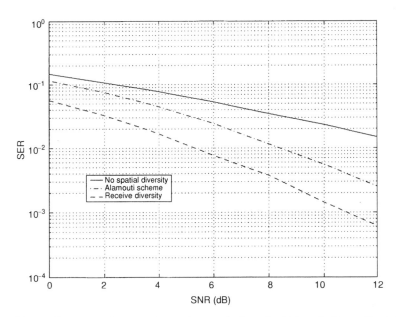

Figure 5.5: Comparison of Alamouti transmit diversity ($M_T = 2$, $M_R = 1$) with receive diversity ($M_T = 1$, $M_R = 2$). Both schemes have the same diversity order of 2, but receive diversity has an additional 3 dB receive array gain.

5.4.2   Channel known to the transmitter: MISO

Consider a MISO system with M_T transmit antennas and a frequency flat fading channel. The vector channel **h** is given by

$$\mathbf{h} = \begin{bmatrix} h_1 & h_2 & \cdots & h_{M_T} \end{bmatrix}. \tag{5.27}$$

To exploit spatial diversity, the signal is transmitted from each transmit antenna after being weighted appropriately, so that the signals arrive in phase at the receive antenna and add coherently. The signal at the receiver is given by

$$y = \sqrt{\frac{E_s}{M_T}}\mathbf{h}\mathbf{w}s + n, \tag{5.28}$$

where y is the received signal, **w** is a weight vector of dimension $M_T \times 1$ and n is ZMCSCG noise. The weight vector **w** must be chosen subject to $\|\mathbf{w}\|_F^2 = M_T$, to ensure that the average total power of the transmitted signal is E_s. Clearly, the weight vector **w** that maximizes the received SNR is given by [Lo, 1999]

$$\mathbf{w} = \sqrt{M_T}\frac{\mathbf{h}^H}{\sqrt{\|\mathbf{h}\|_F^2}}. \tag{5.29}$$

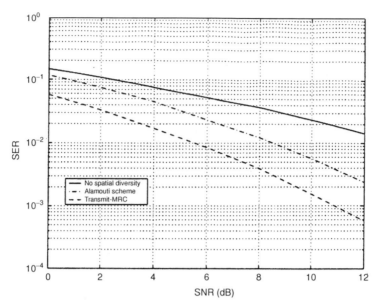

Figure 5.6: Comparison of Alamouti transmit diversity with transmit-MRC diversity for $M_T = 2$ and $M_R = 1$. Again note the difference due to transmit array gain.

This scheme is known as transmit-maximal ratio combining (transmit-MRC) [Lo, 1999; Feng and Leung, 2001]. The SNR at the receiver η is given by

$$\eta = \|\mathbf{h}\|_F^2 \rho. \tag{5.30}$$

If $\mathbf{h} = \mathbf{h}_w$, it follows from Eqs. (5.4)–(5.6) that the average probability of symbol error in the high SNR regime is upper-bounded by

$$P_e \leq N_e \left(\frac{\rho d_{min}^2}{4} \right)^{-M_T}. \tag{5.31}$$

Thus, in the presence of IID Rayleigh fading, transmit-MRC delivers a diversity order of M_T and since $\mathcal{E}\{\|\mathbf{h}_w\|_F^2\} = M_T$, the average received SNR, $\bar{\eta}$, equals

$$\bar{\eta} = M_T \rho. \tag{5.32}$$

The average SNR at the receiver is improved by a factor of M_T (the number of transmit antennas) over a SISO link, and is the transmit array gain. Hence, if perfect channel knowledge is available to the transmitter, transmit-MRC will deliver array gain and diversity gain. Figure 5.6 compares the performance of the Alamouti scheme with transmit-MRC for BPSK modulation with $M_T = 2$. Transmit-MRC outperforms the Alamouti scheme on account of array gain while providing the same diversity gain. Also, MISO systems with M transmit antennas, using transmit-MRC, will have the same performance as SIMO systems with the same number of receive antennas employing receive-MRC.

5.4.3   Channel unknown to the transmitter: MIMO

Consider a MIMO system with two transmit antennas and two receive antennas. The Alamouti scheme described in Section 5.4.1 may be used to extract diversity in such a system. As in the MISO system described previously, two different symbols s_1 and s_2 are transmitted simultaneously from antennas 1 and 2 respectively during the first symbol period, following which symbols $-s_2^*$ and s_1^* are launched from antennas 1 and 2 respectively. We assume that the channel remains constant over consecutive symbol periods. Assume also a frequency flat channel. Let the 2×2 channel matrix $\mathbf{H}$ be

$$\mathbf{H} = \begin{bmatrix} h_{1,1} & h_{1,2} \\ h_{2,1} & h_{2,2} \end{bmatrix}, \tag{5.33}$$

and the signals received at the receive antenna array over consecutive symbol periods be $\mathbf{y}_1$ and $\mathbf{y}_2$. It follows that

$$\mathbf{y}_1 = \sqrt{\frac{E_s}{2}} \begin{bmatrix} h_{1,1} & h_{1,2} \\ h_{2,1} & h_{2,2} \end{bmatrix} \begin{bmatrix} s_1 \\ s_2 \end{bmatrix} + \begin{bmatrix} n_1 \\ n_2 \end{bmatrix}, \tag{5.34}$$

$$\mathbf{y}_2 = \sqrt{\frac{E_s}{2}} \begin{bmatrix} h_{1,1} & h_{1,2} \\ h_{2,1} & h_{2,2} \end{bmatrix} \begin{bmatrix} -s_2^* \\ s_1^* \end{bmatrix} + \begin{bmatrix} n_3 \\ n_4 \end{bmatrix}, \tag{5.35}$$

where n_1, n_2, n_3 and n_4 are uncorrelated ZMCSCG noise samples with $\mathcal{E}\{|n_i|^2\} = N_o$ ($i = 1, 2, \ldots, 4$). As in the MISO channel (with the channel unknown to transmitter), the energy available at the transmitter is evenly divided between the transmit antennas. The receiver now forms a signal vector $\mathbf{y}$ according to

$$\mathbf{y} = \begin{bmatrix} \mathbf{y}_1 \\ \mathbf{y}_2^* \end{bmatrix}. \tag{5.36}$$

It follows that $\mathbf{y}$ may be expressed as

$$\mathbf{y} = \sqrt{\frac{E_s}{2}} \begin{bmatrix} h_{1,1} & h_{1,2} \\ h_{2,1} & h_{2,2} \\ h_{1,2}^* & -h_{1,1}^* \\ h_{2,2}^* & -h_{2,1}^* \end{bmatrix} \begin{bmatrix} s_1 \\ s_2 \end{bmatrix} + \begin{bmatrix} n_1 \\ n_2 \\ n_3^* \\ n_4^* \end{bmatrix}$$

$$= \sqrt{\frac{E_s}{2}} \mathbf{H}_{eff} \mathbf{s} + \mathbf{n}, \tag{5.37}$$

where $\mathbf{s} = [s_1 \ s_2]^T$ and $\mathbf{n} = [n_1 \ n_2 \ n_3^* \ n_4^*]^T$. Furthermore, we note that $\mathbf{H}_{eff}$ is orthogonal irrespective of the channel realization (i.e., $\mathbf{H}_{eff}^H \mathbf{H}_{eff} = \|\mathbf{H}\|_F^2 \mathbf{I}_2$). If $\mathbf{z} = \mathbf{H}_{eff}^H \mathbf{y}$, we get

$$\mathbf{z} = \sqrt{\frac{E_s}{2}} \|\mathbf{H}\|_F^2 \mathbf{I}_2 \mathbf{s} + \tilde{\mathbf{n}}, \tag{5.38}$$

where $\mathcal{E}\{\tilde{\mathbf{n}}\} = \mathbf{0}_{2,1}$ and $\mathcal{E}\{\tilde{\mathbf{n}}\tilde{\mathbf{n}}^H\} = \|\mathbf{H}\|_F^2 N_o \mathbf{I}_2$. Hence, the effective channel for either data symbol s_i $(i = 1, 2)$ becomes

$$z_i = \sqrt{\frac{E_s}{2}} \|\mathbf{H}\|_F^2 s_i + \tilde{n}_i, \quad i = 1, 2, \tag{5.39}$$

with the corresponding received SNR given by

$$\eta = \frac{\|\mathbf{H}\|_F^2 \rho}{2}. \tag{5.40}$$

From Eqs. (5.4)–(5.6) and assuming $\mathbf{H} = \mathbf{H}_w$, it follows that the average probability of symbol error in the high SNR regime is upper-bounded by

$$\overline{P}_e \leq \overline{N}_e \left(\frac{\rho d_{min}^2}{8} \right)^{-4}. \tag{5.41}$$

Therefore, the Alamouti scheme extracts order $M_T M_R$ diversity (fourth order in this case), though channel knowledge is not available to the transmitter. Since $\mathcal{E}\{\|\mathbf{H}\|_F^2\} = 4$ for $\mathbf{H} = \mathbf{H}_w$, the average SNR at the receiver, $\overline{\eta}$, is

$$\overline{\eta} = 2\rho. \tag{5.42}$$

Therefore, in the absence of channel knowledge at the transmitter, the Alamouti scheme is capable of extracting only receive array gain.

The Alamouti technique may be used to extract diversity in MIMO systems with two transmit antennas and any number (M_R) of receive antennas – we get $2M_R$ order diversity (full diversity) and an array gain of M_R. Signal design criteria for exploiting spatial diversity in MIMO systems with more than two transmit antennas are studied in Chapter 6.

5.4.4   Channel known to the transmitter: MIMO

We consider a system with M_R receive antennas and M_T transmit antennas. When the channel is known to the transmitter, spatial diversity may be extracted through a technique known as dominant eigenmode transmission. Here, as with transmit-MRC for MISO systems, the same signal is transmitted from all antennas in the transmit array with weight vector $\mathbf{w}$. The received signal vector is then given by

$$\mathbf{y} = \sqrt{\frac{E_s}{M_T}} \mathbf{H}\mathbf{w}s + \mathbf{n}, \tag{5.43}$$

where $\mathbf{y}$ is the $M_R \times 1$ received signal vector, $\mathbf{H}$ is the $M_R \times M_T$ channel transfer function, $\mathbf{w}$ is the $M_T \times 1$ complex weight vector and $\mathbf{n}$ is spatially white ZMCSCG noise. We note that $\mathbf{w}$ must satisfy $\|\mathbf{w}\|_F^2 = M_T$ to maintain total average transmitted energy. Let the receiver form a weighted sum of antenna outputs according to

$$z = \mathbf{g}^H \mathbf{y}, \tag{5.44}$$

where $\mathbf{g}$ is a $M_R \times 1$ vector of complex weights. The SNR at the receiver, η, is given by

$$\eta = \frac{\|\mathbf{g}^H \mathbf{H} \mathbf{w}\|_F^2}{M_T \|\mathbf{g}\|_F^2} \rho. \tag{5.45}$$

Hence, maximizing the SNR at the receiver is equivalent to maximizing $\|\mathbf{g}^H \mathbf{H} \mathbf{w}\|_F^2 / \|\mathbf{g}\|_F^2$. We recall from Chapter 3 that the SVD of $\mathbf{H}$ is given by

$$\mathbf{H} = \mathbf{U} \mathbf{\Sigma} \mathbf{V}^H. \tag{5.46}$$

It can be verified that η is maximized when $\mathbf{w}/\sqrt{M_T}$ and $\mathbf{g}$ are the input and output singular vectors respectively, corresponding to the maximum singular value σ_{max} of $\mathbf{H}$. With the appropriate (SNR maximizing) choice of $\mathbf{w}$ and $\mathbf{g}$, the effective input–output relation for the channel reduces to

$$z = \sqrt{E_s} \sigma_{max} s + n, \tag{5.47}$$

where n is ZMCSCG noise with variance N_o. Recalling from Chapter 3 that $\sigma_{max}^2 = \lambda_{max}$, where λ_{max} is the maximum eigenvalue of $\mathbf{H} \mathbf{H}^H$, the SNR at the receiver is given by

$$\eta = \lambda_{max} \rho. \tag{5.48}$$

Therefore, the array gain in dominant eigenmode transmission is given by $\mathcal{E}\{\lambda_{max}\}$. It is obvious that the transmit-MRC technique for MISO systems described earlier is essentially dominant eigenmode transmission over a channel with a single mode (rank 1 channel). Since $\sum_{i=1}^r \lambda_i = \|\mathbf{H}\|_F^2$ (recall from Chapter 3 that λ_i ($i = 1, 2, \ldots, M_R$) are the eigenvalues of $\mathbf{H} \mathbf{H}^H$), λ_{max} may be upper- and lower-bounded according to

$$\frac{\|\mathbf{H}\|_F^2}{r} \leq \lambda_{max} \leq \|\mathbf{H}\|_F^2, \tag{5.49}$$

where r is the rank of $\mathbf{H}$. Since the channel $\mathbf{H} = \mathbf{H}_w$ is full-rank with probability 1, $r = \min(M_T, M_R)$. Hence, the average SNR at the receiver may be upper- and lower-bounded as follows:

$$\rho \max(M_T, M_R) \leq \overline{\eta} \leq \rho M_T M_R. \tag{5.50}$$

Equation (5.50) implies that array gain when the channel is known to the transmitter is greater than or equal to the array gain when the channel is unknown.

From Eqs. (5.47) and (5.49), following the derivation for SER in Section 5.2, it follows that $\overline{P}_e$ for a system employing dominant eigenmode transmission may be upper- and lower-bounded (allowing for the approximation that the Chernoff upper

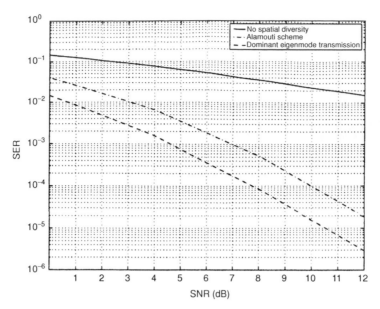

Figure 5.7: Comparison of the Alamouti scheme with dominant eigenmode transmission for $M_T = M_R = 2$. Dominant eigenmode transmission outperforms the Alamouti scheme due to array gain.

bound is a close estimate of the SER at high SNR) in the high SNR regime by

$$\overline{N}_e \left(\frac{\rho d_{min}^2}{4 \min(M_T, M_R)} \right)^{-M_T M_R} \geq \overline{P}_e \geq \overline{N}_e \left(\frac{\rho d_{min}^2}{4} \right)^{-M_T M_R}. \tag{5.51}$$

Equation (5.51) implies that the SER must maintain a slope of magnitude $M_T M_R$, as a function of SNR (on a log–log scale). Hence, we can conclude that dominant eigenmode transmission extracts a full diversity order of $M_T M_R$ [Paulraj, 2002].

Figure 5.7 compares the performance of the Alamouti scheme with dominant eigenmode transmission for a system with two transmit antennas, two receive antennas and BPSK modulation. As expected, both schemes extract the same order of diversity. However, dominant eigenmode transmission outperforms the Alamouti scheme due to the higher array gain.

The main conclusions of Sections 5.3 and 5.4 are summarized in Table 5.1. In the following section we quantify the effect of diversity on channel gain variability.

5.5   Diversity order and channel variability

Sections 5.2 and 5.3 showed how the effective channel depends on the squared Frobenius norm of the channel, $\|\mathbf{H}\|_F^2$. Figure 5.8 shows the evolution over time of $\|\mathbf{H}\|_F^2$, assuming $\mathbf{H} = \mathbf{H}_w$ for varying diversity order.

Table 5.1. *Array gain and diversity order for different multiple antenna configurations*

(CU = channel unknown to the transmitter and CK = channel known to the transmitter.)

Configuration	Expected array gain	Diversity order
SIMO (CU)	M_R	M_R
SIMO (CK)	M_R	M_R
MISO (CU)	1	M_T
MISO (CK)	M_T	M_T
MIMO (CU)	M_R	$M_R M_T$
MIMO (CK)	$\mathcal{E}\{\lambda_{max}\}$	$M_R M_T$

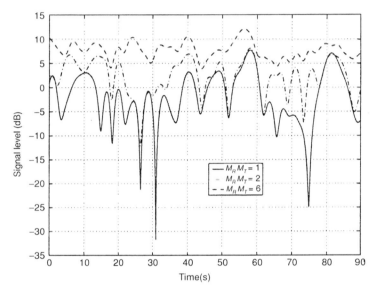

Figure 5.8: Link stability induced with increasing orders of spatial diversity. In the limit, as $M_T M_R \to \infty$, the channel is perfectly stabilized and approaches an AWGN link.

In the absence of spatial diversity ($M_T = M_R = 1$) the signal suffers deep fades. With increasing $M_T M_R$ (spatial diversity), the depth of the fades reduces considerably and the effective channel tightens. This effect has been amply demonstrated in real channels in [Bölcskei *et al.*, 2001]. Furthermore, we observe a progressive increase in the mean signal level with increasing diversity order. This comes from the array gain at the receiver and transmitter antennas depending on whether the channel is known at the receiver (as generally assumed) and the transmitter (optional) respectively.

The degree of tightening (also referred to sometimes as "hardening") of the channel may be quantified by its *coefficient of variation*, μ_{var}, [Sokal and Rohlf, 1995; Nabar *et al.*, 2002b]. The coefficient of variation of a random variable is defined as the ratio of the standard deviation of the random variable to its mean. Assuming an $\mathbf{H}_w$ MIMO channel with M_T transmit antennas and M_R receive antennas, the coefficient of variation for such a channel is given by

$$\mu_{var} = \frac{1}{\sqrt{M_T M_R}}. \qquad (5.52)$$

The degree of tightening of the effective channel is inversely proportional to the square root of the diversity order. Furthermore, asymptotically as the number of degrees of freedom in the channel increases, i.e., $M_R M_T \to \infty$, the coefficient of variation approaches 0. This verifies that the channel is perfectly stabilized with infinite diversity order.

5.6   Diversity performance in extended channels

So far we have considered the IID ($\mathbf{H} = \mathbf{H}_w$) MIMO channel. In practice, we know that several departures from $\mathbf{H}_w$ are possible. For example, the elements of $\mathbf{H}$ may be correlated, or may have gain imbalances or Ricean statistics. In this section, we shall study the influence of these conditions on the performance of diversity techniques. For clarity, we consider a MIMO system with $M_T = M_R = 2$ employing the Alamouti scheme. First, we study the case where the elements of $\mathbf{H}$ are correlated and have gain imbalances.

5.6.1   Influence of signal correlation and gain imbalance

Assuming a Rayleigh fading MIMO channel, correlation and gain imbalances between the elements of $\mathbf{H}$ are captured by the 4×4 covariance matrix $\mathbf{R} = \mathcal{E}\{\text{vec}(\mathbf{H})\text{vec}(\mathbf{H})^H\}$. We recall from Section 5.4.3 that the received signal with Alamouti coding for any channel realization $\mathbf{H}$ is given by

$$y = \sqrt{\frac{E_s}{2}} \|\mathbf{H}\|_F^2 s + n. \qquad (5.53)$$

Using the Chernoff upper-bound described in Section 5.2, the probability of symbol error for a sample realization of the channel $\mathbf{H}$ with Alamouti coding may be upper-bounded as

$$P_e \leq \overline{N}_e e^{-\frac{\rho d_{min}^2 \|\mathbf{H}\|_F^2}{8}}. \qquad (5.54)$$

From Eq. (3.44) the average probability of symbol error may be upper-bounded by

$$\overline{P}_e \leq \overline{N}_e \prod_{i=1}^{4} \frac{1}{1 + (\rho d_{min}^2/8)\lambda_i(\mathbf{R})}, \qquad (5.55)$$

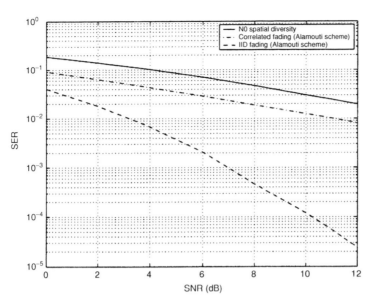

Figure 5.9: Impact of spatial fading correlation on the performance of the Alamouti scheme with $M_T = M_R = 2$. IID fading is optimal for diversity.

where $\lambda_i(\mathbf{R})$ ($i = 1, 2, \ldots, 4$) is the ith eigenvalue of $\mathbf{R}$. Since the logarithm is a strictly monotonic function, minimizing $\overline{P}_e$ is equivalent to maximizing

$$\sum_{i=1}^{4} \log \left(1 + \frac{\rho d_{min}^2}{8} \lambda_i(\mathbf{R}) \right). \tag{5.56}$$

A total average power constraint on the channel translates to constraining $\text{Tr}(\mathbf{R}) = \sum_{i=1}^{4} \lambda_i(\mathbf{R})$. Constraining $\text{Tr}(\mathbf{R}) = 4$, Eq. (5.56) is maximized when all eigenvalues of $\mathbf{R}$ are equal, i.e., $\lambda_i(\mathbf{R}) = \lambda_j(\mathbf{R}) = 1$ ($i, j = 1, \ldots, 4$). This condition is satisfied in the absence of fading signal correlation and gain imbalances, and corresponds to the $\mathbf{H}_w$ MIMO channel model. Consequently, gain imbalance and fading signal correlation are detrimental to the performance of diversity schemes. Further, in the high SNR regime Eq. (5.55) simplifies to

$$\overline{P}_e \leq \overline{N}_e \left(\frac{\rho d_{min}^2}{8} \right)^{-r(\mathbf{R})} \prod_{i=1}^{r(\mathbf{R})} (\lambda_i(\mathbf{R}))^{-1}. \tag{5.57}$$

Hence, the diversity order extracted by diversity transmission in the presence of correlated fading or gain imbalance is related to the rank of the covariance matrix $\mathbf{R}$. Figure 5.9 compares the SER performance of the Alamouti scheme, with BPSK modulation for $\mathbf{H} = \mathbf{H}_w$, with a channel where the elements of $\mathbf{H}$ are fully correlated with no gain imbalances ($\mathbf{R}$ is an all 1s matrix). Fully correlated fading destroys diversity gain as expected from Eq. (5.57). However, array gain is still available.

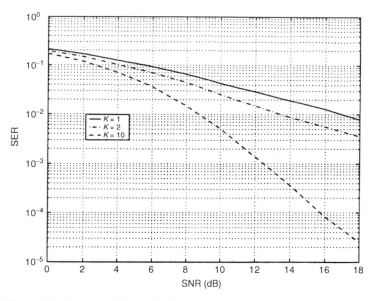

Figure 5.10: Impact of Ricean fading on the performance of the Alamouti scheme. The presence of an invariant component in the channel stabilizes the link and improves performance at high K-factor.

5.6.2   Influence of Ricean fading

The presence of a mean (or LOS) component in the channel $\mathcal{E}\{\mathbf{H}\} = \overline{\mathbf{H}}$ leads to Ricean fading. We know from Chapter 3 that the MIMO channel may be written as

$$\mathbf{H} = \sqrt{\frac{K}{1+K}}\,\overline{\mathbf{H}} + \sqrt{\frac{1}{1+K}}\,\mathbf{H}_w, \tag{5.58}$$

where $\sqrt{K/(1+K)}\,\overline{\mathbf{H}}$ represents the mean component of the channel matrix and $\sqrt{1/(1+K)}\,\mathbf{H}_w$ is the fading component of the channel matrix and K is the Ricean factor. The probability of symbol error for a particular channel realization of $\mathbf{H}$ is upper-bounded according to Eq. (5.54). The SER may then be upper-bounded as [Nabar *et al.*, 2002b]

$$\overline{P}_e \leq \overline{N}_e \left(\frac{1+K}{1+K+\rho d_{min}^2/8} \right)^4 e^{-\left(\frac{\frac{\rho d_{min}^2}{8} K \|\overline{\mathbf{H}}\|_F^2}{1+K+\frac{\rho d_{min}^2}{8}} \right)}. \tag{5.59}$$

Asymptotically, as $K \to \infty$

$$\overline{P}_e \leq \overline{N}_e\, e^{-\frac{\rho d_{min}^2}{8} \|\overline{\mathbf{H}}\|_F^2}, \tag{5.60}$$

which is the Chernoff upper bound on the probability of symbol error for a non-fading (AWGN) SISO channel with channel transfer function $(1/2)\|\overline{\mathbf{H}}\|_F^2$. Ricean fading stabilizes the link due to the presence of a mean (LOS) component in the channel. Figure 5.10

shows the performance of the Alamouti scheme with BPSK modulation for varying degrees of Ricean fading. As expected, the SER performance improves with K, the Ricean factor.

The impact of Ricean fading on diversity gain for more complex environments that include spatial fading correlation and gain imbalance between channel elements has been studied in [Nabar et al., 2002b, c, d]. Straightforward analysis derives the corresponding results for SIMO and MISO channels.

5.6.3   Degenerate MIMO channels

Consider a frequency flat pin-hole degenerate MIMO channel with M_T transmit and M_R receive antennas. Such channels may be modeled as (see Chapter 3)

$$\mathbf{H} = \mathbf{h}_r \mathbf{h}_t^T,$$
(5.61)

where $\mathbf{h}_r$ and $\mathbf{h}_t$ are both $\mathbf{h}_w$ vectors of dimension $M_R \times 1$ and $M_T \times 1$ respectively. In Chapter 4 we saw that degenerate channels are rank one and have much lower capacity than the corresponding $\mathbf{H}_w$ channel. In this section we show that the diversity order is also severely limited.

From Eq. (5.54) the probability of error for a sample channel realization with Alamouti encoding is upper-bounded as per

$$P_e \leq \overline{N_e} e^{-\frac{\rho d_{min}^2 \|\mathbf{H}\|_F^2}{4 M_T}}.$$
(5.62)

Note that for the degenerate channel we have $\|\mathbf{H}\|_F^2 = \|\mathbf{h}_r\|_F^2 \|\mathbf{h}_t\|_F^2$ and that both $\|\mathbf{h}_r\|_F^2$ and $\|\mathbf{h}_t\|_F^2$ are Chi-squared random variables. The average probability of error is upper-bounded by

$$\overline{P_e} \leq \mathcal{E} \left\{ \overline{N_e} e^{-\frac{\rho d_{min}^2 \|\mathbf{H}\|_F^2}{4 M_T}} \right\}.$$
(5.63)

For $M_T = 1$ the average SER is upper-bounded by

$$\overline{P_e} \leq \overline{N_e} \frac{e^{\frac{1}{\gamma}} \Gamma(1 - M_R, 1/\gamma)}{\gamma^{M_R}},$$
(5.64)

and for $M_T = 2$ we have

$$\overline{P_e} \leq \overline{N_e} \frac{e^{\frac{1}{\gamma}}}{\gamma^{M_R}} \left[\Gamma\left(2 - M_R, \frac{1}{\gamma}\right) - \frac{1}{\gamma} \Gamma\left(1 - M_R, \frac{1}{\gamma}\right) \right],$$
(5.65)

where $\gamma = (\rho/4 M_T) d_{min}^2$, and $\Gamma(a, b) = \int_b^\infty t^{a-1} e^{-t} dt$ is the incomplete Gamma function.

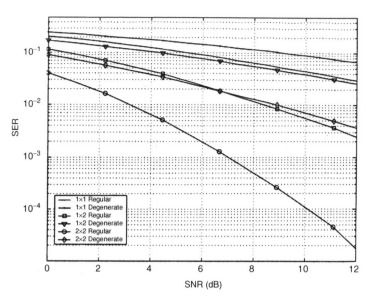

Figure 5.11: SER vs SNR in degenerate and $\mathbf{H}_w$ channels. The diversity order for degenerate channels is $\min(M_T, M_R)$ compared with $M_T M_R$ for $\mathbf{H}_w$ channels.

Figure 5.11 compares the SER curves for several antenna configurations in both $\mathbf{H}_w$ and degenerate channels. The diversity order for the degenerate channel is much lower than the corresponding $\mathbf{H}_w$ channel and is equal to $\min(M_T, M_R)$. For more details refer to [Venkatesh *et al.*, 2002].

5.7   Combined space and path diversity

In situations where there is significant delay spread in the channel, diversity can also be extracted from the multipath to yield both space and path (or frequency) diversity. We now analyze the diversity performance of a frequency selective SIMO channel, and later summarize the results for MISO and MIMO channels. The SIMO channel with delay spread (see Section 3.7.3) is characterized by $h_i[l]$ ($l = 0, 1, 2, \ldots, L - 1$), where i ($i = 1, 2, \ldots, M_R$) is the receive antenna index and l is the tap index. Equivalently, the SIMO channel is an $M_R \times 1$ vector $\mathbf{h}[l]$, whose elements are $h_i[l]$.

We note that even though the underlying multipaths are uncorrelated (the US assumption in Chapter 2), the time taps of the sampled channel $h_i[l]$ at a given antenna are in general correlated due to pulse-shaping and matched-filtering. The sampled channel taps will be uncorrelated only if the underlying multipath is symbol spaced, the combination of pulse-shaping and matched-filtering satisfies the Nyquist criterion and the sampling phase is correctly aligned. For a given tap, the channel at the antennas will have a correlation depending on angle spread and antenna spacing (see Chapter 3).

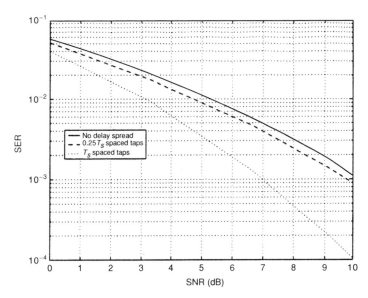

Figure 5.12: Impact of frequency selective fading on the diversity performance of a SIMO ($M_R = 2$) channel. The diversity performance improves when the spacing of the physical channel taps increases from $T_s/4$ to T_s.

To eliminate the effect of ISI, we consider the matched-filter (MF) at the receiver for a single transmitted pulse and study its SER assuming ISI is eliminated. This is the well-known matched-filter bound (MFB) that can be used to upper-bound performance. We temporarily redefine **h** to be the $M_R L \times 1$ vector given by

$$\mathbf{h}^T = \left[\mathbf{h}[0]^T \cdots \mathbf{h}[L-1]^T \right]. \tag{5.66}$$

The diversity performance of the MF receiver is governed by the eigenvalues $\lambda_i(\mathbf{R})$, where $\mathbf{R} = \mathcal{E}\{\mathbf{hh}^H\}$ is similar to the analysis in Section 5.6.1. Clearly the maximum diversity order possible is $M_R L$.

Figure 5.12 plots the SER vs SNR for the MFB case. We assume $M_R = 2$, BPSK modulation, raised cosine pulse shaping with 30% excess bandwidth and that the channel is uncorrelated across the antennas. Three cases are considered: (a) no delay spread, (b) a two-tap physical channel, equal power with $T_s/4$ tap spacing, (c) a two-tap physical channel, equal power with T_s tap spacing. Clearly, the diversity order improves with delay spread. With T_s tap spacing we reach $M_R L = 4$ order diversity.

The diversity order in space and frequency selective channels for the MISO case is bounded by $M_T L$. However, we lose the array gain if the channel is unknown at the transmitter [Lindskog and Paulraj, 2000]. Likewise, in the MIMO case, the achievable diversity order is $M_T M_R L$ and we lose the transmit array gain if the transmitter has no channel knowledge.

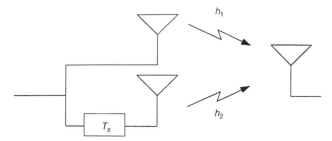

Figure 5.13: Schematic of delay diversity – a space selective channel at the transmitter is converted into a frequency selective channel at the receiver.

5.8   Indirect transmit diversity

So far we have discussed diversity schemes that directly extract the spatial diversity inherent in MIMO systems. Diversity schemes for SISO data transmission over time or frequency selective channels (time/frequency diversity) have been well studied [Biglieri *et al.*, 1991]. In this section we see how spatial diversity may be converted to time/frequency diversity which can be readily exploited using standard techniques (such as forward error correction (FEC) and interleaving). We discuss two cases: delay diversity, which converts space to frequency diversity, and phase-roll diversity, which converts space to time diversity.

5.8.1   Delay diversity

We assume $M_T = 2$ and $M_R = 1$ (i.e., a MISO channel). Delay diversity techniques [Seshadri and Winters, 1994] convert the available spatial diversity into frequency diversity by transmitting the data signal from the first antenna and a delayed replica of the same from the second antenna (see Fig. 5.13).

Assuming that the delay is one symbol interval, the effective channel seen by the transmitted data signal is a SISO channel given by

$$h[i] = h_1\delta[i] + h_2\delta[i-1], \ i = 0, 1, 2, \ldots, \tag{5.67}$$

where h_1 and h_2 are the channel gains between transmit antennas 1 and 2 and the receive antenna respectively. We assume h_1 and h_2 are IID ZMCSCG random variables with unit variance. Such a channel looks exactly like a two-path (symbol spaced) SISO channel with independent path fading and equal average path energy. An ML detector can capture full second-order diversity at the receiver (see Section 6.3.3).

5.8.2   Phase-roll diversity

An indirect transmit diversity technique for converting spatial diversity into temporal diversity was first described in [Hiroike *et al.*, 1992]. Assume that $M_T = 2$ and $M_R = 1$.

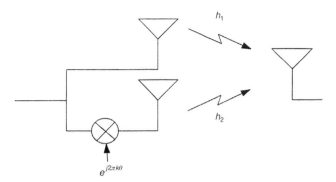

Figure 5.14: Schematic of phase-roll diversity – a space selective channel at the transmitter is converted into a time selective channel at the receiver.

After coding and modulation, the data signal is transmitted from the first antenna and a frequency shifted (phase rotated) version of the same is transmitted simultaneously from the second antenna (see Fig. 5.14).

The effective channel in time is a SISO channel given by

$$h[k] = h_1 + h_2 e^{j2\pi k\theta}, \quad k = 0, 1, 2, \ldots, \tag{5.68}$$

where θ ($|\theta| < 1/2$) is the frequency offset introduced at the second antenna. Once again, we assume that h_1 and h_2 are IID complex Gaussian random variables. To see how spatial diversity is converted into temporal diversity, we compute the temporal correlation function of the channel $h[k]$, i.e., $R_k[\Delta k] = \frac{1}{2}\mathcal{E}\{h[k]h^*[k + \Delta k]\}$. Using Eq. (5.68) we get

$$R_k[\Delta k] = \frac{1}{2}(1 + e^{-j2\pi \Delta k\theta}). \tag{5.69}$$

If $\Delta k\theta = \frac{1}{2}, \frac{3}{2}, \frac{5}{2}, \ldots$ then $R_k[\Delta k] = 0$ and symbols spaced Δk symbol intervals apart undergo independent fading. The resulting temporal diversity can be exploited by using FEC in combination with time interleaving, just as we do in time selective fading channels.

5.9   Diversity of a space-time-frequency selective fading channel

Channels exhibiting space(transmit–receive)-time-frequency selective fading offer diversity in all four dimensions that can be exploited by the wireless link. The actual diversity captured by the modem depends on the available (or inherent) diversity in the link, the coding and modulation scheme, and the receiver design. The available diversity in the channel depends on the codeword dimensions and the coherence parameters. Codeword dimensions are the number of transmit (M_T) and receive (M_R) antennas, the duration (T) of the codeword and the signal bandwidth (B). The coherence parameters

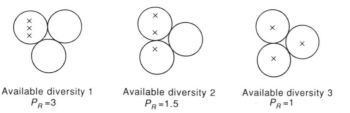

Available diversity 1   Available diversity 2   Available diversity 3
$P_R=3$ $P_R=1.5$ $P_R=1$

Figure 5.15: Packing factor P_R and available diversity in a three-element array. The diameter of the circles is equal to the coherence distance D_C and × represents an antenna location.

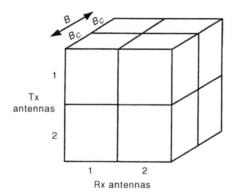

Figure 5.16: Schematic of the diversity composition of a ST channel with $M_T = M_R = 2$, $B/B_C = 2$. Each inner-cube represents one diversity dimension.

of the channel are the coherence time (T_C), the coherence bandwidth (B_C) and the coherence distance (D_C), defined in Chapter 2. The number of independent diversity branches (the available diversity) in time the codeword can capture in duration T is T/T_C. Likewise B/B_C is the number of independent frequency diversity branches. In space, the available diversity branches depend on the number of antennas and the topology of the transmit–receive antenna array. In turn, this depends on the packing factor or the number of coherence distances (or areas) occupied by at least one antenna. Figure 5.15 shows how a three-element receive array can have an available diversity order of one, two or three. Note that the coherence circles need not be circular as shown (true in cylindrically isotropic channels) and will in general be elliptical or even less regular. If P_R represents the packing factor for the receive array, then the available diversity is M_R/P_R. Similar comments apply to the available transmit diversity and packing factor P_T. Furthermore, even if $P_T = P_R = 1$, i.e., the antennas are independently fading, in order to extract all $M_T M_R$ order diversity, two conditions must be satisfied: (a) for every transmit antenna, the M_R receive antennas must show independent fading, i.e., the receive antennas must be spaced D_C (at receiver) apart and (b) for every receive antenna the M_T transmit antennas must show independent fading, i.e., the transmit antennas must be spaced D_C (at transmitter) apart. The available diversity (see Fig. 5.16) offered by the channel for a codeword spread across duration T, bandwidth B, M_T transmit and M_R receive antennas is then given by

$$\text{available diversity} \leq \frac{T}{T_C} \times \frac{B}{B_C} \times \frac{M_T}{P_T} \times \frac{M_R}{P_R}. \tag{5.70}$$

Equality in Eq. (5.70) is reached only when there is full independence of fading across the space, time and frequency dimensions (see [Sayeed and Veeravalli, 2002] for details).

The actual diversity realized by the modem depends on the codeword design and the receiver design and is clearly upper-bounded by the available diversity. Further discussion on codeword and receiver design to maximize actual diversity follows in Chapters 6 and 7.

6 ST coding without channel knowledge at transmitter

6.1 Introduction

In this chapter we address how coding can be used across space and time to maximize link performance. Our broad goals are, of course, to maximize link throughput and minimize error. These goals can be translated to supporting performance criteria such as the signaling rate (in bps/Hz or bits per transmission), the diversity gain (or diversity order, which is the slope of the error vs SNR curve), the coding gain (from code design that increases effective SNR), and the array gain (from antenna combining that also increases effective SNR). In this chapter we assume block fading with no channel knowledge at the transmitter. We are interested in: (a) improving error performance – this implies maximizing diversity which we know from Chapter 5 is upper-bounded by $M_T M_R$ in MIMO $\mathbf{H}_w$ channels; (b) increasing coding gain, which depends on the minimum distance of the code; and (c) increasing array gain, which is upper-bounded by M_R. The chapter discusses a variety of ST coding schemes that support different tradeoffs between rate, diversity and coding/array gain. Our formulation focuses on optimizing the average error rate performance with SC modulation.

We begin this chapter with a brief discussion on coding architectures and introduce the rate definitions used to classify various coding schemes that will be used in the remainder of the book. Next, we introduce ST coding techniques that focus on schemes extracting $M_T M_R$ order diversity. We formulate code construction criteria that extract diversity gain and coding gain [Tarokh *et al.*, 1998]. We introduce two forms of ST diversity coding: (a) ST trellis coding (STTC), where the codes have a trellis definition; and (b) ST block coding (STBC), where a block description is more appropriate. Several examples are provided to help to clarify concepts and develop intuition. Next we discuss spatial multiplexing (SM) with the full spatial rate (M_T) as a special case of ST coding and introduce some popular SM encoding architectures. We generalize the ST coding design to allow any spatial rate up to M_T and discuss a ST block coding scheme [Heath *et al.*, 2001; Heath and Paulraj, 2002] as an example.

We then focus on ST coding in frequency selective channels using SC modulation. We show that the code design criterion to guarantee full spatio-temporal diversity is

Figure 6.1: Coding architecture. The signaling rate is the product of the logarithm of the modulation order (q), the temporal coding rate (r_t) and the spatial rate (r_s).

similar to that in the flat fading case. We discuss the generalized delay diversity scheme for frequency selective channels and the Lindskog–Paulraj scheme that extends the Alamouti ST code to frequency selective channels.

6.2   Coding and interleaving architecture

We now propose a general coding architecture for transmission over multiple antennas. Consider Fig. 6.1. A block of qK bits is input to a block that performs the functions of temporal coding, interleaving and symbol mapping. In the process $q(N - K)$ parity bits are added and N symbols are output. 2^q is the modulation order (e.g., 4 if 4-QAM modulation is used). The N symbols are now input to a ST coder that adds an additional $M_T T - N$ parity symbols and packs the resulting $M_T T$ symbols into an $M_T \times T$ frame of length T. The block/frame is then transmitted over T symbol periods and is referred to in this chapter as the ST codeword. The signaling (data) rate on the channel is clearly qK/T bits/transmission and should not exceed the channel capacity if we wish to signal with zero error. Note that we can rewrite the signaling rate as

$$\frac{qK}{T} = q \left(\frac{qK}{qN} \right) \left(\frac{N}{T} \right)$$
$$= q\, r_t\, r_s, \tag{6.1}$$

where $r_t = qK/qN$ is the temporal code rate of the outer encoder and $r_s = N/T$ is the spatial code rate defined as the average number of independent symbols (constructed from the N input symbols) transmitted from the M_T antennas over T symbol periods. When all transmit antennas send one symbol per symbol period we get $r_s = 1$. On the other hand, in SM we send M_T independent symbols per symbol period to get $r_s = M_T$. Depending on the choice of ST coding, the spatial rate r_s varies between 0 to M_T. For certain classes of codes discussed below such as STTC, the functions of the symbol mapper and ST encoder are combined into a single block.

The generic coding architecture is clearly block oriented, but it applies equally to convolutional or trellis coding schemes, where the blocks are appropriately defined in terms of rates and other appropriate metrics, not in terms of block sizes.

ST codes can usefully exploit concatenated coding ideas to efficiently exploit the error correcting behavior of different codes (see [Sampath *et al.*, 2002] for details of use of concatenated coding in practical MIMO systems). Generic multiple antenna coding (such as V-BLAST) often uses parallel scalar (single input stream and single output stream) coding within the overall coding framework and each stream is accessible separately at the receiver for decoding. In principle, any scalar coding/decoding, including turbo codes, can be employed in such schemes. A deeper understanding of code construction in fading MIMO channels is an area of active study. See the comments at the end of Section 12.5.

Interleaving

Interleaving is used to spread burst errors due to fades across codewords or constraint lengths to improve error correction performance. Ignoring time selective and frequency selective fading (for the SC modulation case discussed here), interleaving in ST coding is motivated by the need to mitigate space selective fading across transmit antennas and is absolutely necessary to exploit all available spatial diversity. The ST diversity codes discussed in Section 6.3.3 have an inherent form of spatial interleaving built into the code framework, i.e., the ST codes have a spatial interleaver built into the code structure to exploit full $M_T M_R$ order diversity. Spatial interleaving can also be used with SM through stream rotation and is discussed in Section 6.3.5.

6.3   ST coding for frequency flat channels

This section derives the well-known rank and determinant criteria [Tarokh *et al.*, 1998] for ST diversity coding (we assume the channel is unknown to the transmitter).

6.3.1   Signal model

Consider a MIMO system with M_T transmit antennas and M_R receive antennas. Assume the information bit stream to be transmitted is encoded into a ST codeword of dimension $M_T \times T$, where T is the block length, comprising data symbols drawn from unit average energy constellations. We denote the ST codeword by $\mathbf{S} = [\mathbf{s}[1]\ \mathbf{s}[2]\ \cdots\ \mathbf{s}[T]]$, where $\mathbf{s}[k] = [s_1[k]\ \ldots\ s_{M_T}[k]]^T$, is the transmitted vector symbol over the kth symbol period. The channel is assumed to be IID Gaussian with quasi-static flat fading, i.e., the channel remains constant over the length of the ST codeword. The signal model from

Chapter 3 is

$$\mathbf{y}[k] = \sqrt{\frac{E_s}{M_T}} \mathbf{H}\mathbf{s}[k] + \mathbf{n}[k], \quad k = 1, 2, \ldots, T, \tag{6.2}$$

where E_s is the total average energy available at the transmitter over a symbol period, $\mathbf{H}$ is the $M_R \times M_T$ channel transfer function and $\mathbf{n}[k]$ is the ZMCSCG noise vector with covariance matrix $N_o \mathbf{I}_{M_R}$. All T received vector symbols in the codeword may be stacked together so that

$$\mathbf{Y} = \sqrt{\frac{E_s}{M_T}} \mathbf{H}\mathbf{S} + \mathbf{N}, \tag{6.3}$$

where $\mathbf{Y} = [\mathbf{y}[1] \; \mathbf{y}[2] \; \cdots \; \mathbf{y}[T]]$ and $\mathbf{N} = [\mathbf{n}[1] \; \mathbf{n}[2] \; \cdots \; \mathbf{n}[T]]$ are matrices of size $M_R \times T$.

Decoding

Assume the receiver uses a ML detection criterion based on perfect channel knowledge. The estimated codeword is

$$\widehat{\mathbf{S}} = \arg \min_{\mathbf{S}} \|\mathbf{Y} - \sqrt{\frac{E_s}{M_T}} \mathbf{H}\mathbf{S}\|_F^2$$

$$= \arg \min_{\mathbf{S}} \sum_{k=1}^{T} \left\| \mathbf{y}[k] - \sqrt{\frac{E_s}{M_T}} \mathbf{H}\mathbf{s}[k] \right\|_F^2, \tag{6.4}$$

where the minimization is performed over all admissible codewords $\mathbf{S}$. An error occurs when the receiver mistakes a transmitted codeword for another codeword from the set of possible codewords.

6.3.2   ST codeword design criteria

Assume that a codeword $\mathbf{S}^{(i)}$ is transmitted. Given that the receiver constructs a ML estimate of the transmitted codeword according to Eq. (6.4), the probability that the receiver mistakes the transmitted codeword $\mathbf{S}^{(i)}$ for another codeword $\mathbf{S}^{(j)}$, given knowledge of the channel realization at the receiver (also referred to as the pairwise error probability (PEP)), is

$$P(\mathbf{S}^{(i)} \to \mathbf{S}^{(j)}|\mathbf{H}) = Q\left(\sqrt{\frac{E_s \|\mathbf{H}(\mathbf{S}^{(i)} - \mathbf{S}^{(j)})\|_F^2}{2 M_T N_o}} \right)$$

$$= Q\left(\sqrt{\frac{\rho \|\mathbf{H}\mathbf{E}_{i,j}\|_F^2}{2 M_T}} \right), \tag{6.5}$$

where $\mathbf{E}_{i,j} = \mathbf{S}^{(i)} - \mathbf{S}^{(j)}$ is the $M_T \times T$ codeword difference matrix and $\rho = E_s/N_o$ is the SNR. Applying the Chernoff bound, we get

$$P(\mathbf{S}^{(i)} \to \mathbf{S}^{(j)}|\mathbf{H}) \leq e^{-\frac{\rho\|\mathbf{H}\mathbf{E}_{i,j}\|_F^2}{4M_T}}. \tag{6.6}$$

Let $\mathbf{h}_i$ $(m = 1, 2, \ldots, M_R)$ be the rows of $\mathbf{H}$. Then, $\|\mathbf{H}\mathbf{E}_{i,j}\|_F^2$ may be rewritten as

$$\|\mathbf{H}\mathbf{E}_{i,j}\|_F^2 = \sum_{m=1}^{M_R} \mathbf{h}_m \mathbf{E}_{i,j} \mathbf{E}_{i,j}^H \mathbf{h}_m^H$$
$$= \|\mathcal{H}\mathcal{E}_{i,j}\|_F^2, \tag{6.7}$$

where $\mathcal{E}_{i,j} = \mathbf{I}_{M_R} \otimes \mathbf{E}_{i,j}$ and $\mathcal{H} = \text{vec}(\mathbf{H}^T)^T$. Combining Eqs. (6.6) and (6.7), the PEP averaged over all channel realizations may be upper-bounded as

$$P(\mathbf{S}^{(i)} \to \mathbf{S}^{(j)}) \leq \frac{1}{\det\left(\mathbf{I}_{M_R M_T} + \frac{\rho}{4M_T}\mathcal{E}\{\mathcal{E}_{i,j}^H \mathcal{H}^H \mathcal{H} \mathcal{E}_{i,j}\}\right)}, \tag{6.8}$$

which follows from $\mathcal{H}$ being a Gaussian vector (of dimension $1 \times M_R M_T$). Assuming $\mathbf{H} = \mathbf{H}_w$, straightforward manipulations lead to

$$P(\mathbf{S}^{(i)} \to \mathbf{S}^{(j)}) \leq \left(\frac{1}{\det\left(\mathbf{I}_{M_T} + \frac{\rho}{4M_T}\mathbf{E}_{i,j}\mathbf{E}_{i,j}^H\right)}\right)^{M_R}. \tag{6.9}$$

This can be rewritten as

$$P(\mathbf{S}^{(i)} \to \mathbf{S}^{(j)}) \leq \prod_{k=1}^{r(\mathbf{G}_{i,j})} \left(\frac{1}{1 + \rho\lambda_k(\mathbf{G}_{i,j})/4M_T}\right)^{M_R}, \tag{6.10}$$

where $\lambda_k(\mathbf{G}_{i,j})$ $(k = 1, 2, \ldots, r(\mathbf{G}_{i,j}))$ are the non-zero eigenvalues of $\mathbf{G}_{i,j} = \mathbf{E}_{i,j}\mathbf{E}_{i,j}^H$. In the high SNR regime $(\rho \gg 1)$, Eq. (6.10) may be further simplified as

$$P(\mathbf{S}^{(i)} \to \mathbf{S}^{(j)}) \leq \frac{1}{\left(\prod_{k=1}^{r(\mathbf{G}_{i,j})} \lambda_k(\mathbf{G}_{i,j})\right)^{M_R}} \left(\frac{\rho}{4M_T}\right)^{-r(\mathbf{G}_{i,j})M_R}. \tag{6.11}$$

Equation (6.11) leads us to the two well-known criteria for ST codeword construction, namely the "rank criterion" and the "determinant criterion" [Tarokh et al., 1998]. We examine these criteria more closely in the following discussion.

Rank criterion

The rank criterion optimizes the spatial diversity extracted by a ST code. From Eq. (6.11) note that the ST code extracts $r(\mathbf{G}_{i,j})M_R$ order diversity ($r(\mathbf{G}_{i,j})$ being the rank of $\mathbf{G}_{i,j}$). Clearly, to extract the full spatial diversity gain of $M_T M_R$, the code design should be such that the difference matrix between any pair of codeword matrices $\mathbf{S}^{(i)}$ and $\mathbf{S}^{(j)}$ is full-rank ($r(\mathbf{G}_{i,j}) = M_T$).

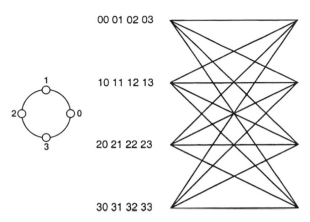

Figure 6.2: Trellis diagram for a 4-QAM, four-state trellis code for $M_T = 2$ with a rate of 2 bps/Hz.

Determinant criterion

The determinant criterion optimizes the coding gain. From Eq. (6.11), it is clear that the coding gain depends on the term $\prod_{k=1}^{r(\mathbf{G}_{i,j})} \lambda_k(\mathbf{G}_{i,j})$. Hence, for high coding gain, one should maximize the minimum of the determinant of $\mathbf{E}_{i,j}\mathbf{E}_{i,j}^H$ over all possible pairs of codeword matrices $\mathbf{S}^{(i)}$ and $\mathbf{S}^{(j)}$. The determinant criterion provides rules for maximizing coding gain. It does not, however, provide an accurate estimate of the true coding gain.

The design criteria for arbitrary diversity order can be found in [Tarokh *et al.*, 1998]. Extracting full diversity order through ST coding requires only the rank criterion to be satisfied.

6.3.3   ST diversity coding ($r_s \leq 1$)

In this section we discuss two flavors of ST diversity codes – STTC and STBC, which extract full diversity order ($M_T M_R$) with spatial rate $r_s \leq 1$.

STTC

STTC are an extension of conventional trellis codes [Biglieri *et al.*, 1991] to multi-antenna systems. These codes may be designed to extract diversity gain and coding gain using the criteria described in the previous section. Binary criteria to identify ST trellis codes that yield full spatial diversity have been discussed in [Hammons and El Gamal, 2000]. Each STTC can be described using a trellis. The number of nodes in the trellis diagram corresponds to the number of states in the trellis. Each node has A groups of symbols to the left (A being the constellation size), with each group consisting of M_T entries. Each group corresponds to the output for a given input symbol. The M_T entries in each group correspond to the symbols to be transmitted from the M_T antennas. This description is clarified further through the illustration in Fig. 6.2.

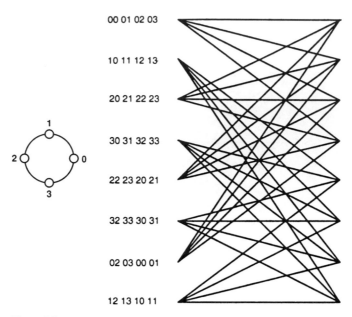

Figure 6.3: Trellis diagram for 4-QAM, eight-state, trellis code for $M_T = 2$ with a rate of 2 bps/Hz.

Figure 6.2 shows the trellis diagram for a simple 4-QAM, four-state trellis code, for $M_T = 2$ with rate 2 bps/Hz. The trellis has four nodes corresponding to four states. There are four groups of symbols at the left of every node since there are four possible inputs (4-QAM constellation). Each group has two entries corresponding to the symbols to be output through the two transmit antennas. For example, 02 (third entry for first node) corresponds to symbols 2 and 0 being output from antennas 1 and 2 respectively, when the input is the third symbol. The outputs 0, 1, 2 and 3 are mapped to data symbols 1, j, -1 and $-j$ respectively. The encoder is required to be in the zero state at the beginning and at the end of each frame (block). Beginning in state 0, if the incoming two bits are 10, the encoder outputs a 2 on antenna 1 and a 0 on antenna 2 and changes state to state 2.

Figure 6.3 shows the trellis digram for a 4-QAM ST trellis code with eight states. The trellis is to be read in a similar fashion to the trellis described in Fig. 6.2. The codes have been designed by hand so as to maximize the coding gain for a given rate, diversity order, constellation and ease of decoding [Tarokh *et al.*, 1998].

The transmitted frame is decoded at the receiver using ML sequence estimation (MLSE) using the Viterbi algorithm. Figures 6.4 and 6.5 compare the frame error rate (each frame comprises 130 transmissions) as a function of SNR for the two trellis codes described in Figs. 6.2 and 6.3, and for systems with one and two receive antennas respectively. These show that increasing the number of states increases the coding gain advantage. The coding gain also seems to increase with the number of receive antennas (part of which is due to the array gain through multiple receive antennas). Further, the diversity order realized by a system with two receive antennas is double that of a system with a single receive antenna.

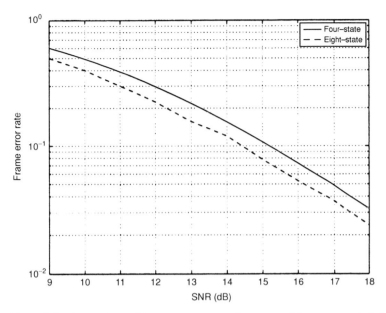

Figure 6.4: Comparison of frame error rate performance of four-state and eight-state trellis codes for $M_T = 2$, $M_R = 1$. Increasing the number of states increases the coding gain.

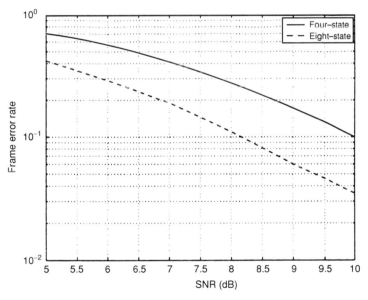

Figure 6.5: Comparison of the frame error rate performance of four-state and eight-state trellis codes for $M_T = 2$, $M_R = 2$. Fourth-order diversity is achieved in both codes.

Delay diversity as a STTC

The delay diversity scheme discussed in Chapter 5 can be recast as a STTC. Assume a system with two transmit antennas and one receive antenna. Delay diversity transmission involves transmitting the symbol stream over one antenna and a one-symbol

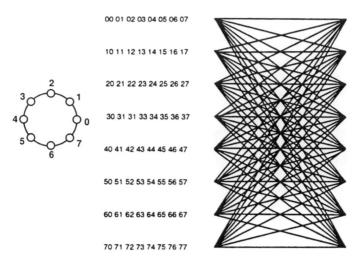

Figure 6.6: Trellis diagram for delay diversity code with 8-PSK transmission and $M_T = 2$.

delayed replica of the symbol stream over the second antenna. Figure 6.6 shows the trellis diagram for a delay diversity code for 8-PSK transmission over a system with two transmit antennas.

Observe that the ST codeword for delay diversity transmission of T data symbols s_i $(i = 1, 2, \ldots, T)$ over two transmit antennas may be expressed as

$$\mathbf{S} = \begin{bmatrix} s_1 & s_2 & s_3 & s_4 & \cdots & s_T & 0 \\ 0 & s_1 & s_2 & s_3 & \cdots & s_{T-1} & s_T \end{bmatrix}. \tag{6.12}$$

It is easily verified that all possible codeword difference matrices described in Eq. (6.12) have rank 2. Therefore, applying the rank criterion for ST codeword design described in Section 6.3.2, the delay diversity transmission extracts the full diversity order of $2M_R$.

STTC are an effective means of capturing diversity. However, the computational complexity for decoding a STTC increases exponentially with the number of states. In contrast, the encoding/decoding associated with the Alamouti scheme described in Chapter 5 is simpler than that for STTC discussed above. The Alamouti scheme is therefore appealing in terms of simplicity and performance and we next discuss extending Alamouti type schemes to systems with more than two transmit antennas.

STBC from orthogonal designs

We study the performance of the Alamouti scheme [Alamouti, 1998] from the rank and determinant criteria discussed previously. Recall from Chapter 5 that given symbols s_1 and s_2 to be transmitted, the Alamouti scheme transmits symbols s_1 and s_2 from antennas 1 and 2 respectively, during the first symbol period, followed by $-s_2^*$ and s_1^* from antennas 1 and 2 respectively during the following symbol period. Hence, the transmitted ST codeword may be expressed as

$$\mathbf{S} = \begin{bmatrix} s_1 & -s_2^* \\ s_2 & s_1^* \end{bmatrix}. \tag{6.13}$$

It is easy to verify that the codeword difference matrix between any pair of codewords, say $\mathbf{S}^{(i)}$ and $\mathbf{S}^{(j)}$, is of the form

$$\mathbf{E}_{i,j} = \begin{bmatrix} e_1 & -e_2^* \\ e_2 & e_1^* \end{bmatrix}. \tag{6.14}$$

Clearly, $\mathbf{E}_{i,j}$ is an orthogonal matrix with two non-zero eigenvalues (rank 2) of equal magnitude. The Alamouti scheme therefore delivers full $2M_R$ order diversity, where M_R is the number of receive antennas. More importantly, the structure of the transmitted signal is such that the effective channel is rendered orthogonal regardless of the channel realization, thus decoupling the otherwise complex vector ML detection problem into simpler scalar detection problems as discussed in Chapter 5. Recall that the receiver output is

$$y_i = \sqrt{\frac{E_s}{2}} \|\mathbf{H}\|_F^2 s_i + n_i, \quad i = 1, 2, \tag{6.15}$$

where y_i is the scalar processed received signal corresponding to transmitted symbol s_i and n_i is ZMCSCG noise with variance $\|\mathbf{H}\|_F^2 N_o$.

ST code construction for Alamouti type schemes ($M_T = 2$) can be generalized using orthogonal designs for $M_T > 2$. In fact orthogonal ST codewords for real constellations may be designed [Tarokh et al., 1999b] for systems with any number of transmit antennas using the solution to the Hurwitz–Radon problem [Radon, 1922]. An example of an orthogonal design for $M_T = 4$ is

$$\mathbf{S} = \begin{bmatrix} s_1 & -s_2 & -s_3 & -s_4 \\ s_2 & s_1 & s_4 & -s_3 \\ s_3 & -s_4 & s_1 & s_2 \\ s_4 & s_3 & -s_2 & s_1 \end{bmatrix}, \tag{6.16}$$

where symbols s_1, s_2, s_3 and s_4 are all drawn from a real constellation. It is easy to verify that the difference between any two such codewords, say $\mathbf{S}^{(i)}$ and $\mathbf{S}^{(j)}$, is an orthogonal matrix, $\mathbf{E}_{i,j}$. From Eq. (6.11) the average PEP in the high SNR regime for an orthogonal STBC (OSTBC) is

$$P(\mathbf{S}^{(i)} \to \mathbf{S}^{(j)}) \le \left(\frac{M_T}{\|\mathbf{E}_{i,j}\|_F^2} \right)^{M_T M_R} \left(\frac{\rho}{4M_T} \right)^{-M_T M_R}. \tag{6.17}$$

Clearly, OSTBC extract the full diversity gain of $M_T M_R$. Further, simple linear processing at the receiver decouples the vector detection problem into simpler scalar detection problems resulting in a simple input–output relation similar to that for the Alamouti scheme described in Eq. (6.15). The spatial rate for the ST codes discussed so far is 1.

In the case of complex constellations, we know [Tarokh *et al.*, 1999b; Ganesan and Stoica, 2001; Wang and Xia, 2002] that an orthogonal design with spatial rate 1 does not exist for systems with more than two transmit antennas. The Alamouti scheme is a rate 1 design for a system with two transmit antennas. However, orthogonal designs for rates less than or equal to $\frac{1}{2}$ have been shown to exist for systems with any number of transmit antennas. A rate $\frac{1}{2}$ orthogonal design for a system with three antennas is shown below:

$$\mathbf{S} = \begin{bmatrix} s_1 & -s_2 & -s_3 & -s_4 & s_1^* & -s_2^* & -s_3^* & -s_4^* \\ s_2 & s_1 & s_4 & -s_3 & s_2^* & s_1^* & s_4^* & -s_3^* \\ s_3 & -s_4 & s_1 & s_2 & s_3^* & -s_4^* & s_1^* & s_2^* \end{bmatrix}. \tag{6.18}$$

OSTBC with rates greater than $\frac{1}{2}$ are known to exist for systems with three or four transmit antennas. One such code with $r_s = 3/4$ for a three-transmit-antenna system is

$$\mathbf{S} = \begin{bmatrix} s_1 & -s_2^* & \dfrac{s_3^*}{\sqrt{2}} & \dfrac{s_3^*}{\sqrt{2}} \\ s_2 & s_1^* & \dfrac{s_3^*}{\sqrt{2}} & -\dfrac{s_3^*}{\sqrt{2}} \\ \dfrac{s_3}{\sqrt{2}} & \dfrac{s_3}{\sqrt{2}} & \dfrac{-s_1 - s_1^* + s_2 - s_2^*}{2} & \dfrac{s_2 + s_2^* + s_1 - s_1^*}{2} \end{bmatrix}. \tag{6.19}$$

A simpler form of the above code has been presented in [Ganesan and Stoica, 2001]. Higher rate quasi-orthogonal ST codes have been proposed in [Papadias and Foschini, 2001; Jafarkhani, 2001; Sharma and Papadias, 2002]. Though OSTBC are attractive due to their low implementation and decoding complexity, they will be outperformed by STTC designed to optimize the rank and determinant criteria. However, OSTBC concatenated with standard AWGN codes can outperform some of the best-known STTC (with the same transmit power and signaling rate) in terms of error performance [Sandhu *et al.*, 2001].

Thus far we have discussed ST coding for frequency flat Rayleigh fading channels. Their performance in correlated or Ricean channels or in presence of channel estimation errors can be found in [Naguib *et al.*, 1998b; Fitz *et al.*, 1999; Tarokh *et al.*, 1999a, d; Bölcskei and Paulraj, 2000a; Uysal and Georghiades, 2001; Nabar *et al.*, 2002a, b]. Additionally, design criteria for ST codeword construction for optimizing outage error rate performance rather than average error rate performance can be found in [Gorokhov, 2001]. ST coding when the transmit antennas are grouped for decoding is discussed in [Tarokh *et al.*, 1999c].

6.3.4   Performance issues

As stated in the introduction to this chapter, the principal performance metrics are error performance and signaling rate. The error performance depends on the diversity order and coding gain. The rank criterion provides a useful tool for designing codes that provide full spatial diversity. Both STTC and STBC, if designed correctly, achieve $M_T M_R$ diversity. However, OSTBC do not provide coding gain. On the contrary, a properly designed STTC will have both diversity gain and coding gain. The OSTBC error rate curve may be used as a baseline to estimate the coding gain of various STTC. Care must be taken to subtract out the array gain if using multiple receive antennas since the effect (parallel shift of the curves) is indistinguishable from the effect of coding gain. Our code design criteria has focused on the PEP, i.e., the error probability between a pair of codewords. Typically we must use the PEP of the worst-case nearest neighbor codeword pair to design the ST code. However, the true error rate must take into account other types of error events (not necessarily nearest neighbor) and also their relative frequencies. PEP based designs may not therefore truly optimize link performance. A better criterion would be to use the union bound which is an upper bound on the error rate. Analysis of ST code design based on union bound and alternative criteria can be found in [Bouzekri and Miller, 2001; Sandhu, 2002; Biglieri, 2002].

6.3.5   Spatial multiplexing as a ST code ($r_s = M_T$)

Thus far we have considered codes with spatial rate $r_s \leq 1$ and diversity order $M_T M_R$, where there was one or less independent symbol transmitted per symbol period over the M_T transmit antennas. In this section we consider SM, where we transmit M_T independent symbols per symbol period. We begin with an uncoded SM scheme where the input data stream is $1 : M_T$ demultiplexed and the resultant sub-streams are transmitted over M_T antennas.

In uncoded SM (as above), $r_t = 1$ and $r_s = M_T$, resulting in a signaling rate of $q M_T$ bits/transmission. The receiver treats each received signal vector as a codeword, i.e., $T = 1$, and performs ML decoding over every vector symbol. The performance analysis follows easily once we notice that the codeword difference matrix, $\mathbf{E}_{i,j}$, is now an $M_T \times 1$ vector defined as $\mathbf{E}_{i,j} = \mathbf{s}^{(i)} - \mathbf{s}^{(j)}$, where $\mathbf{s}^{(i)}$ and $\mathbf{s}^{(j)}$ are two possible transmitted vector codewords. Furthermore, $\mathbf{E}_{i,j}\mathbf{E}_{i,j}^H$ is now a rank 1 matrix. Applying the above observations to Eq. (6.11), we can write the average PEP as

$$P(\mathbf{s}^{(i)} \rightarrow \mathbf{s}^{(j)}) \leq \frac{1}{\lambda(\mathbf{G}_{i,j})^{M_R}} \left(\frac{\rho}{4M_T} \right)^{-M_R}, \tag{6.20}$$

where $\lambda(\mathbf{G}_{i,j}) = \mathbf{E}_{i,j}^H \mathbf{E}_{i,j}$ since $\mathbf{G}_{i,j}$ is rank 1. Notice that the diversity order is M_R. To summarize, SM with no coding may be considered as a ST code with spatial rate

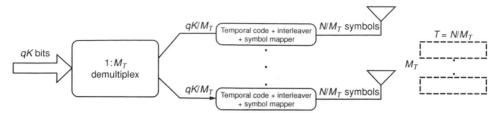

Figure 6.7: Horizontal encoding. This is a sub-optimal encoding technique that captures at most M_R order diversity.

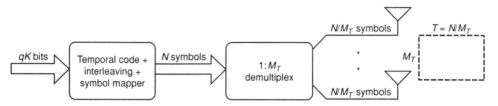

Figure 6.8: Vertical encoding allows spreading of information bits across all antennas. It usually requires complex decoding techniques.

M_T with M_R order diversity. The performance of SM with ML decoding is revisited in Chapter 7 with an additional discussion on sub-optimal receivers. Next, we briefly discuss popular encoder structures used for SM.

Horizontal encoding (HE)

In HE the bit stream is first demultiplexed into M_T separate streams (see Fig. 6.7). Each stream undergoes independent temporal coding, interleaving and symbol mapping, and is transmitted from one antenna. The spatial rate is clearly $r_s = M_T$. The signaling rate is therefore $q r_t M_T$ bits/transmission. The HE scheme (like uncoded SM) can at most achieve M_R order diversity, since any given symbol is transmitted from only one transmit antenna and received by M_R receive antennas. This is a source of the sub-optimality of this particular encoding architecture but it does simplify receiver design. Coding gain depends on the strength of the temporal code. Array gain of M_R is achievable.

Vertical encoding (VE)

In VE the bit stream undergoes temporal coding, interleaving and symbol mapping after which it is demultiplexed into M_T streams that are transmitted over the antennas (see Fig. 6.8). This form of encoding can reach optimality since potentially each information bit can be spread across all antennas. However, VE requires joint decoding of the sub-streams at the receiver and can be very complex.

The spatial rate is $r_s = M_T$ and the signaling rate is $q r_t M_T$ bits/transmission. Since the information symbols are spread over more than one antenna, VE can achieve a

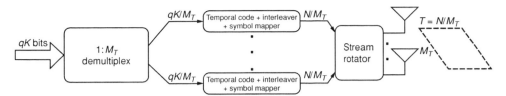

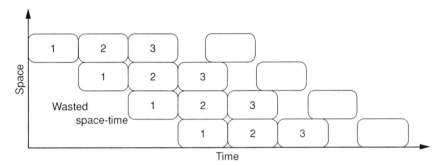

Figure 6.9: Diagonal encoding is HE with stream rotation. Stream rotation enables information bits to be spread across all antennas. D-BLAST transmission uses same encoding.

Figure 6.10: D-BLAST encoding – numerals represent layers belonging to the same codeword.

diversity order greater than M_R. For example, if we use a trivial repetition code with $r_t = 1/2$ with the repeated bits/symbol transmitted on the second antenna delayed by one symbol period, the scheme becomes similar to delay diversity (STTC), and would achieve $2M_R$ (for $M_T = 2$) diversity. Coding gain will depend on the temporal code design and array gain of M_R is achievable.

Various combinations/variations of the above schemes are possible. We discuss one such transmission technique (diagonal encoding – DE) below.

Diagonal encoding (DE)

In DE the incoming data stream first undergoes HE encoding (see Fig. 6.9) after which each codeword is split into frames/slots. These frames pass through a stream rotator that rotates the frames in a round robin fashion so that the bit stream–antenna association is periodically cycled. Making the codeword large enough ensures that the codeword from any one demultiplexed stream is transmitted over all M_T antennas. The D-BLAST [Foschini, 1996] transmission technique follows such an encoding strategy (includes an initial wasted triangular block (see Fig. 6.10) where no transmission takes place). This initial wastage is required to enable optimal decoding (see Chapter 11). The spatial rate is M_T and the signaling rate is $q r_t M_T$ bits/transmission.

D-BLAST like schemes can achieve full $M_T M_R$ diversity if the temporal coding with stream rotation is optimal (Gaussian code books with infinite block size). Coding gain will depend on the temporal code design and an array gain of M_R is achievable. A variation of the diagonal layered ST coding approach is the threaded ST code proposed by [El Gamal and Hammons, 2001]. The codeword in this case extends beyond one diagonal stripe to wrap around and span over multiple stripes. Threaded codes can potentially offer improved temporal diversity. This does not have the advantage of one codeword at a time layered decoding and needs joint decoding of multiple threads so that it is more complex to implement.

6.3.6 ST coding for intermediate rates ($1 < r_s < M_T$)

We have discussed ST coding schemes of $r_s \leq 1$ and $r_s = M_T$. A natural question is whether ST codes with $1 < r_s < M_T$ make sense, and how they may be motivated. Another point worth noting is that the design metrics (rank and determinant criteria) used so far to develop ST codes are not directly related to the capacity of the encoding scheme. In fact, when a particular choice of coding is used, the encoding scheme can be viewed as an operator on the channel, to yield a new effective channel whose capacity can be computed. Including both metrics in the design procedure is an area of active research. The linear dispersion framework proposed in [Hassibi and Hochwald, 2001] spreads the symbols to be transmitted across space and time through matrix modulation and superposition with the objective of ergodic capacity maximization. Similar schemes, designed for both diversity and capacity optimization, are discussed in [Heath *et al.*, 2001; Heath and Paulraj, 2002; Sandhu, 2002].

Below, we discuss a ST hybrid code similar in nature to the linear dispersion framework of [Hassibi and Hochwald, 2001] that offers spatial rates that range from 1 to M_T and that explicitly includes both capacity efficiency and diversity/coding gain metrics in its design.

Signal model
An input $N \times 1$ vector $\mathbf{s}$ of N complex data symbols is modulated by a code matrix of dimension $M_T \times N$, and transmitted over the $M_R \times M_T$ channel $\mathbf{H}$ for each symbol period. Assume that there are T such distinct matrices (code matrices), i.e., at time $1 \leq k \leq T$, signal $\mathbf{X}[k]\mathbf{s}$ is transmitted, where $\mathbf{X}[k]$ is the kth code matrix. The received symbol vector at time instant k is

$$\mathbf{y}[k] = \sqrt{\frac{E_s}{M_T}} \mathbf{H}\mathbf{X}[k]\mathbf{s} + \mathbf{n}[k], \tag{6.21}$$

where $\mathbf{n}[k]$ is the $M_R \times 1$ ZMCSCG noise vector and all other parameters remain as defined earlier. If we stack the T received vectors, we have a block signal model defined by

$$
\begin{bmatrix} \mathbf{y}[1] \\ \vdots \\ \mathbf{y}[T] \end{bmatrix} = \sqrt{\frac{E_s}{M_T}} \mathcal{H} \begin{bmatrix} \mathbf{X}[1] \\ \vdots \\ \mathbf{X}[T] \end{bmatrix} \mathbf{s} + \begin{bmatrix} \mathbf{n}[1] \\ \vdots \\ \mathbf{n}[T] \end{bmatrix},
\tag{6.22}
$$

or equivalently,

$$
\mathcal{Y} = \sqrt{\frac{E_s}{M_T}} \mathcal{H} \mathcal{X} \mathbf{s} + \mathcal{N},
\tag{6.23}
$$

where $\mathcal{Y} = [\mathbf{y}[1]^T \cdots \mathbf{y}[T]^T]^T$ is a vector of dimension $(M_R T \times 1)$, $\mathcal{H} = \mathbf{I}_T \otimes \mathbf{H}$ is a matrix of dimension $M_R T \times M_T T$, $\mathcal{X} = [\mathbf{X}[1]^T \cdots \mathbf{X}[T]^T]^T$ is a matrix of dimension $M_T T \times N$ and $\mathcal{N} = [\mathbf{n}[1]^T \cdots \mathbf{n}[T]^T]^T$ is the stacked noise vector of dimension $M_R T \times 1$.

Spatial rate

In every block of T symbol periods, N independent symbols are transmitted, so that the spatial rate is given by $r_s = N/T$. If $N = T$ we have a spatial rate of 1. The case of $N = T M_T$ corresponds to a spatial rate of M_T (SM). For $T < N < T M_T$ we have $1 < r_s < M_T$ corresponding to a signaling framework that "goes between" the extremes of ST diversity coding and SM.

Code design

The code design involves identifying the matrices $\mathbf{X}[k]$ $(k = 1, 2, \ldots, T)$ that constitute the code. The matrices are computed so as to maximize both diversity and ergodic capacity. Employing analysis techniques similar to those presented earlier, it can be shown that the average PEP, given perfect knowledge of the channel at the receiver, is upper-bounded as

$$
P(\mathbf{s}^{(i)} \to \mathbf{s}^{(j)}) \leq \mathcal{E} \left\{ e^{-\frac{\rho}{4M_T} \| \mathcal{H} \mathcal{X} \mathbf{e}_{i,j} \|_F^2} \right\},
\tag{6.24}
$$

where $\mathbf{e}_{i,j} = \mathbf{s}^{(i)} - \mathbf{s}^{(j)}$. Simplification yields

$$
P(\mathbf{s}^{(i)} \to \mathbf{s}^{(j)}) \leq \frac{1}{\det \left(\mathbf{I}_{M_T M_R} + (\rho/4M_T) \mathbf{I}_{M_R} \otimes \mathbf{R} \right)},
\tag{6.25}
$$

where $\mathbf{R} = \sum_{k=1}^{T} \mathbf{X}_k \mathbf{e}_{i,j} \mathbf{e}_{i,j}^H \mathbf{X}_k^H$. The diversity order of the code is clearly

$$
M_R \min \, r \left(\sum_{k=1}^{T} \mathbf{X}_k \mathbf{e}_{i,j} \mathbf{e}_{i,j}^H \mathbf{X}_k^H \right),
\tag{6.26}
$$

where the minimization is performed over all possible codeword error vectors $\mathbf{e}_{i,j}$. The diversity order ranges between $M_R M_T$ and M_R depending upon the choice of spatial signaling rate. Further, the ergodic capacity of this signaling scheme may be optimized according to

$$\overline{C} = \max_{\mathrm{Tr}(\mathcal{X}^H \mathcal{X}) = M_T T} \frac{1}{T} \mathcal{E} \left\{ \log_2 \det \left(\mathbf{I}_{M_R T} + \frac{\rho}{M_T} \mathcal{H} \mathcal{X} \mathcal{X}^H \mathcal{H}^H \right) \right\}. \tag{6.27}$$

This analysis is greatly simplified by working with an approximation (upper bound) instead. Applying Jensen's inequality, the code matrices may be chosen according to

$$\max_{\mathrm{Tr}(\mathcal{X}^H \mathcal{X}) = M_T T} \frac{1}{T} \log_2 \det \left(\mathbf{I}_N + \frac{\rho}{M_T} M_R \mathcal{X}^H \mathcal{X} \right). \tag{6.28}$$

The solution to Eq. (6.28) satisfies $\mathcal{X}^H \mathcal{X} = (M_T T / N) \mathbf{I}_N$, implying $\mathcal{X}$ is a tight frame. The code is designed by searching among this class of matrices for the matrix that minimizes Eq. (6.26) thereby maximizing diversity. Hence, the code is designed with the twin metrics of diversity and capacity. Note that several matrices may have the same (upper-bound) ergodic capacity. The diversity metric allows us to select a code that has good throughput as well as error rate performance. An example of one such code with parameters $M_T = 3$, $M_R = 2$, $T = 2$ and $N = 4$, corresponding to a spatial rate of $r_s = 2$ is presented below:

$$\mathbf{X}[1] = \begin{bmatrix} -0.0250 + 0.0991j & -0.3543 + 0.2061j & -0.1027 + 0.1441j & 0.1830 - 0.1459j \\ 0.3338 - 0.2626j & 0.1479 - 0.1324j & -0.0419 + 0.0670j & 0.0191 - 0.2380j \\ -0.0904 + 0.3446j & 0.1578 - 0.1835j & 0.2134 - 0.1749j & 0.3154 - 0.1131j \end{bmatrix},$$

$$\tag{6.29}$$

$$\mathbf{X}[2] = \begin{bmatrix} -0.2202 + 0.2648j & -0.0701 - 0.1779j & -0.3267 + 0.2913j & -0.1392 - 0.1645j \\ 0.0059 - 0.1472j & 0.0458 - 0.2895j & 0.1988 + 0.3879j & 0.2581 - 0.1384j \\ -0.2047 + 0.0073j & 0.2387 + 0.2336j & -0.0031 + 0.0690j & -0.0556 - 0.3750j \end{bmatrix}.$$

$$\tag{6.30}$$

Figure 6.11 compares the performance of this code with various other temporally un-coded signaling schemes such as the Alamouti coding scheme and SM. The spatial rate and signaling rate have been maintained at a constant rate of 2 and 4 bps/Hz respectively, for all schemes. The optimized hybrid code ($M_T = 3$, $M_R = 2$) outperforms all other schemes and has approximately fourth-order diversity (a slope similar to that achieved by the Alamouti code, $M_T = 2$, $M_R = 2$). The SM (uncoded) along with the ML receiver only reaches second-order diversity and will underperform Alamouti at higher SNRs.

The hybrid code achieves a tradeoff between spatial rate and diversity. For a spatial rate defined by N/T, $(N \leq M_T T)$, the code achieves a diversity order of $M_R \min(M_T, T)$.

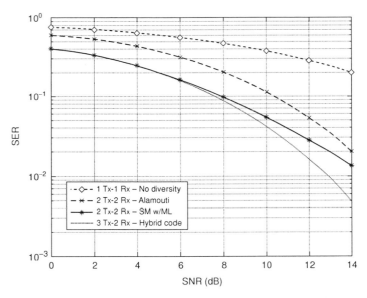

Figure 6.11: Performance of various signaling schemes. The rate is normalized to 4 bps/Hz.

6.4   ST coding for frequency selective channels

We now extend the ST code construction criteria to a delay spread channel with SC modulation [Gong and Letaief, 2000; Liu *et al.*, 2001a, b; Gore and Paulraj, 2001; Zhou and Giannakis, 2001]. We treat the OFDM case in Chapter 9 and show that up to $M_T M_R L_{eff}$ order diversity may be realized.

6.4.1   Signal model

Assume that the channel between the ith receive antenna and jth transmit antenna is frequency selective. The symbol-sampled baseband impulse response is denoted by $h_{i,j}[l]$, $(l = 0, \ldots, L - 1)$ and assumed to be complex circular Gaussian random variables with zero mean and correlations depending on the baseband pulse, the RF channel time response and the sampling frequency. As in the flat fading case, it is assumed that there is no channel knowledge at the transmitter and full channel knowledge with ML decoding at the receiver. In the following we assume $L_{eff} = L$. As described in Chapter 3, the signal model is as follows:

$$\mathbf{y}[k] = \sqrt{\frac{E_s}{M_T}} \begin{bmatrix} \mathbf{h}_{1,1} & \cdots & \mathbf{h}_{1,M_T} \\ \vdots & \vdots & \vdots \\ \mathbf{h}_{M_R,1} & \cdots & \mathbf{h}_{M_R,M_T} \end{bmatrix} \begin{bmatrix} \mathbf{s}_1[k] \\ \vdots \\ \mathbf{s}_{M_T}[k] \end{bmatrix} + \mathbf{n}[k], \tag{6.31}$$

where

$$\mathbf{h}_{i,j} = \left[h_{i,j}[L-1] \cdots h_{i,j}[0] \right], \; \mathbf{s}_j[k] = \begin{bmatrix} s_j[k-L+1] \\ \vdots \\ s_j[k] \end{bmatrix}. \tag{6.32}$$

T contiguous received samples may be stacked and written as

$$\left[\mathbf{y}[k] \cdots \mathbf{y}[k+T-1] \right] = \sqrt{\frac{E_s}{M_T}} \begin{bmatrix} \mathbf{h}_{1,1} & \cdots & \mathbf{h}_{1,M_T} \\ \vdots & \vdots & \vdots \\ \mathbf{h}_{M_R,1} & \cdots & \mathbf{h}_{M_R,M_T} \end{bmatrix} \begin{bmatrix} \mathcal{S}_1 \\ \vdots \\ \mathcal{S}_{M_T} \end{bmatrix} + \mathcal{N}, \tag{6.33}$$

or equivalently

$$\mathcal{Y} = \sqrt{\frac{E_s}{M_T}} \mathcal{H} \mathcal{S} + \mathcal{N}, \tag{6.34}$$

where $\mathcal{S}_j$ and $\mathcal{N}$ are Hankel blocks as defined earlier in Chapter 3. We may rewrite Eq. (6.34) as

$$\overline{\mathcal{Y}} = \sqrt{\frac{E_s}{M_T}} \overline{\mathcal{H}} \overline{\mathcal{S}} + \overline{\mathcal{N}}, \tag{6.35}$$

where $\overline{\mathcal{Y}} = \mathrm{vec}(\mathcal{Y}^T)^T$, $\overline{\mathcal{H}} = \mathrm{vec}(\mathcal{H}^T)^T$, $\overline{\mathcal{S}} = \mathbf{I}_{M_R} \otimes \mathcal{S}$ and $\overline{\mathcal{N}} = \mathrm{vec}(\mathcal{N}^T)^T$.

Decoding

The matrix $\mathcal{S}$ is just a rearranged version of the actual transmitted codeword $\mathbf{S}$ (as defined in Section 6.3 for a flat fading channel). In fact, $\mathcal{S}$ collapses to $\mathbf{S}$ for $L=1$, i.e., when there is no delay spread. From these observations and the fact that $\overline{\mathcal{S}}^{(i)} = \mathbf{I}_{M_R} \otimes \mathcal{S}^{(i)}$ we can conclude that the probability of mistaking $\overline{\mathcal{S}}^{(i)}$ for $\overline{\mathcal{S}}^{(j)}$, $j \neq i$ is the same as the probability of mistaking $\mathcal{S}^{(i)}$ for $\mathcal{S}^{(j)}$ which in turn is the same as the probability of mistaking $\mathbf{S}^{(i)}$ for $\mathbf{S}^{(j)}$.

Assuming ML decoding at the receiver and applying the Chernoff bound, the probability of decoding $\overline{\mathcal{S}}^{(i)}$ for $\overline{\mathcal{S}}^{(j)}$ may be upper-bounded as follows:

$$P(\overline{\mathcal{S}}^{(i)} \to \overline{\mathcal{S}}^{(j)} | \overline{\mathcal{H}}) \leq e^{-\frac{\rho}{4M_T} D_{i,j}}, \tag{6.36}$$

where

$$\begin{aligned} D_{i,j} &= \| \overline{\mathcal{H}} (\overline{\mathcal{S}}^{(i)} - \overline{\mathcal{S}}^{(j)}) \|_F^2 \\ &= \| \overline{\mathcal{H}} \left(\mathbf{I}_{M_R} \otimes (\mathcal{S}^{(i)} - \mathcal{S}^{(j)}) \right) \|_F^2 \\ &= \| \overline{\mathcal{H}} (\mathbf{I}_{M_R} \otimes \mathcal{E}_{i,j}) \|_F^2, \end{aligned} \tag{6.37}$$

and $\mathcal{E}_{i,j} = \mathcal{S}^{(i)} - \mathcal{S}^{(j)}$ is the modified codeword difference matrix. Assuming that the channel is spatially and temporally white, the average PEP may be upper-bounded as

$$P(\mathcal{S}^{(i)} \to \mathcal{S}^{(j)}) \leq \left(\frac{1}{\det \left(\mathbf{I} + \frac{\rho}{4M_T} \mathcal{E}_{i,j} \mathcal{E}_{i,j}^H \right)} \right)^{M_R}, \tag{6.38}$$

where $\mathcal{E}_{i,j}$ has dimensions $M_T L \times T$. At high SNR we may approximate Eq. (6.38) as

$$P(\mathcal{S}^{(i)} \to \mathcal{S}^{(j)}) \leq \frac{1}{\left(\prod_{k=1}^{r(\mathcal{G}_{i,j})} \lambda_k(\mathcal{G}_{i,j}) \right)^{M_R}} \left(\frac{\rho}{4M_T} \right)^{-r(\mathcal{G}_{i,j})M_R}, \tag{6.39}$$

where $\lambda_k(\mathcal{G}_{i,j})$ $(k = 1, 2, \ldots, r(\mathcal{G}_{i,j}))$ with $1 \leq r(\mathcal{G}_{i,j}) \leq M_T L$ is the kth eigenvalue of $\mathcal{G}_{i,j} = \mathcal{E}_{i,j} \mathcal{E}_{i,j}^H$. The analysis indicates that the diversity order is $r(\mathcal{G}_{i,j})M_R$. Therefore, the maximum achievable diversity order is $M_T M_R L$, where we have assumed that $L = L_{eff}$.

6.4.2 ST codeword design criteria

Note the similarity between Eq. (6.39) and Eq. (6.11). The codeword construction criterion for obtaining full system diversity is very similar to the criterion when the channel is frequency flat, i.e., it ensures that the difference matrix between every pair of codewords is full rank. The codeword has in effect $\mathcal{M} = M_T L$ virtual antennas [Gong and Letaief, 2000]. More precisely, in the flat fading case, the error matrices depend only on the $M_T \times T$ codeword, whereas in the frequency selective case the $\mathcal{M} \times T$ equivalent codewords consist of stacked Hankel versions of the actual transmitted $M_T \times T$ codewords.

This concludes the discussion on ST code design for frequency selective channels with SC modulation. The performance of codes designed for flat fading scenarios in frequency selective channels is discussed next.

Performance and example codes

Although there are $M_T L$ virtual antennas, there is a certain structure imposed on the codewords in a frequency selective environment. Codes designed for flat fading environments when used in delay spread environments are guaranteed to exploit $M_T M_R$ order diversity. However, in a frequency selective channel, the additional structure imposed may prevent these codes from exploiting full spatio-temporal diversity equal to $M_T M_R L_{eff}$. This fact has also been shown for multicarrier modulation [Bölcskei and Paulraj, 2000b]. We demonstrate this below by considering the delay diversity code.

The delay diversity code delays the data stream by one symbol on the second transmit antenna, two symbols on the third antenna and so on. We now show that this code fails to exploit full spatio-temporal diversity in the presence of delay spread. Consider the case of $M_T = 2$, $M_R = 1$ and $L = 2$. The channel taps are independent across space and time, implying a potential fourth-order spatio-temporal diversity order. The transmitted symbol sequences for the standard delay diversity (SDD) code assuming four independent symbols transmitted are

$$\mathbf{S}_{SDD} = \begin{bmatrix} 0 & s_1 & s_2 & s_3 & s_4 & 0 & 0 \\ 0 & 0 & s_1 & s_2 & s_3 & s_4 & 0 \end{bmatrix}, \tag{6.40}$$

where one extra zero has been added at the beginning and the end assuming a guard period of length $L - 1$. In effect, we have $\mathcal{M} = 4$ transmit antennas with the equivalent codeword $\mathcal{S}_{SDD}$ as follows:

$$\mathcal{S}_{SDD} = \begin{bmatrix} 0 & s_1 & s_2 & s_3 & s_4 & 0 \\ s_1 & s_2 & s_3 & s_4 & 0 & 0 \\ 0 & 0 & s_1 & s_2 & s_3 & s_4 \\ 0 & s_1 & s_2 & s_3 & s_4 & 0 \end{bmatrix}. \tag{6.41}$$

The first and fourth rows of $\mathcal{S}_{SDD}$ are identical and as a result all error matrices will be of a low rank. This code only extracts a diversity order of 3 instead of a possible 4.

It is worth mentioning that codes designed for flat fading scenarios are guaranteed to extract at least $M_T M_R$ order diversity if used in a frequency selective environment [Gong and Letaief, 2000].

Generalized delay diversity (GDD)

A simple example that extends the delay diversity code to frequency selective environments is discussed in this section. Examples of ST codes with SC modulation in delay spread environments can be found in [Liu *et al.*, 2001b].

The improvement in performance of multiple symbol delay diversity over single symbol delay diversity in frequency selective channels has been demonstrated through simulations in [Mogensen, 1993; Ostling, 1993; Winters, 1998]. The code design analysis developed in the previous section can be used to show that such GDD codes exploit full spatio-temporal diversity.

Consider a GDD scheme for a channel of length L in which the data stream is delayed on the second transmit antenna by L symbols, on the third antenna by $2L$ symbols and so on. The symbol sequences for the GDD code, assuming four independent symbols are transmitted and for the case of $M_T = 2$, $M_R = 1$, $L = 2$, are

$$\mathbf{S}_{GDD} = \begin{bmatrix} 0 & s_1 & s_2 & s_3 & s_4 & 0 & 0 & 0 \\ 0 & 0 & 0 & s_1 & s_2 & s_3 & s_4 & 0 \end{bmatrix}. \tag{6.42}$$

The equivalent codeword has $\mathcal{M} = 4$ transmit antennas and is represented by

$$\mathcal{S}_{GDD} = \begin{bmatrix} 0 & s_1 & s_2 & s_3 & s_4 & 0 & 0 \\ s_1 & s_2 & s_3 & s_4 & 0 & 0 & 0 \\ 0 & 0 & 0 & s_1 & s_2 & s_3 & s_4 \\ 0 & 0 & s_1 & s_2 & s_3 & s_4 & 0 \end{bmatrix}. \tag{6.43}$$

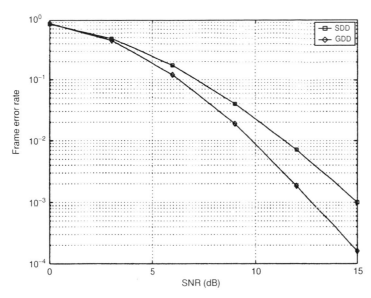

Figure 6.12: Comparison of the performance of GDD and SDD, $M_T = 2$, $L = 2$. Increased delay for GDD allows full fourth-order spatio-temporal diversity as compared to second-order for SDD.

Due to the structure of the GDD codewords all error matrices will be full rank (i.e., of rank $\mathcal{M}$) guaranteeing that the code will extract full $M_T M_R L$ order diversity. Figure 6.12 confirms this observation. The slope of the error rate curve for GDD is clearly greater than that for SDD. Further extensions of this technique for OFDM may be found in [Gore *et al.*, 2002c].

Lindskog–Paulraj (LP) scheme

The LP transmit diversity scheme [Lindskog and Paulraj, 2000] is a ST block coding technique that captures the spatio-temporal diversity in frequency selective channels. It may be regarded as an extension of the Alamouti code to a delay spread channel. We shall show that the LP scheme for MISO channels (which can be extended to MIMO) with no channel knowledge at the transmitter extracts all available spatio-temporal diversity order.

For ease of presentation, we introduce notation that will be used exclusively in the remaining part of this chapter. A SISO channel (which is a discrete-time filter) may be modeled as a polynomial according to

$$h(D) = h[0] + h[1]D + \cdots + h[L-1]D^{L-1}, \tag{6.44}$$

where D refers to a delay of one symbol period. The complex conjugate of the discrete-time channel, $h(D)^H$, is defined as

$$h(D)^H = h[0]^H + h[1]^H D^{-1} + \cdots + h[L-1]^H D^{-(L-1)}. \tag{6.45}$$

The operation $h(D)u[k]$ is defined as

$$h(D)u[k] = h[0]u[k] + h[1]u[k-1] + \cdots + h[L-1]u[k-L+1], \tag{6.46}$$

and represents discrete-time convolution. Similarly, SIMO/MISO channels are represented as polynomial columns and row vectors respectively, and MIMO channels as polynomial matrices. The complex conjugate transpose of a vector/matrix filter is the transpose of the vector/matrix filter with all elements complex conjugated as described above.

Consider a frequency selective SIMO system with M_R receive antennas. The channel impulse response can be expressed as $\mathbf{h}(D)$, where $\mathbf{h}(D)$ is a $M_R \times 1$ polynomial column vector. Assume that we transmit a discrete time sequence $s[k]$ of N symbols. The received sequence $\mathbf{y}[k]$ may be represented as

$$\mathbf{y}[k] = \sqrt{E_s}\mathbf{h}(D)s[k] + \mathbf{n}[k], \tag{6.47}$$

where $\mathbf{n}[k]$ represents ZMCSCG noise. The ML detection for $s[k]$ is given by

$$\arg\min_{s[k]} \sum_{k=1}^{N} \left\| \mathbf{y}[k] - \sqrt{E_s}\mathbf{h}(D)s[k] \right\|_F^2. \tag{6.48}$$

Standard MLSE techniques may be used for detection. However, the computational complexity of MLSE may be high due to a vector channel. The MLSE may be reformulated approximately (with low error for $N \gg L$) as [Lindskog, 1999; Stoica and Lindskog, 2001]

$$\arg\min_{s[k]} \Re \left\{ \sum_{k=1}^{N} s[k]^H \left[\sqrt{E_s}\mathbf{h}(D^{-1})^H \mathbf{h}(D)s[k] - 2\mathbf{h}(D^{-1})^H \mathbf{y}[k] \right] \right\}. \tag{6.49}$$

MLSE extracts all available spatio-temporal diversity order in the channel.

Next consider a MISO system with two transmit antennas. The input–output relation is

$$y[k] = \sqrt{\frac{E_s}{2}}\mathbf{h}(D)^T \mathbf{s}[k] + n[k], \tag{6.50}$$

where $\mathbf{h}(D) = [h_1(D)\ h_2(D)]^T$ is as defined earlier for the SIMO channel. The transmission strategy for the LP scheme is as follows. The symbol stream to be transmitted (say of length $2N$) is split into two sub-streams $s_1[k]$ and $s_2[k]$, each of length N, that are transmitted simultaneously from antennas 1 and 2 respectively. This is followed by a guard interval of L symbol periods. Following the guard period, the sub-stream $-s_2[N-k+1]^H$ obtained by complex conjugating, negating and time-reversing stream $s_2[k]$ is launched from antenna 1, while the sub-stream $s_1[N-t+1]^H$ obtained by time-reversing and complex conjugating stream $s_1[k]$ is launched from antenna 2.

The output of the MISO system for the first burst may be represented as

$$y_1[k] = \sqrt{\frac{E_s}{2}}\,[h_1(D)\ h_2(D)] \begin{bmatrix} s_1[k] \\ s_2[k] \end{bmatrix} + n_1[k], \tag{6.51}$$

while the received signal for the second burst may be represented as

$$y_2[k] = \sqrt{\frac{E_s}{2}} \, [h_1(D) \ h_2(D)] \begin{bmatrix} -s_2[N-k+1]^H \\ s_1[N-k+1]^H \end{bmatrix} + n_2[k]. \tag{6.52}$$

Observe that after time reversal and complex conjugation we may rewrite the second received burst as

$$y_2[N-k+1]^H = \sqrt{\frac{E_s}{2}} \, [h_2(D^{-1})^H \ -h_1(D^{-1})] \begin{bmatrix} s_1[k] \\ s_2[k] \end{bmatrix} + n_2[k]. \tag{6.53}$$

The noise has not been changed since complex conjugation followed by time reversal does not change the noise statistics. We may combine the two bursts as

$$y_2[N-k+1]^H = \sqrt{\frac{E_s}{2}} \begin{bmatrix} h_1(D) & h_2(D) \\ h_2(D^{-1})^H & h_1(D^{-1})^H \end{bmatrix} \begin{bmatrix} s_1[k] \\ s_2[k] \end{bmatrix} + \begin{bmatrix} n_1[k] \\ n_2[k] \end{bmatrix}$$

$$= \sqrt{\frac{E_s}{2}} \mathbf{H}(D)\mathbf{s}[k] + \mathbf{n}[k]. \tag{6.54}$$

The ML receiver is formulated as

$$\arg \min_{s_1[k],s_2[k]} \sum_{k=1}^{N} \left\| \mathbf{y}[k] - \sqrt{\frac{E_s}{2}} \mathbf{H}(D)\mathbf{s}[k] \right\|_F^2. \tag{6.55}$$

With minor approximations ($N \gg L$) and manipulations (see [Stoica and Lindskog, 2001]), MLSE may be reformulated as

$$\arg \min_{s_1[k],s_2[k]} \Re \left\{ \sum_{k=1}^{N} \mathbf{s}[k]^H \left[\sqrt{\frac{E_s}{2}} \mathbf{H}(D^{-1})^H \mathbf{H}(D)\mathbf{s}[k] - 2\mathbf{H}(D^{-1})^H \mathbf{y}[k] \right] \right\}. \tag{6.56}$$

Next, note that $\mathbf{H}(D^{-1})^H \mathbf{H}(D) = \mathbf{h}(D^{-1})^H \mathbf{h}(D)\mathbf{I}_2$, i.e., the matrix filter decouples after multiplying with a complex conjugated filter (matched-filtering). This decouples the vector MLSE into two independent MLSE operations on scalar sequences as follows:

$$\arg \min_{s_i[k]} \Re \left\{ \sum_{k=1}^{N} s_i[k]^H \left[\sqrt{\frac{E_s}{2}} \mathbf{h}(D^{-1})^H \mathbf{h}(D)s_i[k] - 2r_i[k] \right] \right\}, \ i = 1, 2, \tag{6.57}$$

where $\mathbf{r}[k] = [r_1[k] \ r_2[k]]^T = \mathbf{H}(D^{-1})^H \mathbf{y}[k]$. In practice, this decoupling is achieved by filtering the received vector sequence $\mathbf{y}[k]$ with the matched filter $\mathbf{H}(D^{-1})^H$. Note that the decoupled form of the MLSE is identical to that for the SIMO channel, i.e., the MLSE is posed as though $s_i[k]$ ($i = 1, 2$) were transmitted through a SIMO channel with two receive antennas. To see this note that $\mathbf{r}[k]$ satisfies

$$\mathbf{r}[k] = \mathbf{H}(D^{-1})^H \mathbf{y}[k] = \sqrt{\frac{E_s}{2}} \mathbf{h}(D^{-1})^H \mathbf{h}(D)\mathbf{s}[k] + \mathbf{H}(D^{-1})^H \begin{bmatrix} n_1[k] \\ n_2[k] \end{bmatrix}, \tag{6.58}$$

and may be decomposed as

$$\mathbf{r}_i[k] = \mathbf{h}(D^{-1})^H \widetilde{\mathbf{y}}_i[k], \tag{6.59}$$

where $\widetilde{\mathbf{y}}_i[k]$ is the output of a fictitious SIMO system defined by

$$\widetilde{\mathbf{y}}_i[k] = \sqrt{\frac{E_s}{2}}\mathbf{h}(D)s_i[k] + \mathbf{e}_i[k], \tag{6.60}$$

with $\mathbf{e}_i[k]$ representing the fictitious noise. To summarize, each component sequence of the transmit vector sequence sees a fictitious channel without array gain. The LP scheme has been extended in [Stoica and Lindskog, 2001] for two or more transmit antennas.

Remarks

The case of ST coding for multicarrier modulation over frequency selective MIMO channels is briefly discussed in Chapter 9. The interested reader is referred to Kim *et al.*, 1998; Lang *et al.*, 1999; Li *et al.*, 1999; Lin *et al.*, 2000; Bölcskei and Paulraj, 2000a, b; Liu *et al.*, 2001c; Blum *et al.*, 2001; Lu *et al.*, 2002] for further details.

When the channel is known to the transmitter, the problem decouples (see Fig. 4.2) so that all channel modes are accessible through linear processing at the transmitter and receiver. Error-free transmission is then possible through SISO capacity achieving codes.

7 ST receivers

7.1 Introduction

In Chapter 6 we introduced ST coding techniques for maximizing rate, diversity gain and array gain by exploiting the spatial dimension in multiple antenna systems when the channel is unknown to the transmitter. This chapter studies receiver structures for ST wireless channels. We introduce SISO receivers and focus on receivers for SIMO and MIMO channels. The optimal receiver for a MISO ($M_R = 1$) channel has already been covered in Chapters 5 and 6.

We begin with a brief review of SISO receivers for frequency flat and frequency selective channels. We then extend these to SIMO channels. The performance of the receivers is characterized in terms of SER vs SNR and the corresponding diversity gain and array gain. We discuss different receivers ranging from optimal ML techniques to more practical linear and non-linear receivers. Finally, we discuss different receivers for MIMO channels with diversity and SM signaling. In SM we cover ML, linear and successive cancellation receivers. Our discussion is primarily for the temporally uncoded ($r_t = 1$) case. The asymptotic performance with temporal (HE and DE) coding is also addressed.

7.2 Receivers: SISO

We begin with receivers for SISO channels to develop the basic theory before going on to SIMO and MIMO receivers.

7.2.1 Frequency flat channel

Recall from Chapter 3 that the signal model for a SISO flat fading channel is given by

$$y = \sqrt{E_s}hs + n, \tag{7.1}$$

where y is the received signal, h is the scalar channel gain, s is the complex data symbol drawn from a unit average energy constellation, n is ZMCSCG noise with variance N_o and E_s is the average energy at the transmitter per symbol period. The principal impairments are fading and additive noise.

Receivers

In a frequency flat SISO channel, symbol by symbol detection is optimal (in the absence of channel coding). The optimal ML estimate of the transmitted signal assuming all data symbols are equally likely to be transmitted is given by

$$\widehat{s} = \arg\min_{s} |y - \sqrt{E_s} hs|^2, \tag{7.2}$$

where $\widehat{s}$ is the estimated data symbol and the minimization is performed over all possible scalar constellation points from which the symbol s is drawn. The ML receiver searches for the symbol s that after being scaled by the channel is closest in Euclidean distance to the received signal y.

Alternatively, the received signal can be multiplied by a weight g prior to detection, and the problem restated as

$$\widehat{s} = \arg\min_{s} |gy - g\sqrt{E_s} hs|^2. \tag{7.3}$$

$g = (\sqrt{E_s} h)^{-1}$ is the channel inverting or the ZF solution, while $g = E_s^{-1/2}(|h|^2 + \rho^{-1})^{-1} h^*$ with $\rho = E_s/N_o$ is the MMSE solution. Both schemes reduce to the optimal ML detection formulated in Eq. (7.2).

Error performance

The probability of symbol error, P_e, for a given instance of the SISO flat fading channel was derived in Chapter 5. Assuming Rayleigh flat fading, the average SER $\overline{P}_e$ at high SNR is upper-bounded by (see Eq. (5.7))

$$\overline{P}_e \leq \overline{N}_e \left(\frac{\rho d_{min}^2}{4} \right)^{-1}. \tag{7.4}$$

As expected, the diversity order of the Rayleigh fading SISO channel is 1.

7.2.2   Frequency selective channel

The signal model for a SISO frequency selective channel is (see Eq. (3.55))

$$\mathbf{Y}[k] = \sqrt{E_s}\,\mathcal{H}\mathbf{S}[k] + \mathbf{N}[k], \tag{7.5}$$

where

$$\mathbf{Y}[k] = \begin{bmatrix} y[k] \\ \vdots \\ y[k+T-1] \end{bmatrix}, \quad \mathbf{S}[k] = \begin{bmatrix} s[k-L+1] \\ \vdots \\ s[k+T-1] \end{bmatrix}, \quad \mathbf{N}[k] = \begin{bmatrix} n[k] \\ \vdots \\ n[k+T-1] \end{bmatrix}$$

$$\tag{7.6}$$

and

$$
\mathcal{H} = \begin{bmatrix} h[L-1] & \cdots & h[0] & 0 & \cdots & 0 \\ \vdots & \ddots & \ddots & \ddots & \ddots & \vdots \\ 0 & \cdots & 0 & h[L-1] & \cdots & h[0] \end{bmatrix}. \tag{7.7}
$$

The vectors $\mathbf{Y}[k]$ and $\mathbf{N}[k]$ are $T \times 1$, while the vector $\mathbf{S}[k]$ has dimension $(T + L - 1) \times 1$ and $\mathcal{H}$ is $T \times (T + L - 1)$.

The impairments are channel fading, ISI and additive noise.

MLSE

The MLSE receiver identifies the symbol sequence $\mathbf{S}[k]$ that is most likely to have been transmitted given the received sequence $\mathbf{Y}[k]$. The MLSE receiver, assuming all symbol sequences are equally likely to be transmitted, may be formulated as

$$
\widehat{\mathbf{S}}[k] = \arg\min_{\mathbf{S}[k]} \|\mathbf{Y}[k] - \sqrt{E_s}\mathcal{H}\mathbf{S}[k]\|_F^2, \tag{7.8}
$$

where $\widehat{\mathbf{S}}[k]$ represents the estimated data symbol sequence and the minimization is performed over all possible sequences $\mathbf{S}[k]$. The MLSE receiver searches for the data sequence that after convolution with the channel is closest in Euclidean distance to the received signal sequence. Sequence detection techniques such as the Viterbi algorithm [Omura, 1971; Forney, 1972] may be used to simplify this search. Although optimal, the complexity of the MLSE receiver increases exponentially with channel length L making its use prohibitive in environments with large delay spread. A discussion of sub-optimal low complexity receivers follows.

ZF receiver

The ZF receiver aims to invert the channel and eliminate ISI. In the following we consider the finite impulse response (FIR) ZF receiver. It may not be possible to perfectly invert the channel with FIR filtering as the channel itself is an FIR filter and perfect inversion (elimination of ISI) will require infinite impulse response (IIR) filtering at the receiver. However, with oversampling, a polyphase FIR filter of sufficient length can perfectly equalize the channel.

The received signal sequence, $\mathbf{Y}[k]$, is filtered by $\mathbf{g}_{ZF}$ $(1 \times T)$, which is a linear FIR filter that inverts the channel $\mathcal{H}$ to yield the desired symbol stream free of ISI and is given by

$$
\mathbf{g}_{ZF} = \frac{1}{\sqrt{E_s}}\mathbf{1}_{\Delta_D,T+L-1}\mathcal{H}^{\dagger}, \tag{7.9}
$$

where $\mathbf{1}_{\Delta_D,T+L-1}$ is a $1 \times (T + L - 1)$ vector whose Δ_Dth $(1 \le \Delta_D \le T + L - 1)$ element is 1, all other elements are 0, and $\mathcal{H}^{\dagger}$ is the pseudo-inverse of $\mathcal{H}$. Without oversampling $\mathbf{g}_{ZF}$ cannot cancel all ISI. With oversampling and sufficient equalizer length, the channel matrix $\mathcal{H}$ is tall, allowing perfect equalization with no residual ISI.

The delay, Δ_D ($1 \leq \Delta_D \leq T + L - 1$), corresponds to $s[k - L + \Delta_D]$, the target data symbol that is to be detected. Δ_D can be optimized to minimize noise and residual ISI energy at the output.

MMSE receiver

The ZF equalizer discussed earlier seeks to maximally eliminate ISI at the expense of noise enhancement. A more sophisticated approach is the MMSE equalizer that balances noise enhancement with ISI mitigation. The received signal sequence is multiplied by a $1 \times T$ vector of weight $\mathbf{g}_{MMSE}$, given by

$$\mathbf{g}_{MMSE} = \arg \min_{\mathbf{g}} \mathcal{E}\{|\mathbf{g}\mathbf{Y}[k] - s[k - L + \Delta_D]|^2\}, \tag{7.10}$$

where Δ_D ($1 \leq \Delta_D \leq T + L - 1$) is again the delay parameter. The MMSE filter $\mathbf{g}_{MMSE}$ follows from the orthogonality principle

$$\mathcal{E}\{(\mathbf{g}\mathbf{Y}[k] - s[k - L + \Delta_D])\mathbf{Y}[k]^H\} = \mathbf{0}_{1,T}, \tag{7.11}$$

to yield

$$\mathbf{g}_{MMSE} = \mathbf{R}_{s[k-L+\Delta_D]\mathbf{Y}}\mathbf{R}_{\mathbf{Y}\mathbf{Y}}^{-1}, \tag{7.12}$$

where

$$\begin{aligned} \mathbf{R}_{\mathbf{Y}\mathbf{Y}} &= \mathcal{E}\{\mathbf{Y}[k]\mathbf{Y}[k]^H\} \\ &= E_s \mathcal{H}\mathcal{H}^H + N_o \mathbf{I}_T, \end{aligned} \tag{7.13}$$

is the received signal covariance matrix and $\mathbf{R}_{s[k-L+\Delta_D]\mathbf{Y}}$ (of dimension $1 \times T$) is given by

$$\begin{aligned} \mathbf{R}_{s[k-L+\Delta_D]\mathbf{Y}} &= \mathcal{E}\{s[k - L + \Delta_D]\mathbf{Y}^H\} \\ &= \sqrt{E_s}\mathbf{1}_{\Delta_D,T+L-1}\mathcal{H}^H. \end{aligned} \tag{7.14}$$

Decision feedback equalizer (DFE)

The DFE [Austin, 1967] is a non-linear receiver consisting of two filters, a feedforward filter and a feedback filter (see Fig. 7.1). The input to the feedforward filter is the received signal sequence. The feedback filter has as its input the sequence of decisions on previously detected symbols, $\hat{s}[i]$. The feedback filter removes the portion of ISI from the present signal caused by previously detected symbols.

The feedforward filter $\mathbf{g}_{FF}$ ($1 \times T$) and feedback filter $\mathbf{b}_{FB}$ ($1 \times F$) (T and F are the number of feedforward and feedback filter taps respectively) are designed based on the MMSE criterion according to

$$\begin{aligned} \mathbf{g}_{FF}, \mathbf{b}_{FB} &= \arg \min_{\mathbf{g},\mathbf{b}} \mathcal{E}\{|\mathbf{g}\mathbf{Y}[k] - \mathbf{b}\, \mathbf{s}_{F,\Delta_D}[k] - s[k - L + \Delta_D]|^2\} \\ &= \arg \min_{\mathbf{g},\mathbf{b}} \mathcal{E}\{|\widetilde{\mathbf{g}}\widetilde{\mathbf{Y}}[k] - s[k - L + \Delta_D]|^2\}, \end{aligned} \tag{7.15}$$

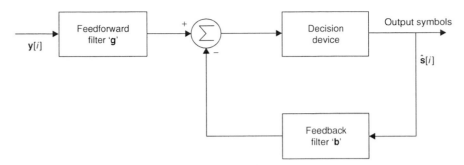

Figure 7.1: Schematic of DFE equalization for SISO channels. The feedback filter subtracts trailing ISI from the current symbol to be detected.

where $\widetilde{\mathbf{g}} = [\mathbf{g} \; -\mathbf{b}]$ is of dimension $1 \times (F + T)$ and $\widetilde{\mathbf{Y}}[k] = [\mathbf{Y}[k]^T \, \mathbf{s}_{F,\Delta_D}[k]^T]^T$ is $(F + T) \times 1$ with $\mathbf{s}_{F,\Delta_D}[k] = [s[k - L + \Delta_D - F] \cdots s[k - L + \Delta_D - 2] \, s[k - L + \Delta_D - 1]]^T$. The $F \times 1$ vector $\mathbf{s}_{F,\Delta_D}[k]$ comprises previously detected symbols (assuming the decisions are correct). In reality, errors in previously detected symbols will lead to error propagation resulting in a loss in performance. Optimal feedforward and feedback filter weights are obtained by solving Eq. (7.15) using the orthogonality principle

$$\widetilde{\mathbf{g}} = \mathbf{R}_{s[k-L+\Delta_D]\widetilde{\mathbf{Y}}} \mathbf{R}_{\widetilde{\mathbf{Y}}\widetilde{\mathbf{Y}}}^{-1}. \tag{7.16}$$

$\mathbf{R}_{\widetilde{\mathbf{Y}}\widetilde{\mathbf{Y}}} = \mathcal{E}\{\widetilde{\mathbf{Y}}[k]\widetilde{\mathbf{Y}}[k]^H\}$ (of dimension $(T + F) \times (T + F)$) in the above equation is given by[1]

$$\mathbf{R}_{\widetilde{\mathbf{Y}}\widetilde{\mathbf{Y}}} = \begin{bmatrix} E_s \mathcal{H}\mathcal{H}^H + N_o \mathbf{I}_T & \sqrt{E_s}\mathcal{H}\mathbf{J}_{\Delta_D} \\ \sqrt{E_s}\mathbf{J}_{\Delta_D}^H \mathcal{H}^H & \mathbf{I}_F \end{bmatrix}, \tag{7.17}$$

where $\mathbf{J}_{\Delta_D} = \mathcal{E}\{\mathbf{S}[k]\mathbf{s}_{F,\Delta_D}[k]^H\}$ is of dimension $(T + L - 1) \times F$. Furthermore, $\mathbf{R}_{s[k-L+\Delta_D]\widetilde{\mathbf{Y}}} = \mathcal{E}\{s[k - L + \Delta_D]\widetilde{\mathbf{Y}}[k]^H\}$ (of size $1 \times (T + F)$) in Eq. (7.16) is given by

$$\mathbf{R}_{s[k-L+\Delta_D]\widetilde{\mathbf{Y}}} = \begin{bmatrix} \sqrt{E_s}\mathbf{1}_{\Delta_D,T+L-1}\mathcal{H}^H & \mathbf{0}_{1,F} \end{bmatrix}. \tag{7.18}$$

As for the case of the ZF and MMSE receivers discussed earlier, the delay Δ_D can be adjusted for optimal performance. The DFE receiver can also be designed based on a ZF criterion. For a discussion on DFE implementation, the reader is referred to [Proakis, 1995; Stüber, 1996; Cioffi, 2002].

Comparative performance
The performance (SER) of all the receivers discussed above may be lower-bounded by the MFB discussed in Chapter 5. The output SNR of the receive MF for a given

---

[1] $\mathbf{J}_{\Delta_D}$ is composed of only 0s and 1s, the exact arrangement of which depends on Δ_D, F and T. For more details see [Cioffi, 2002].

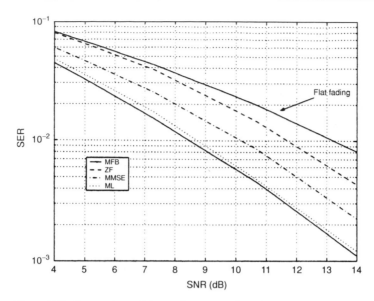

Figure 7.2: Comparison of the performance of MLSE, ZF and MMSE receivers for a two-path SISO channel with T_s path delay. The MLSE receiver performs close to MFB.

realization of the SISO frequency selective channel is given by

$$\eta \leq \|\mathbf{h}\|_F^2 \rho, \tag{7.19}$$

where $\mathbf{h} = [h[0]\ h[1]\ \cdots\ h[L-1]]^T$ is the SISO channel (Eq. (3.50)). Assuming that the taps $h[l]$ are equi-power and fade independently (this is usually not the case in practice, see Chapter 5 for a detailed discussion), the diversity order of the channel with the MF receiver is given by $L_{eff} = L$.

Figure 7.2 compares the performance of the MLSE, ZF and MMSE receivers with BPSK modulation for a channel with two independent and equi-powered physical taps, spaced T_s apart, with root raised-cosine pulse-shaping and 40% excess bandwidth. The total average power in the channel is normalized to 1, i.e., $\mathcal{E}\{\|\mathbf{h}\|_F^2\} = 1$. It is clear from the figure that the ML receiver performs almost as well as the MFB and captures all the temporal diversity in the channel ($L_{eff} = 2$). As expected, the MMSE receiver outperforms the ZF receiver. The MMSE and ZF receivers have the same diversity slope as ML, but have a SNR loss. Although DFE performance has not been plotted, the SER for the DFE will lie between that of the ML and MMSE receivers [Balaban and Salz, 1992].

As discussed in Chapter 4, the channel taps $\{h[l]\}$ will typically not fade independently. Thus, all the receivers will deliver diversity $L_{eff} < L$ in practice, maintaining the same relative performance characteristics. Figure 7.3 plots the performance of the receivers for a channel with two independent and equi-powered physical paths with a path delay of $0.25T_s$. Again, we assume BPSK modulation with root raised-cosine

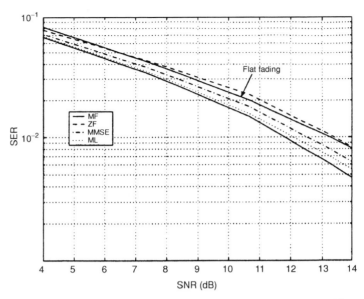

Figure 7.3: Comparison of the performance of MLSE, ZF and MMSE receivers for a SISO channel with $0.25T_s$ path delay. There is very little diversity to be extracted.

pulse-shaping and 40% excess bandwidth. The degradation in performance is clear – the temporal diversity in the channel is mostly lost and all four receivers perform somewhat similarly, with the receivers maintaining the relative order. The performance of the SISO receivers depends on the spacing of the physical multipaths, excess bandwidth and sampling delay (phase).

7.3 Receivers: SIMO

7.3.1 Frequency flat channel

The signal model for a SIMO flat fading channel Eq. (3.58) is given by

$$\mathbf{y} = \sqrt{E_s}\mathbf{h}s + \mathbf{n}, \tag{7.20}$$

where $\mathbf{y}$ is the received signal vector of dimension $M_R \times 1$, $\mathbf{h}$ is the channel vector of dimension $M_R \times 1$, s is the transmitted signal and $\mathbf{n}$ is the $M_R \times 1$ ZMCSCG noise vector with covariance matrix $N_o\mathbf{I}_{M_R}$.

Receiver
The optimal receiver strategy is to perform MRC (see Section 5.3). The received signal post-MRC z is given by

$$z = \sqrt{E_s}\mathbf{h}^H\mathbf{h}s + \mathbf{h}^H\mathbf{n}. \tag{7.21}$$

Effectively, this is a SISO channel and symbol by symbol decoding is optimal. In Chapter 5 we saw that MRC provides diversity gain and array gain proportional to M_R.

7.3.2   Frequency selective channels

Using Eq. (7.5) for a SISO channel, the SIMO channel can be written as

$$\mathbf{Y}_i[k] = \sqrt{E_s}\mathcal{H}_i\mathbf{S}[k] + \mathbf{N}_i[k], \tag{7.22}$$

where $\mathcal{H}_i(i = 1, \ldots, M_R)$ is the stacked channel at the ith receive antenna. The input–output relation for the SIMO channel is given by

$$\mathcal{Y}[k] = \sqrt{E_s}\overline{\mathcal{H}}\mathbf{S}[k] + \mathcal{N}[k], \tag{7.23}$$

where

$$\mathcal{Y}[k] = \begin{bmatrix} \mathbf{Y}_1[k] \\ \vdots \\ \mathbf{Y}_{M_R}[k] \end{bmatrix}, \quad \overline{\mathcal{H}} = \begin{bmatrix} \mathcal{H}_1 \\ \vdots \\ \mathcal{H}_{M_R} \end{bmatrix}, \quad \mathbf{N}[k] = \begin{bmatrix} \mathbf{N}_1[k] \\ \vdots \\ \mathbf{N}_{M_R}[k] \end{bmatrix}. \tag{7.24}$$

The vectors $\mathcal{Y}[k]$ and $\mathbf{N}[k]$ have dimension $M_R T \times 1$, while $\overline{\mathcal{H}}$ is $M_R T \times (T + L - 1)$. Note that the signal model in Eq. (7.24) is similar to a SISO model with a temporal oversampling factor of M_R. Oversampled channels have different correlation properties. Therefore, multiple antenna systems also offer advantages for equalization and blind estimation that are similar to those offered by oversampling in conventional SISO channels.

MLSE receiver
The MLSE estimate of the transmitted sequence $\widehat{\mathbf{S}}[k]$ is given by

$$\widehat{\mathbf{S}}[k] = \arg\min_{\mathbf{S}[k]} \|\mathcal{Y}[k] - \sqrt{E_s}\overline{\mathcal{H}}\mathbf{S}[k]\|_F^2. \tag{7.25}$$

The sequence of symbols $\mathbf{S}[k]$ that meets this criterion can be determined by the vector Viterbi algorithm. Note that the SIMO channel increases the computational requirements of direct Viterbi decoding [Lindskog, 1997]. The complexity can be mitigated through a multi-dimensional MF that reduces the detection problem to standard scalar MLSE detection. To summarize the technique, the received signal on the ith $(i = 1, \ldots, M_R)$ antenna is filtered with a time-reversed conjugated version of the channel corresponding to that receive antenna, $h_i[-l]^*$ $(l = (L - 1), \ldots, 1, 0)$. The M_R signal streams are then added together to generate a scalar sequence. Thus, the effective scalar channel post-combining at the receiver is given by $q[l]$ $(l = -(L - 1), \ldots, -1, 0, 1, \ldots, L - 1)$, where

$$q[l] = \sum_{i=1}^{M_R} \sum_{k=0}^{L-1} h_i[k]h_i[k - l]^*. \tag{7.26}$$

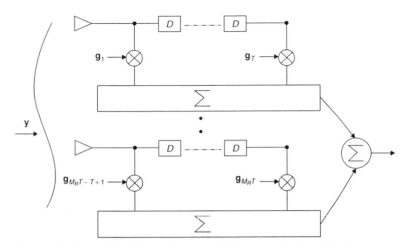

Figure 7.4: ZF and MMSE equalizers in SIMO use an $M_R T$ tap FIR filter.

However, the equivalent channel doubles in length and a brute force implementation of the Viterbi algorithm would require A^{2L-1} states (A is the alphabet size of the transmitted symbols) thereby forgoing any reduction in computational complexity that the scalar processing attains. However, noting that the effective channel has conjugate symmetry (i.e., $q[l] = q[-l]^*$), the number of states in the Viterbi trellis can be maintained at A^{L-1}. See [Krenz and Wesolowski, 1997; Lindskog, 1997; Larsson, 2001; Larsson *et al.*, 2002] for more details.

ZF receiver

The ZF equalizer (see Fig. 7.4) for the SIMO channel can be derived in a similar fashion to that for the SISO channel. The ST ZF weight vector $\mathbf{g}_{ZF}$ ($1 \times M_R T$) inverts the channel to yield the desired symbol stream and is given by

$$\mathbf{g}_{ZF} = \frac{1}{\sqrt{E_s}} \mathbf{1}_{\Delta_D . T + L - 1} \overline{\mathcal{H}}^\dagger. \tag{7.27}$$

As for the SISO channel, the delay Δ_D can be selected to optimize performance. The spatial dimension due to multiple receive antennas will, in general, allow perfect equalization of the SIMO channel, resulting in zero residual ISI.

MMSE receiver

The ST ZF receiver discussed earlier eliminates ISI at the expense of noise enhancement. A ST MMSE receiver can be designed that balances noise enhancement with ISI mitigation. Following the development of the MMSE receiver for SISO channels, the ST MMSE filter $\mathbf{g}_{MMSE}$ ($1 \times M_R T$), is given by

$$\mathbf{g}_{MMSE} = \mathbf{R}_{s[k-L+\Delta_D]y} \mathbf{R}_{yy}^{-1}, \tag{7.28}$$

where

$$\mathbf{R}_{yy} = \mathcal{E}\{\mathcal{Y}[k]\mathcal{Y}[k]^H\}$$
$$\mathbf{R}_{yy} = E_s \overline{\mathcal{H}\mathcal{H}}^H + N_o \mathbf{I}_{TM_R}, \tag{7.29}$$

has dimension $M_R T \times M_R T$ and

$$\mathbf{R}_{s[k-L+\Delta_D]y} = \mathcal{E}\{s[k - L + \Delta_D]\mathcal{Y}[k]^H\}$$
$$= \sqrt{E_s}\mathbf{1}_{\Delta_D, T+L-1}\overline{\mathcal{H}}^H \tag{7.30}$$

is of dimension $1 \times M_R T$. The delay Δ_D must be adjusted for optimal performance.

DFE

Analogous to the DFE for SISO channels, ST DFE receivers have been designed for SIMO channels and operate on the same general principle. The SIMO DFE comprises ST feedforward and feedback filters. The ST feedback filter subtracts the influence of previously detected symbols from the current symbol to be detected [Lindskog, 1999; Hwang et al., 2002]. We briefly review the SIMO DFE below.

The feedforward filter $\mathbf{g}_{FF}$ $(1 \times M_R T)$ and feedback filter $\mathbf{b}_{FB}$ $(1 \times F)$ are designed based on the MMSE criterion according to

$$\mathbf{g}_{FF}, \mathbf{b}_{FB} = \arg\min_{\mathbf{g},\mathbf{b}} \mathcal{E}\{|\mathbf{g}\mathcal{Y}[k] - \mathbf{b}\mathbf{s}_{F,\Delta_D}[k] - s[k - L + \Delta_D]|^2\}$$
$$= \arg\min_{\mathbf{g},\mathbf{b}} \mathcal{E}\{|\widetilde{\mathbf{g}}\widetilde{\mathcal{Y}}[k] - s[k - L + \Delta_D]|^2\}, \tag{7.31}$$

where $\widetilde{\mathbf{g}} = [\mathbf{g} \ -\mathbf{b}]$ is of dimension $1 \times (F + M_R T)$ and $\widetilde{\mathcal{Y}}[k] = [\mathcal{Y}[k]^T \mathbf{s}_{F,\Delta_D}[k]^T]^T$ is $(F + M_R T) \times 1$ with $\mathbf{s}_{F,\Delta_D}[k] = [s[k - L + \Delta_D - F] \cdots s[k - L + \Delta_D - 2] \ s[k - L + \Delta_D - 1]]^T$. As for the SISO DFE, $\mathbf{s}_{F,\Delta_D}[k]$ is the $F \times 1$ vector of previously detected symbols (assuming the decisions are correct). Optimal feedforward and feedback filter weights are given by

$$\widetilde{\mathbf{g}} = \mathbf{R}_{s[k-L+\Delta_D]\widetilde{y}}\mathbf{R}_{\widetilde{y}\widetilde{y}}^{-1}. \tag{7.32}$$

$\mathbf{R}_{\widetilde{y}\widetilde{y}}$ (of dimension $(M_R T + F) \times (M_R T + F)$) in the above equation is given by

$$\mathbf{R}_{\widetilde{y}\widetilde{y}} = \mathcal{E}\{\widetilde{\mathcal{Y}}[k]\widetilde{\mathcal{Y}}[k]^H\}$$
$$= \begin{bmatrix} E_s \overline{\mathcal{H}\mathcal{H}}^H + N_o \mathbf{I}_T & \sqrt{E_s} \ \overline{\mathcal{H}}\mathbf{J}_{\Delta_D} \\ \sqrt{E_s}\mathbf{J}_{\Delta_D}^H \overline{\mathcal{H}}^H & \mathbf{I}_F \end{bmatrix}, \tag{7.33}$$

where $\mathbf{J}_{\Delta_D} = \mathcal{E}\{\mathbf{S}[k]\mathbf{s}_{F,\Delta_D}[k]^H\}$ is of dimension $M_R(T + L - 1) \times F$, and is composed of 0s and 1s. Furthermore, $\mathbf{R}_{s[k-L+\Delta_D]\widetilde{y}}$ $(1 \times (M_R T + F))$ in Eq. (7.32) is given by

$$\mathbf{R}_{s[k-L+\Delta_D]\widetilde{y}} = \mathcal{E}\{s[k - L + \Delta_D]\widetilde{\mathcal{Y}}^H\}$$
$$= [\sqrt{E_s}\mathbf{1}_{\Delta_D, T+L-1}\overline{\mathcal{H}}^H \ \ \mathbf{0}_{1,F}]. \tag{7.34}$$

The delay Δ_D must be adjusted for optimal performance.

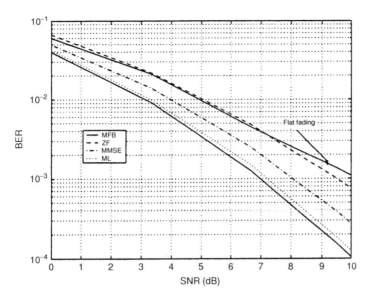

Figure 7.5: Comparison of the performance of MLSE, ZF and MMSE receivers for a SIMO channel with $M_R = 2$ and T_s spaced physical channel taps. The MLSE receiver extracts all available spatio-temporal diversity.

Comparative performance

The SNR assuming a MF at the receiver is upper-bounded by

$$\eta \leq \|\mathbf{h}\|_F^2 \rho, \tag{7.35}$$

where $\mathbf{h} = [\mathbf{h}[0]^T \ \mathbf{h}[1]^T \cdots \mathbf{h}[L-1]^T]^T$ is the SIMO channel stacked into a vector of dimension $M_R L \times 1$. The corresponding maximum realizable diversity order is $M_R L$. MFB assumes one-shot transmission.

Figure 7.5 compares the performance of the MLSE, ZF and MMSE receivers for a SIMO channel with $M_R = 2$ and a physical channel with T_s spaced independent and equi-powered paths. We assume BPSK modulation with root raised-cosine pulse-shaping and 40% excess bandwidth and constrain the average total power in each of the SISO channels to unity. Once again, we see that the ML receiver extracts all the spatio-temporal diversity in the channel (order 4). Further, it is easily verified that the MLSE receiver also extracts array gain proportional to the number of receive antennas. As expected, the MMSE receiver outperforms ZF, but is clearly sub-optimal compared with MLSE and suffers a SNR loss.

Figure 7.6 compares the performance of the receivers for $M_R = 2$ with $0.25T_s$ spaced independent equi-powered physical taps. With the use of the ZF or MMSE receivers, the loss of diversity is evident. As expected, the MLSE receiver extracts all available spatio-temporal diversity.

The remaining part of this chapter is dedicated to receiver design for MIMO receivers. We treat only the flat fading case in detail.

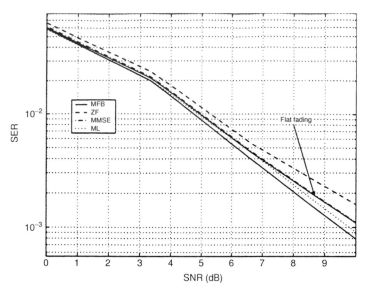

Figure 7.6: Comparison of the performance of ML, ZF and MMSE receivers for a SIMO channel with $0.25T_s$ spaced physical channel taps. The loss in temporal diversity is evident.

7.4   Receivers: MIMO

We begin with a brief discussion of MIMO receivers for ST diversity coding ($r_s \leq 1$) and then focus on receivers for SM ($r_s = M_T$).

7.4.1   ST diversity schemes

Recall that the Alamouti coding in a frequency flat $M_R \times 2$ channel collapses the MIMO channel into a SISO channel (Eq. (6.15)) with

$$y = \sqrt{\frac{E_s}{2}} \|\mathbf{H}\|_F^2 s + n. \tag{7.36}$$

This reduction can be extended to $M_T \geq 2$, if we use OSTBC. Similar extensions can be made to frequency selective channels [Stoica and Lindskog, 2001]. Therefore, SISO receiver techniques discussed in the previous sections are applicable.

There is generally no equivalent SISO/SIMO channel model for STTC. Nevertheless there may be certain codes that allow such a representation. Consider, for example, the delay diversity code. As discussed in Chapter 5, this code converts spatial diversity to frequency diversity. For a MIMO system with delay diversity coding, the equivalent channel is frequency selective (SIMO) with M_T temporal taps [Seshadri and Winters, 1994] per receive antenna. The techniques developed in Section 7.3.2 are now directly applicable. For a general STTC, we will need scalar or vector Viterbi decoding based on the encoding state transition maps.

7.4.2   SM schemes

The rest of the chapter overviews MIMO receivers for SM schemes. The new problem faced by a MIMO receiver is the presence of multistream interference (MSI), since the multiple transmit streams interfere with each other.

Our discussion usually assumes uncoded SM, i.e., the transmitter demultiplexes the uncoded data stream to the M_T antennas. Cases where temporal coding ($r_t \leq 1$) is used with HE or DE are also briefly discussed.

We assume a frequency flat MIMO channel, recalling from Chapter 3 that the signal model is

$$\mathbf{y} = \sqrt{\frac{E_s}{M_T}} \mathbf{H s} + \mathbf{n}, \tag{7.37}$$

where $\mathbf{y}$ is the received $M_R \times 1$ signal vector, $\mathbf{H}$ is the $M_R \times M_T$ channel matrix, $\mathbf{s}$ is the $M_T \times 1$ transmit signal vector and $\mathbf{n}$ is the $M_R \times 1$ ZMCSCG noise vector with covariance matrix $N_o \mathbf{I}_{M_R}$.

The impairments are MSI, channel fading and additive noise.

ML receiver

The ML receiver performs vector decoding and is the optimal receiver. Assuming equally likely, temporally uncoded transmit symbols, the ML receiver chooses the vector $\mathbf{s}$ that solves

$$\hat{\mathbf{s}} = \arg \min_{\mathbf{s}} \left\| \mathbf{y} - \sqrt{\frac{E_s}{M_T}} \mathbf{H s} \right\|_F^2, \tag{7.38}$$

where the optimization is performed through an exhaustive search over all candidate vector symbols $\mathbf{s}$. The ML receiver searches through the entire vector constellation for the most probable transmitted signal vector. A brute force implementation requires a search through a total of A^{M_T} vector symbols and for a naive implementation the decoding complexity of the ML receiver is exponential in M_T. However, the development of fast algorithms [Hassibi and Vikalo, 2001] for sphere decoding techniques [Fincke and Pohst, 1985; Viterbo and Boutros, 1999; Damen *et al.*, 2000] promises to reduce complexity significantly and we briefly discuss it below.

Sphere decoding

The ML detection problem can be posed as an integer least-squares problem that can be solved via sophisticated methods such as Kannan's algorithm [Kannan, 1983] and the sphere decoding algorithm of Fincke–Pohst [Fincke and Pohst, 1985].

The main idea behind sphere decoding is to reduce computational complexity by searching over only those lattice points (defined as $\mathbf{Hs}$) that lie within a hypersphere of

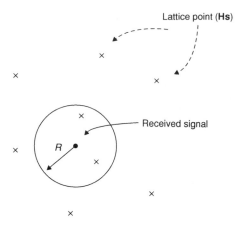

Figure 7.7: Schematic of the sphere decoding principle. The choice of the decoding radius R is critical to the performance.

radius R around the received signal $\mathbf{y}$, rather than searching over the entire lattice (see Fig. 7.7). Clearly, the point in the hypersphere closest to the received signal is also the closest lattice point for the whole lattice. The two immediate questions are:

(a) How do we choose R? If R is too large, then there are too many points to be searched and if R is too small there is the possibility of having no point.

(b) How do we determine which of the lattice points lies inside the given sphere?

A natural answer for the first question is the covering radius which is defined as the radius of the spheres centered at the lattice points that cover the entire space most economically. The Fincke–Pohst algorithm provides an efficient means of solving the second question, thereby avoiding an exhaustive search that would otherwise negate the benefits of sphere decoding. Though the worst-case complexity of sphere decoding is exponential, its expected complexity (averaged over noise and the lattice) has been shown to be polynomial (often cubic or even sub-cubic) [Hassibi and Vikalo, 2001]. Thus ML estimation can be applied to practical scenarios through sphere decoding with significant complexity reduction [Viterbo and Boutros, 1999; Brutel and Boutros, 1999; Damen *et al.*, 2000].

Performance   The ML receiver performance lower-bounds the error rate performance of other sub-optimal receivers. Unfortunately, computation of the exact average error probability is difficult in closed form. Therefore, we study the performance of the ML receiver through the PEP criterion (introduced in Chapter 6) that determines the probability that the vector symbol $\mathbf{s}^{(j)}$ is detected when $\mathbf{s}^{(i)}$ is transmitted with $j \neq i$. Assuming $\mathbf{H} = \mathbf{H}_w$, the average PEP at high SNR is upper-bounded by (see Eq. (6.20))

$$P(\mathbf{s}^{(i)} \to \mathbf{s}^{(j)}) \leq \frac{1}{\left(\frac{\rho}{4M_T} \|\mathbf{d}_{i,j}\|_F^2\right)^{M_R}}, \tag{7.39}$$

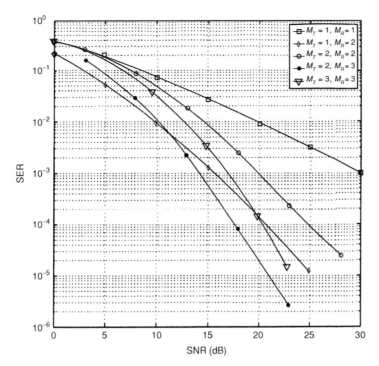

Figure 7.8: Average vector SER performance of the ML receiver over an $\mathbf{H}_w$ MIMO channel, uncoded SM for $M_T > 1$. The ML receiver extracts M_R order spatial diversity on each stream.

where $\mathbf{d}_{i,j} = \mathbf{s}^{(i)} - \mathbf{s}^{(j)}$. The ML decoding process outputs the most likely transmit signal vector. The individual symbols are extracted from this decoded signal vector. An error (in decoding the transmit vector) is declared whenever there is an error in any one of the symbols.

From Eq. (7.39) it is clear that the diversity order is M_R. Figure 7.8 plots the vector SER rate for 4-QAM modulation, $\mathbf{H} = \mathbf{H}_w$ channel, for different antenna configurations. In all cases, the SER curve for a MIMO receiver is parallel to that of a SIMO receiver for the same number of receive antennas, confirming M_R order diversity. However, the MIMO receiver suffers a SNR penalty compared with SISO with an increasing number of transmit antennas. One reason for this is power sharing between antennas in SM – since the normalization keeps the total transmit energy constant, in SM the power per stream is reduced by a factor of M_T. Another reason is the increase in the number of nearest neighbors in a vector constellation. This penalty becomes larger as the number of transmit antennas increase and as the constellation size increases.

The M_R order diversity achieved by the ML receiver is based on uncoded SM transmission. Even if we use HE (Section 6.3.5), the ML receiver will still only have M_R order diversity since the HE scheme does not capture transmit diversity – for a given symbol both systematic and parity bits are transmitted over the same antenna. "Optimal"

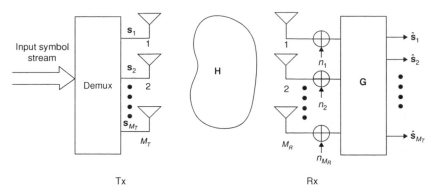

Figure 7.9: Schematic of a linear receiver for separating the transmitted data streams over a MIMO channel.

encoding should transmit the information bits over all the antennas through "spreading" and "mapping". In that case ML receivers would reach $M_T M_R$ order diversity.

The PEP considers only one of the $A^{M_T} \times (A^{M_T} - 1)$ error vectors: $\mathbf{d}_{i,j} = \mathbf{s}^{(i)} - \mathbf{s}^{(j)}$. This is generally not sufficient to correctly characterize the actual error probability. A more accurate characterization is the union bound [Sandhu and Paulraj, 2001; Marzetta *et al.*, 2001], which averages the effects of all error vectors. Applying the union bound, the average vector SER $\overline{P}_v$ is upper-bounded at high SNR by

$$\overline{P}_v \le \frac{1}{A^{M_T}} \sum_i \sum_{j,j \ne i} \frac{1}{\left(\frac{\rho}{4M_T} \|\mathbf{d}_{i,j}\|_F^2 \right)^{M_R}}. \tag{7.40}$$

Inclusion of all error vectors does not change our observations regarding the diversity order though it does show that systems with a higher number of transmit antennas have a larger error vector space (assuming the same constellation on each transmit antenna).

Linear receivers for SM

We can reduce the decoding complexity of the ML receiver by using a linear filter to separate the transmitted data streams, and then independently decode each stream (see Fig. 7.9). We discuss the ZF and MMSE linear receivers below. Whereas ZF receivers for SISO and SIMO channels were constructed to cancel ISI, here they are used to cancel MSI.

ZF receiver

The ZF matrix filter that separates the received signal into its component transmitted streams is given by

$$\mathbf{G}_{ZF} = \sqrt{\frac{M_T}{E_s}} \mathbf{H}^\dagger, \tag{7.41}$$

where $\mathbf{G}$ is an $M_T \times M_R$ matrix that simply inverts the channel.

The output of the ZF receiver is given by

$$\mathbf{z} = \mathbf{s} + \sqrt{\frac{M_T}{E_s}}\mathbf{H}^\dagger\mathbf{n}, \tag{7.42}$$

where we assume that $M_R \geq M_T$ and $\mathbf{H}$ has full column rank. The ZF receiver decouples the matrix channel into M_T parallel scalar channels with additive noise. Clearly, the noise is enhanced by $\mathbf{G}_{ZF}$. Furthermore, the noise is correlated across the channels. Each scalar channel is decoded independently ignoring noise correlation. ZF reduces complexity, but the receiver is sub-optimal and results in significant performance degradation.

Performance   From Eq. (7.42) the noise power on the kth ($k = 1, 2, \ldots, M_T$) output data stream is $(M_T/\rho)[(\mathbf{H}^H\mathbf{H})^{-1}]_{k,k}$ with the corresponding SNR, η_k, equal to

$$\eta_k = \frac{\rho}{M_T}\frac{1}{[(\mathbf{H}^H\mathbf{H})^{-1}]_{k,k}}. \tag{7.43}$$

It has been shown in [Winters *et al.*, 1994; Gore *et al.*, 2002b] that the SNR on each of the M_T streams (assuming $\mathbf{H} = \mathbf{H}_w$) is distributed as

$$f(x) = \frac{M_T}{\rho(M_R - M_T)!}e^{-\frac{M_T}{\rho}x}\left(\frac{M_T}{\rho}x\right)^{M_R - M_T}u(x). \tag{7.44}$$

η_k is a Chi-squared random variable with $2(M_R - M_T + 1)$ degrees of freedom. Since the noise in the separated M_T streams is correlated, the SNRs are not independent. The SER on any one channel averaged over all channel instances is upper-bounded by

$$\overline{P}_e \leq \overline{N}_e\left(\frac{\rho d_{min}^2}{4M_T}\right)^{-(M_R - M_T + 1)}. \tag{7.45}$$

Equation (7.45) shows that the diversity order of each stream is $M_R - M_T + 1$. Further, we can show that the average SNR on each stream is $(M_R - M_T + 1)(\rho/M_T)$. In effect, the ZF receiver decomposes the MIMO link into M_T parallel streams, each with diversity gain and array gain proportional to $M_R - M_T + 1$. Figure 7.10 plots SER for uncoded SM receivers for various antenna configurations assuming an $\mathbf{H}_w$ channel and 4-QAM modulation. Curves for SIMO channels are also presented for comparison. The slope of the curve with M_T transmit and M_R receive antennas with ZF reception is the same as the slope of the SIMO curve with $M_R - M_T + 1$ receive antennas using MRC. However, there is SNR loss since each stream has only $1/M_T$ of the total power.

To summarize, the ZF receiver with horizontal encoding realizes only $M_R - M_T + 1$ order diversity out of a maximum possible M_R order diversity.

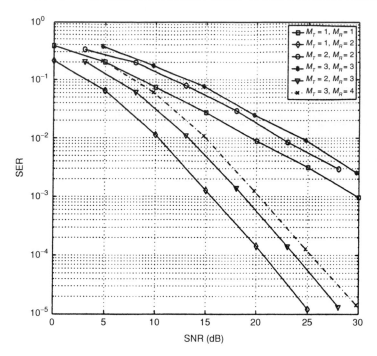

Figure 7.10: SER curves for a ZF receiver over an $\mathbf{H}_w$ channel, uncoded SM for $M_T > 1$. The diversity order extracted per stream equals $M_R - M_T + 1$.

MMSE receiver

The ZF receiver completely eliminates MSI at the expense of noise enhancement. The MMSE receiver balances MSI mitigation with noise enhancement and minimizes the total error, i.e.,

$$\mathbf{G}_{MMSE} = \arg\min_{\mathbf{G}} \mathcal{E}\{\|\mathbf{Gy} - \mathbf{s}\|_F^2\}. \tag{7.46}$$

Utilizing the orthogonality principle,

$$\mathcal{E}\{(\mathbf{Gy} - \mathbf{s})\mathbf{y}^H\} = \mathbf{0}_{M_T, M_R}, \tag{7.47}$$

$\mathbf{G}_{MMSE}$ is easily derived as

$$\mathbf{G}_{MMSE} = \sqrt{\frac{M_T}{E_s}} \left(\mathbf{H}^H\mathbf{H} + \frac{M_T}{\rho}\mathbf{I}_{M_T}\right)^{-1}\mathbf{H}^H. \tag{7.48}$$

Performance The signal to interference and noise ratio (SINR) on the kth $(1, 2, \ldots, M_T)$ decoded stream can be shown to be

$$\eta_k = \frac{1}{\left[\left(\frac{\rho}{M_T}\mathbf{H}^H\mathbf{H} + \mathbf{I}_{M_T}\right)^{-1}\right]_{k,k}} - 1, \tag{7.49}$$

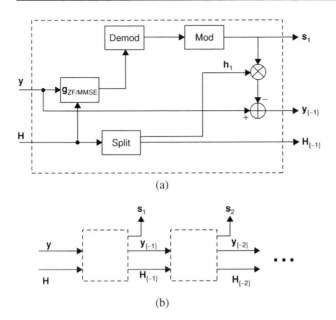

Figure 7.11: The SUC receiver: (a) one stage of SUC; (b) layers "peeled" at each stage to demodulate vector symbol.

or equivalently

$$\eta_k = \frac{\det(\mathbf{R_{yy}})}{\det\left(\mathbf{R_{yy}} - \dfrac{E_s}{M_T} \mathbf{h}_k \mathbf{h}_k^H \right)} - 1, \tag{7.50}$$

where $\mathbf{h}_k$ is the kth column of $\mathbf{H}$. The -1 term in Eqs. (7.49) and (7.50) is to account for bias [Cioffi, 2002]. The exact statistics of η_k are not yet properly characterized. However, some comments can be made for the high and low SNR regions. At low SNR the MMSE approximates a MF ($\mathbf{G}_{MMSE} = N_o^{-1} \sqrt{E_s/M_T} \mathbf{H}^H$) and outperforms the ZF receiver that continues to enhance noise. At high SNR we get

$$\mathbf{G}_{MMSE} = \sqrt{M_T/E_s}\,\mathbf{H}^\dagger, \tag{7.51}$$

i.e., the MMSE receiver converges to a ZF receiver and we can expect it to extract $M_R - M_T + 1$ order diversity. The MMSE receiver is therefore superior to the ZF receiver.

This completes the discussion on linear receivers. The next section focuses on non-linear ST receivers, in particular on successive cancellation techniques.

SUC receivers for SM
The key idea in a SUC receiver (see Fig. 7.11) is layer peeling where the symbol streams are successively decoded and stripped away layer by layer. For a single vector symbol, the algorithm is briefly summarized below.

Step 1: Extract the symbol from the first stream,

$$z = \mathbf{gy},$$ (7.52)

where $\mathbf{g}$ is the first row of the ZF/MMSE receiver $\mathbf{G}$ (see Eqs. (7.41), (7.48)). Slice z to decode s_1.

Step 2: Assume that the decision on s_1 is correct, remodulate to get s_1 and subtract its contribution from the received signal $\mathbf{y}$. The reduced signal model is

$$\mathbf{y}_{\{-1\}} = \mathbf{y} - \mathbf{h}_1 s_1$$
$$= \sqrt{\frac{E_s}{M_T}} \mathbf{H}_{\{-1\}} \mathbf{s}_{\{-1\}} + \mathbf{n},$$ (7.53)

where $\mathbf{y}_{\{-1\}}$ is the $M_R \times 1$ received vector with the contribution of s_1 removed, $\mathbf{h}_1$ is the first column of $\mathbf{H}$ and $\mathbf{H}_{\{-1\}}$ is a reduced channel matrix of dimension $M_R \times (M_T - 1)$ with

$$\mathbf{H}_{\{-1\}} = [\mathbf{h}_2 \cdots \mathbf{h}_{M_T}],$$ (7.54)

and $\mathbf{s}_{\{-1\}}$ is a reduced signal vector of dimension $(M_T - 1) \times 1$ given by

$$\mathbf{s}_{\{-1\}} = [s_2 \cdots s_{M_T}]^T.$$ (7.55)

Step 3: Return to step 1, extract the second stream and repeat until the vector symbol is decoded.

Performance   Provided all the decisions at each layer are correct (no error propagation), the equivalent system model after the kth decoding step has $M_T - k$ transmit antennas and M_R receive antennas. In fact, if we assume a ZF receiver is used at each stage, then the SNR on the kth stream is a Chi-squared random variable with $2k$ degrees of freedom. A MMSE receiver would have better SNR statistics.

SUC without ordering and assuming no error propagation converts the MIMO channel into a set of parallel channels with increasing diversity at each successive stage. In reality, there will be error propagation (analogous to the SISO/SIMO DFE), especially so if there is inadequate temporal coding. The error rate performance is then dominated by the weakest stream which is the first stream decoded by the receiver. Therefore, the improved diversity performance of succeeding layers does not help. Figure 7.12 plots the SER of SUC and MMSE receivers for SM-HE, 4-QAM modulation, with $M_T = M_R = 2$, in a frequency flat channel with $\mathbf{H} = \mathbf{H}_w$. We find that SUC is only slightly better than MMSE.

Ordered successive cancellation
A better receiver is the ordered SUC (OSUC) receiver, which is used in the V-BLAST receiver [Wolniansky *et al.*, 1998; Golden *et al.*, 1999]. The principle behind OSUC

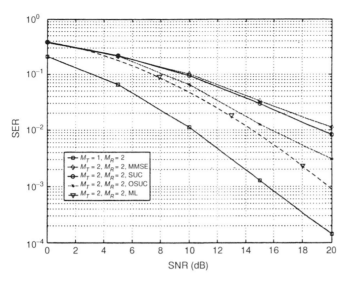

Figure 7.12: Comparison of ML, OSUC, SUC and MMSE receivers over an $\mathbf{H}_w$ MIMO channel, uncoded SM with $M_T > 1$. OSUC is superior to SUC and MMSE.

is that at the beginning of each stage, the stream with the highest SINR is selected for peeling. This improves the quality of the decision and has been shown to be optimal for the SUC approach. The SUC algorithm requires only a small change, wherein the SINRs of the remaining streams are calculated at each stage and the stream with the highest SINR is selected for decoding.

With OSUC, the decoded stream has an inherent form of selection diversity [Eng *et al.*, 1996] since its SINR is the maximum of the SINRs of the remaining streams. Figure 7.12 compares the performance of OSUC and SUC. OSUC is markedly better than SUC but still does not achieve a diversity order of M_R that the ML receiver achieves.

Discussion   Exact analysis of OSUC is difficult and is not attempted here. An upper bound on OSUC performance assuming ideal coding/decoding is presented in [Gore *et al.*, 2002a]. Further, the OSUC algorithm has been shown to be fundamentally similar to the generalized decision feedback equalizer (GDFE) [Cioffi and Forney, 1997; Ginis and Cioffi, 2001], which is an extension of the conventional SISO DFE to vector transmission. One shortcoming of the OSUC receiver is the high computational complexity requirement. The nulling (or alternatively MMSE) and cancellation step in the algorithm is computationally intensive and grows as the fourth power of the number of transmit antennas (for $M_T = M_R = M$). This can be a major limitation in practice. However, efficient square root algorithms [Hassibi, 1999, 2000] reduce this requirement to a complexity that grows as the third power of the number of transmit antennas.

Table 7.1. *Summary of comparative performance of receivers for SM-HE*

Receiver	Diversity order	SNR loss
ZF	$M_R - M_T + 1$	High
MMSE	$\approx M_R - M_T + 1$	Low
SUC	$\approx M_R - M_T + 1$	Low
OSUC	$\geq M_R - M_T + 1 \leq M_R$	Low
ML	M_R	Zero

Table 7.1 summarizes the performance of various receivers with uncoded SM. The ZF, MMSE and SUC receivers provide only $M_R - M_T + 1$ order diversity but have varying SNR loss. The OSUC receiver may have more than $M_R - M_T + 1$ order diversity because of the ordering (selection) process. The ML receiver is optimal and extracts M_R order diversity.

7.4.3   SM with horizontal and diagonal encoding

Thus far we have assumed uncoded SM and focused on vector symbol demodulation. Practical systems would employ coded SM such as SM-HE, SM-DE (D-BLAST) or a variation.

SM with HE

All receivers studied above are applicable with HE. The main difference is that we should decode one block per layer rather than a symbol per layer as discussed above. Since a given information bit and the associated parity bits are transmitted from the same antenna, the diversity order will remain at $M_R - M_T + 1$ for ZF and MMSE receivers. Coding improves the reliability of the decisions and will mitigate the error propagation effects in the successive cancellation receivers. The ML receiver achieves M_R order diversity. Additionally, temporal coding enhances error performance on account of coding gain.

SM with DE (D-BLAST)

This is a SUC receiver where the layers are distributed diagonally across the antennas (hence "diagonal") instead of across a single antenna for each layer as in HE. The initial unused ST triangular block helps initialize the receiver. The diagonal layer peeling is explained in Fig. 7.13. Assume error-free decoding. When we decode and peel layer A (exposing layer B), we are in the same position as we were at the start of decoding layer A. Layer B can now be decoded (peeled) in the same way as layer A. This process can be repeated until all layers are peeled. With a well-designed code, the DE architecture will distribute each information bit uniformly across all antennas. Proper decoding will then guarantee a diversity order of $M_T M_R$.

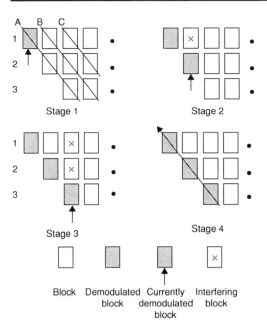

Figure 7.13: Stage 1: MMSE demodulation of A1. Stage 2: MMSE demodulation of A2 (B1 is interferer). Stage 3: MMSE demodulation of A3 (B2 and C1 are interferers). Stage 4: Layer A is decoded and peeled.

DE essentially allows capacity efficient MIMO transmission using scalar channel coding in each layer. In a block fading channel model, this implies that it supports the highest transmission rate for a defined packet error rate and SNR (see Chapter 11).

7.4.4   Frequency selective channel

The ZF, MMSE, DFE and ML receivers studied for SISO and SIMO frequency selective channels can be extended to MIMO frequency selective channels as well. MIMO DFE has received attention [Al-Dhahir and Sayed, 2000; Al-Dhahir *et al.*, 2001; Zhu and Murch, 2001; Lozano and Papadias, 2002]. MIMO receivers have to handle ISI in addition to MSI. In MIMO DFE receivers, we use the general principle of estimating and subtracting ISI from previous symbols and MSI from previous layers. If these estimates are correct, we suffer no penalty from error propagation.

7.5   Iterative MIMO receivers

Following Shannon theory, a capacity achieving scheme should combine appropriate channel encoding with an asymptotically large codeword size followed by an optimal matched (ML) decoder at the receiver. In a practical situation when we have to use

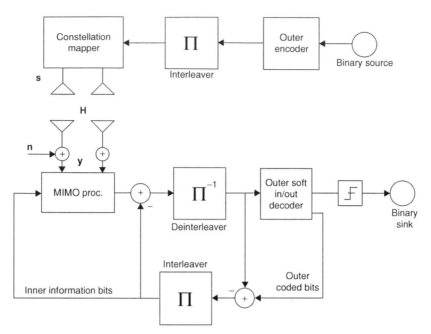

Figure 7.14: Generic block diagram of an iterative receiver.

finite block size codes, an efficient design should have a random codeword structure interleaved over all signal dimensions and, of course, have the largest feasible codeword length to span a maximal message segment. In addition, the receiver should use near-optimal decoding approaching ML performance. While near-optimal performance may be achieved by vertical/diagonal encoding schemes and subsequent mapping of the codeword over the transmit antennas, optimal decoding in such schemes is prohibitively complex. Iterative receiver designs can approach optimal performance with an afford-able receiver complexity. General iterative decoding related ideas have been discussed by several authors: turbo and LDPC coding [Berrou *et al.*, 1993; Benedetto *et al.*, 1998; MacKay, 1995], turbo demodulation [ten Brink *et al.*, 1998] and iterative multiuser detection [Wang and Poor, 1999].

Iterative receivers for MIMO are based on a combination of iterative interference cancellation and decoding. The received signal is regarded as a set of mutually interfering data streams (i.e., with MSI) that arrive from the multiple transmit antennas and are related through an underlying joint encoding scheme. The presence of MSI motivates a multiuser detection framework, whereas the presence of the common underlying code leads to joint iterative decoding of the data streams. A generic block diagram of such a scheme is given in Fig. 7.14. Iterative decoding of the data streams is accomplished by the standard soft-input soft-output decoding [Bahl *et al.*, 1974; Robertson *et al.*, 1974]. The extrinsic information on the code bits, computed as a difference between the inputs and the outputs of soft-input soft-output decoders, is exchanged between the decoders of different streams. Soft iterative interference cancellation consists of subtracting *aposteriori* expected values of the interference from the received signal, where the expectation is based on the preceding soft-input soft-output decoding step.

Layered receivers with iterative interference cancellation have been proposed by [Ariyavisitakul, 2000] and [Sellathurai and Haykin, 2000] and both are based on the D-BLAST scheme [Foschini, 1996] which achieves capacity using asymptotically large diagonally layered one-dimensional codes. While Ariyavisitakul considers a general scalar encoding scheme, Sellathurai and Haykin propose a modified D-BLAST encoding scheme which "fills" the ST wastage and concatenates the whole set of diagonal layers into M_T cycled diagonal layers that are interleaved over M_T antennas. Iterative processing is used at the receiver to approach ML performance. These modifications mitigate the shortcomings of the standard D-BLAST encoder, namely a large ST wastage when optimally large codes are used at each diagonal layer. See Section 7.4.3 for more details. Note that the D-BLAST receiver, which is based on successive peeling of diagonal layers, is not optimal if short codeword blocks are used for each layer.

An iterative receiver such as the one described in Fig. 7.14 combines a number of features such as complexity, which is linear in the number of iterations and in the number of the transmitted data streams, flexibility of the encoder and competitive performance (see [Li *et al.*, 2000]). However, information theoretic analysis of this approach as well as proper statistical performance analysis remains a major open issue.

Another broad family of iterative MIMO receivers is based on the concept of ST coded modulation. MIMO transmission can be regarded as coded modulation (or concatenated coding) where the outer code stands for the actual channel code (or a cascade of parallel codes) while the inner code is given by a map of coded bits onto ST symbols. Hence most ST transmission systems may be cast as a particular case of coded modulation. A generic block diagram of such a system is given in Fig. 7.15.

The standard reception technique is turbo demodulation, which consists of alternating soft-demapping and soft-input soft-output decoding steps with extrinsic information exchange between these steps, see Fig. 7.15. Different versions of ST coded modulation schemes have been proposed [Liu *et al.*, 2000; Tonello, 2000; Su and Geraniotis, 2001; Schlegel and Grant, 2001; van Zelst *et al.*, 2001; Stefanov and Duman, 2001] that make use of various combinations of the outer codes and inner ST maps.

The main drawback of ST coded modulation lies in demapper complexity. The complexity of traditional MAP and bitwise ML demappers scales exponentially with the number of the transmitted (coded) bits per ST symbol. While affordable for most SISO wireless systems, the complexity would be excessive for high data rate MIMO systems. The complexity problem has been addressed through sphere decoding, discussed in more detail in Section 7.4.2. Standard sphere decoding may be extended with list decoding in order to provide a list of candidate ST symbols that ensure accurate computation of soft bits [Hochwald and ten Brink, 2001].

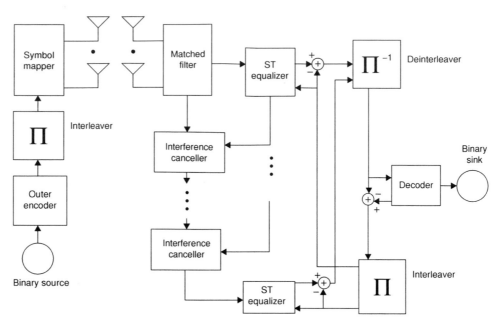

Figure 7.15: Schematic of a ST MIMO receiver based on the concept of ST coded modulation.

The design of outer codes and ST maps is a largely open area of research. A promising design methodology is based on the extrinsic information transfer (EXIT) chart [ten Brink, 1999] which is widely used for analysis and design of turbo codes and turbo demodulation. Although empirical in nature, the analysis based on EXIT charts leads to satisfactory results. Furthermore, results on information-theoretic properties of the EXIT chart [Ashikhmin *et al.*, 2002b] give an insight into the relationship between the shape of the EXIT chart and the maximum achievable information rate. Moreover, these results offer an approach for systematic design of outer codes and ST maps [Ashikhmin *et al.*, 2002a].

8 Exploiting channel knowledge at the transmitter

8.1 Introduction

Thus far we have, in general, assumed channel knowledge is always available at the receiver and not at the transmitter. In this chapter we consider the case where some degree of channel knowledge is also available at the transmitter. This knowledge can be full or partial. Full channel knowledge implies that the channel $\mathbf{H}$ is known to the transmitter. This is not easy to arrange as discussed in Chapter 3. Partial knowledge, on the other hand, can be easier to obtain. Partial channel knowledge might refer to some parameter of the instantaneous channel (such as the matrix channel's condition number) or a statistic (such as transmit or receive correlation).

We shall demonstrate how channel knowledge at the transmitter may be exploited to improve link performance. The scheme to exploit channel knowledge (see Fig. 8.1) at the transmitter depends on a variety of factors:

- The nature of channel knowledge (full or partial)
- The choice of signaling or coding (SM, OSTBC) and receiver (ML, ZF, MMSE, SUC)
- The performance criterion to be optimized (capacity, average error rate, etc.)
- The power constraints at the transmitter (total power, total + peak power, etc.)

The designer may choose from a number of transmit processing schemes. We study two such schemes – linear pre-filtering and selection – and provide representative examples of each. Given a specific transmit scheme, the problem reduces to determining the processor parameters based on the optimization criterion, channel knowledge and signaling and receiver design choices. To simplify our presentation, we assume frequency flat fading and $M_T = M_R = M$ for most of this chapter.

8.2 Linear pre-filtering

A pre-filtering framework for exploiting channel knowledge is shown in Fig. 8.2. The encoded signal to be transmitted, $\mathbf{x}\,(M \times 1)$, is multiplied by a $M \times M$ pre-filter matrix, $\mathbf{W}$, prior to transmission. $\mathbf{W}$ can also be interpreted as a beamformer. The input–output

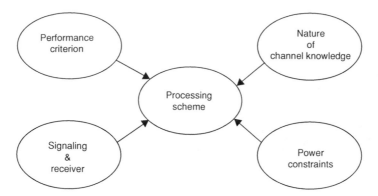

Figure 8.1: Factors that influence transmitter pre-filtering.

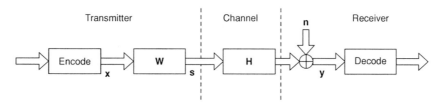

Figure 8.2: A MIMO system with a transmit pre-filter designed by exploiting channel knowledge.

relation for the frequency flat channel is

$$\mathbf{y} = \sqrt{\frac{E_s}{M}} \mathbf{HWx} + \mathbf{n}, \tag{8.1}$$

$$\mathbf{y} = \sqrt{\frac{E_s}{M}} \mathbf{Hs} + \mathbf{n}, \tag{8.2}$$

where $\mathbf{x}$ satisfies $\mathcal{E}\{\mathbf{xx}^H\} = \mathbf{I}_M$ and

$$\mathbf{s} = \mathbf{Wx}. \tag{8.3}$$

It follows that the covariance matrix of the transmitted signal $\mathbf{R_{ss}}$ is given by

$$\mathbf{R_{ss}} = \mathbf{WW}^H. \tag{8.4}$$

The pre-filter matrix $\mathbf{W}$ must satisfy the power constraint at the transmitter. If we require the total average transmit power to be E_s, then

$$\|\mathbf{W}\|_F^2 = \mathrm{Tr}(\mathbf{R_{ss}}) = M. \tag{8.5}$$

Alternatively, if each transmit antenna also has a peak power constraint of E_{peak} in addition to a total average power constraint of E_s, then $\mathbf{W}$ must satisfy a mixed power constraint

$$\|\mathbf{w}_j\|_F^2 = [\mathbf{R}_{ss}]_{j,j} \leq \frac{E_{peak}M}{E_s}, \tag{8.6}$$

$$\|\mathbf{W}\|_F^2 = M, \tag{8.7}$$

where $\mathbf{w}_j$ is the jth ($j = 1, 2, \ldots, M$) row of $\mathbf{W}$.

8.3 Optimal pre-filtering for maximum rate

8.3.1 Full channel knowledge

In the following we assume that the channel $\mathbf{H}$ is full-rank, i.e., $r = M$ and is perfectly known to the transmitter and the receiver. Recall from Chapter 3 that the capacity of the MIMO channel is

$$C = \max_{\mathbf{R}_{ss}} \log_2 \det\left(\mathbf{I}_M + \frac{E_s}{MN_o}\mathbf{H}\mathbf{R}_{ss}\mathbf{H}^H\right) \quad \text{bps/Hz.} \tag{8.8}$$

Since $\mathbf{R}_{ss} = \mathbf{W}\mathbf{W}^H$, capacity is achieved by maximizing RHS over $\mathbf{W}$ with the applicable constraints. Given a total average power constraint, $\mathbf{R}_{ss}^{opt}$ can be found through the waterpouring algorithm described in Chapter 3. Alternatively under the mixed power constraint, we can show that the optimization problem is convex and $\mathbf{R}_{ss}^{opt}$ can be found [Sampath, 2001] by numerical methods. From Eq. (8.4), it follows that once $\mathbf{R}_{ss}^{opt}$ is found, the optimal pre-filtering matrix $\mathbf{W}^{opt}$ is given by

$$\mathbf{W}^{opt} = \mathbf{Q}_{opt}\mathbf{\Lambda}_{opt}^{1/2}, \tag{8.9}$$

where $\mathbf{Q}_{opt}\mathbf{\Lambda}_{opt}\mathbf{Q}_{opt}^H$ is the eigendecomposition of $\mathbf{R}_{ss}^{opt}$.

Total power constraint, $\text{Tr}(\mathbf{R}_{ss}) = M$
From Eq. (4.26) we know that the solution is

$$\mathbf{R}_{ss}^{opt} = \mathbf{V}\mathbf{R}_{\widetilde{ss}}^{opt}\mathbf{V}^H, \tag{8.10}$$

where $\mathbf{R}_{\widetilde{ss}}^{opt}$ is a diagonal matrix found through the waterpouring algorithm. Hence, from Eq. (8.9) we get

$$\mathbf{W}^{opt} = \mathbf{V}\left(\mathbf{R}_{\widetilde{ss}}^{opt}\right)^{1/2}. \tag{8.11}$$

Substituting Eq. (8.11) into Eq. (8.2) and left-multiplying (multiplying with a unitary matrix at the receiver does not alter mutual information) by $\mathbf{U}^H$, where $\mathbf{H} = \mathbf{U}\mathbf{\Sigma}\mathbf{V}^H$, Eq. (8.2) reduces to

$$\widetilde{\mathbf{y}} = \sqrt{\frac{E_s}{M}}\mathbf{\Sigma}\widetilde{\mathbf{s}} + \widetilde{\mathbf{n}}, \tag{8.12}$$

where $\widetilde{\mathbf{n}} = \mathbf{U}^H \mathbf{n}$ and $\widetilde{\mathbf{s}} = \left(\mathbf{R}_{\widetilde{\mathbf{s}}\widetilde{\mathbf{s}}}^{opt}\right)^{1/2} \mathbf{x}$. This is exactly the result found in Chapter 4 using modal decomposition and waterpouring based power allocation for rate maximization.

8.3.2   Partial channel knowledge

We now study $\mathbf{W}^{opt}$ when knowledge of some channel statistics alone is available to the transmitter. We assume that the channel experiences transmit correlation and no receive correlation and that the transmit correlation matrix is known to the transmitter. This correlation model can occur when the transmit antennas are separated by less than the coherence distance, while the receive antennas are fully decorrelated. This channel may be modeled as (see Chapter 3)

$$\mathbf{H} = \mathbf{H}_w \mathbf{R}_t^{1/2}, \tag{8.13}$$

where $\mathbf{R}_t$ is the transmit correlation matrix. We assume the transmitter knows $\mathbf{R}_t$ but not the actual channel realization $\mathbf{H}$.

As only a channel statistic is known to the transmitter, rate optimization is possible only in a statistical sense, leading to the optimization of outage or ergodic capacity optimization. Working with ergodic capacity $\overline{C}$ leads to a tractable analysis and is discussed below. The optimization problem is now given by

$$\overline{C} - \max_{\mathbf{W}} \mathcal{E} \left\{ \log_2 \det \left(\mathbf{I}_{M_R} + \frac{\rho}{M} \mathbf{H} \mathbf{W} \mathbf{W}^H \mathbf{H}^H \right) \right\}. \tag{8.14}$$

The total power constraint requires $\|\mathbf{W}\|_F^2 = M$. It has been shown [Jafar et al., 2001] that capacity is achieved when

$$\mathbf{W}^{opt} = \mathbf{Q}_{\mathbf{R}_t} \Lambda_{\mathbf{W}}^{1/2}, \tag{8.15}$$

where $\mathbf{Q}_{\mathbf{R}_t}$ is the eigenvector matrix of $\mathbf{R}_t$ (i.e., $\mathbf{R}_t = \mathbf{Q}_{\mathbf{R}_t} \Lambda_{\mathbf{R}_t} \mathbf{Q}_{\mathbf{R}_t}^H$) and $\Lambda_{\mathbf{W}}$ is a diagonal power allocation matrix that satisfies

$$\mathrm{Tr}(\Lambda_{\mathbf{W}}) = M. \tag{8.16}$$

The solution therefore suggests we transmit along the eigenvectors of $\mathbf{R}_t$. However, the optimal power allocation, $\Lambda_{\mathbf{W}}$, remains an open problem. Attempts to characterize the solution are made in [Shiu and Kahn, 1998; Jorswieck and Boche, 2002]. For example, if the transmit covariance matrix $\mathbf{R}_t$ approaches rank 1, there is a range of SNR for which the optimal (rate maximizing) strategy is to direct all the power into the dominant eigenmode of $\mathbf{R}_t$. This is analogous to the dominant mode transmission, where the channel is perfectly known to the transmitter and approaches rank 1. A general solution based on "stochastic waterpouring" that maximizes an approximation of the ergodic capacity has been proposed in [Shiu and Kahn, 1998; Gorokhov, 2000]. The solution uses waterpouring on the weighted eigenmodes of $\mathbf{R}_t$ as against the eigenmodes of $\mathbf{H}$.

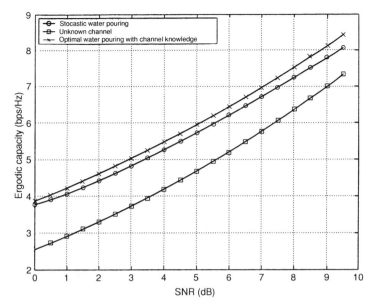

Figure 8.3: Ergodic capacity comparison based on the degree of channel knowledge available to the transmitter.

We can demonstrate the value of statistical knowledge and pre-filtering by a simulation study. Consider a MIMO channel with $M = 4$ and transmit correlation $\mathbf{R}_t$ given by

$$\mathbf{R}_t = \begin{bmatrix} 1 & 0.9 & 0.81 & 0.729 \\ 0.9 & 1 & 0.9 & 0.81 \\ 0.81 & 0.9 & 1 & 0.9 \\ 0.729 & 0.81 & 0.9 & 1 \end{bmatrix}. \tag{8.17}$$

Figure 8.3 plots the ergodic capacity for a channel described by $\mathbf{H} = \mathbf{H}_w \mathbf{R}_t^{1/2}$ in the following scenarios:

(a) No channel knowledge at transmitter and $\mathbf{W} = \mathbf{I}_M$ is chosen, i.e., no pre-filtering is employed.

(b) Partial channel knowledge ($\mathbf{R}_t$) is available and "stochastic waterpouring" as described in [Shiu and Kahn, 1998; Gorokhov, 2000] is used.

(c) Full channel knowledge $\mathbf{H}$ is available and modal transmission with optimal waterpouring power allocation is performed for each channel instance.

Observe that stochastic waterpouring based solely on knowledge of $\mathbf{R}_t$ outperforms the no channel knowledge case but is inferior to the full channel knowledge solution. The performance gap between partial and full channel knowledge reduces with increasing correlation. This follows from earlier comments about full and partial channel knowledge solutions converging for uni-modal (rank 1) channels.

8.4   Optimal pre-filtering for error rate minimization

8.4.1   Full channel knowledge

With full channel knowledge at the transmitter and a sum power constraint, a modal decomposition is optimal and $\mathbf{W}$ is given by Eq. (8.9). The problem of error rate minimization is meaningful only if non-asymptotic (finite block length) coding is used on the decoupled scalar channels. The remaining problem of power allocation can be posed in terms of SNR (i.e., power) in each mode and its influence on SER.

A different approach based on inverse SNR (or equivalently) mean squared estimation error between the input and output of each mode has been studied by several authors [Yang and Roy, 1994; Sampath *et al.*, 2001; Scaglione *et al.*, 2002]. A general solution is a weighted sum of inverse SNRs (or mean squared estimation errors) and leads to a number of well-known solutions depending on the choice of weights [Sampath *et al.*, 2001].

8.4.2   Partial channel knowledge

We now discuss optimizing the pre-filter $\mathbf{W}$ using the probability of error criterion when the transmitter knows the transmit correlation $\mathbf{R}_t$. Specifically, we use the PEP criterion, which leads to a tractable analysis. We can easily absorb $\mathbf{W}$ into the channel itself and for the remainder of this section will refer to $\mathbf{H} = \mathbf{H}_w \mathbf{R}_t^{1/2} \mathbf{W}$ as the effective channel.

As developed in Chapter 6, the PEP, i.e., the probability of mistaking the transmitted codeword $\mathbf{X}^{(i)}$ for $\mathbf{X}^{(j)}$ ($j \neq i$) for a particular realization of the channel $\mathbf{H}$ is upper-bounded by

$$P\left(\mathbf{X}^{(i)} \to \mathbf{X}^{(j)} | \mathbf{H}\right) \leq e^{-\frac{\rho}{4M} \|\mathbf{H}\mathbf{E}_{i,j}\|_F^2}, \tag{8.18}$$

where $\mathbf{E}_{i,j} = \mathbf{X}^{(i)} - \mathbf{X}^{(j)}$ is an error word of dimension $M_T \times T$. The average PEP obtained by averaging over the channel is upper-bounded by

$$P\left(\mathbf{X}^{(i)} \to \mathbf{X}^{(j)}\right) \leq \left(\frac{1}{\det\left(\mathbf{I}_M + (\rho/4M)\mathbf{E}_{i,j}^H \mathbf{W}^H \mathbf{R}_t \mathbf{W}\mathbf{E}_{i,j}\right)}\right)^M. \tag{8.19}$$

Under a total average power constraint, the optimization problem reduces to choosing $\mathbf{W}$ to maximize

$$\max_{\text{Tr}(\mathbf{W}\mathbf{W}^H)=M} \det\left(\mathbf{I}_M + \frac{\rho}{4M}\mathbf{E}_{i,j}^H \mathbf{W}^H \mathbf{R}_t \mathbf{W}\mathbf{E}_{i,j}\right). \tag{8.20}$$

Assuming $\mathbf{R}_t$ and $\mathbf{E}_{i,j}$ are full-rank, the optimal $\mathbf{W}$ satisfies

$$\mathbf{W}^{opt} = \mathbf{Q}_{\mathbf{R}_t} \mathbf{\Lambda}_{\mathbf{W}}^{1/2} \mathbf{Q}_{\mathbf{E}_{i,j}}^H, \tag{8.21}$$

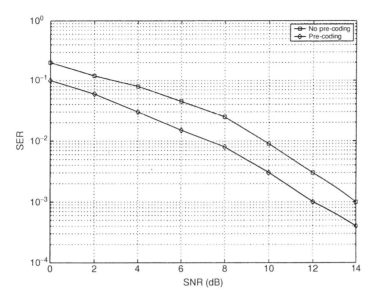

Figure 8.4: Pre-filtering for Alamouti coding based on knowledge of $\mathbf{R}_t$ improves performance.

where $\mathbf{E}_{i,j}\mathbf{E}_{i,j}^{H} = \mathbf{Q}_{\mathbf{E}_{i,j}}\mathbf{\Lambda}_{\mathbf{E}_{i,j}}\mathbf{Q}_{\mathbf{E}_{i,j}}^{H}$ (eigendecomposition of $\mathbf{E}_{i,j}\mathbf{E}_{i,j}^{H}$) and $\mathbf{\Lambda}_{\mathbf{W}}$ is a diagonal matrix whose diagonal elements can be computed using waterpouring [Sampath, 2001]. $\mathbf{W}^{opt}$ depends on the error matrix $\mathbf{E}_{i,j}$. Typically we should choose $\mathbf{E}_{i,j}$ to represent the worst case (maximum PEP). In practice, the worst case error words are difficult to identify, complicating this approach.

The solution simplifies considerably for OSTBC codes which have error matrices $\mathbf{E}_{i,j}$ that satisfy $\mathbf{E}_{i,j}\mathbf{E}_{i,j}^{H} = d_{min}^{2}\mathbf{I}_{M}$, and where d_{min} is the minimum distance of the underlying scalar constellation. Hence, for OSTBC

$$\mathbf{W}_{OSTBC}^{opt} = \mathbf{Q}_{\mathbf{R}_t}\mathbf{\Lambda}_{\mathbf{W}}^{1/2}, \tag{8.22}$$

which is similar to the solution for rate maximization in the sense that the optimal approach is to signal on the modes of $\mathbf{R}_t$.

The gain of pre-filtering for OSTBC is demonstrated via the following example. Consider a MIMO channel with $M = 2$ and Alamouti coding. The channel is correlated with

$$\mathbf{R}_t = \begin{bmatrix} 1 & 0.5 \\ 0.5 & 1 \end{bmatrix}. \tag{8.23}$$

Figure 8.4 plots the SER curves with and without pre-filtering ($\mathbf{W}_{OSTBC}^{opt}$) and demonstrates a 2 dB gain from pre-filtering. For a more detailed treatment refer to [Sampath and Paulraj, 2001]. The above discussion also holds for MISO channels using OSTBC coding.

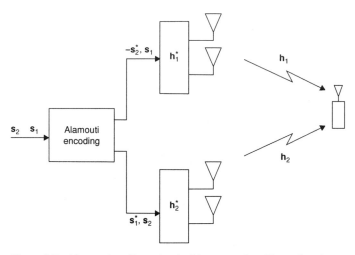

Figure 8.5: Alamouti coding mixed with conventional beamforming.

There is an interesting propagation scenario that combines transmit MRC and Alamouti coding given knowledge of the channel covariance alone. Consider, for example, a MISO system with two compact sub-arrays at the transmitter each with N antennas. The sub-arrays are well separated. We may write the $1 \times 2N$ channel as

$$\mathbf{h}^T = \begin{bmatrix} \mathbf{h}_1^T & \mathbf{h}_2^T \end{bmatrix}, \tag{8.24}$$

where $\mathbf{h}_i^T$ are the $1 \times N$ sub-array channels. Let the channel be coherent (perfectly correlated) across antennas within a sub-array and uncorrelated across sub-arrays. The covariance matrix $\mathbf{R}$ is therefore of the form

$$\mathbf{R} = \mathcal{E}\{\mathbf{h}\mathbf{h}^H\} = \begin{bmatrix} \mathcal{E}\{\mathbf{h}_1\mathbf{h}_1^H\} & \mathbf{0}_{N,N} \\ \mathbf{0}_{N,N} & \mathcal{E}\{\mathbf{h}_2\mathbf{h}_2^H\} \end{bmatrix}. \tag{8.25}$$

Note $r(\mathbf{R}) = 2$. Assuming that $\|\mathbf{h}_1\|_F^2 = \|\mathbf{h}_2\|_F^2 = 1$, the eigenvectors of $\mathbf{R}$ are given by

$$\begin{bmatrix} \mathbf{h}_1 \\ \mathbf{0}_{N,1} \end{bmatrix} \quad \text{and} \quad \begin{bmatrix} \mathbf{0}_{N,1} \\ \mathbf{h}_2 \end{bmatrix}, \tag{8.26}$$

and the two associated non-zero eigenvalues are unity. Therefore, the eigenvectors of $\mathbf{R}$ provide the channel response at each sub-array. An optimal scheme would be beamforming (transmit MRC) at each sub-array with Alamouti coding across sub-arrays to extract available array gain and diversity gain. Figure 8.5 illustrates this scheme.

The unifying approach when performing transmit pre-filtering is to transmit on the eigenmodes of the channel when exact channel knowledge is available and on the eigenmodes of the transmit channel covariance matrix when only partial channel knowledge is available, using appropriate power allocation in both cases.

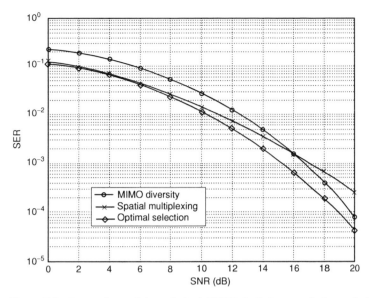

Figure 8.6: Comparison of the switched (OSTBC, SM) transmission technique with fixed OSTBC and fixed SM. The switched scheme outperforms both techniques at all SNRs.

8.5   Selection at the transmitter

We study selection as a type of pre-processing at the transmitter. We consider two approaches: (a) switching between different ST coding techniques and (b) antenna sub-set selection switching.

8.5.1   Selection between SM and diversity coding

Assume that SM and Alamouti encoding schemes are available to the transmitter and that the metric for selection is the probability of error (with the transmission rate held fixed by suitably adjusting the transmit constellation for each scheme). Since the channel is known to the receiver, it can compute the instantaneous error probability (which depends on a realization of the channel) for each scheme, choose the coding scheme with the lower error rate and signal this information to the transmitter.

Figure 8.6 plots the SER in a 2×2 $\mathbf{H}_w$ channel for the uncoded Alamouti scheme, uncoded SM with ML decoding and a scheme that switches optimally between Alamouti and SM. The transmit constellation for SM is BPSK, while the Alamouti code uses the 4-QAM constellation to maintain the data rate constant at 2 bps/Hz for both schemes. A gain of about 1–2 dB is noticeable at SNR = 16 dB for the switched scheme. Of course the selection scheme outperforms both SM and OSTBC at all SNRs.

Instead of computing the exact probability of error, we can derive an approximate criterion as follows. Assuming ML decoding at the receiver, the instantaneous probability of symbol error P_{SM}^{inst} for uncoded SM transmission may be approximated by (see Eq. (6.5))

$$P_{SM}^{inst} \approx Q\left(\sqrt{\frac{\rho}{2M_T}D_{min,SM}^2}\right), \tag{8.27}$$

where $D_{min,SM}^2 = \min_{i,j}\|\mathbf{H}(\mathbf{s}^{(i)} - \mathbf{s}^{(j)})\|_F^2$ is the squared minimum distance of separation of the vector constellation points at the receiver and $\mathbf{s}^{(i)}$ and $\mathbf{s}^{(j)}$ are as defined in Chapter 6. Using the Rayleigh–Ritz criterion we can bound $D_{min,SM}^2$ by

$$D_{min,SM}^2 \geq \lambda_{min}d_{min,SM}^2, \tag{8.28}$$

where λ_{min} is the minimum eigenvalue of $\mathbf{HH}^H$ and $d_{min,SM}^2$ is the squared minimum distance of the transmit scalar constellation.

P_{OSTBC}^{inst}, the instantaneous probability of symbol error with OSTBC, is approximately

$$P_{OSTBC}^{inst} \approx Q\left(\sqrt{\frac{\rho}{2M_T}D_{min,OSTBC}^2}\right), \tag{8.29}$$

where $D_{min,OSTBC}^2 = \|\mathbf{H}\|_F^2 d_{min,OSTBC}^2$, and $d_{min,OSTBC}^2$ is the squared minimum distance of the scalar constellation when OSTBC is used. As stated previously, the scalar constellation used in conjunction with OSTBC is larger than that used with SM to keep the rate constant.

Comparing Eqs. (8.27) and (8.29), P_{OSTBC}^{inst} is lower than P_{SM}^{inst} if

$$D_{min,OSTBC}^2 \geq D_{min,SM}^2, \tag{8.30}$$

or

$$\kappa^2 \geq \frac{d_{min,SM}^2}{d_{min,OSTBC}^2}, \tag{8.31}$$

where $\kappa^2 = \|\mathbf{H}\|_F^2/\lambda_{min}$ and κ is the Demmel condition number [Demmel, 1988; Edelman, 1989] of the matrix $\mathbf{H}$. Equation (8.31) states that it is better to use Alamouti transmission instead of SM when the Demmel condition number of the channel is above a certain threshold (see Eq. (8.31)) and vice versa. The threshold depends on the constellations used for Alamouti coding and SM. Further details can be found in [Heath and Paulraj, 2001b].

8.5.2   Antenna selection

An important practical problem arising with the deployment of multiple transmit antennas is the cost of the hardware (transmit amplifiers, digital-to-analog converters, etc.) associated with every additional antenna. Antenna sub-set selection where transmission is performed through a sub-set of the available antenna elements is a cost-effective solution [Win and Winters, 2001; Molisch et al., 2001; Heath and Paulraj, 2001a; Gorokhov, 2002; Gore et al., 2002a; Gore and Paulraj, 2002;

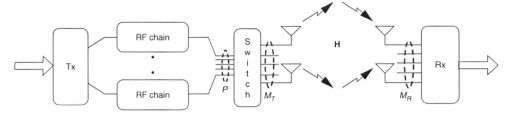

Figure 8.7: Transmit antenna switching schematic.

Blum and Winters, 2002] since the diversity order associated with an optimally selected antenna system is the same as that of the system with all antennas in use.

The antenna selection solution depends, of course, on the signaling scheme, the receiver architecture, the optimization criteria and the nature of channel knowledge available. We very briefly discuss two criteria for antenna selection: (i) maximum information rate and (ii) minimum error (assuming OSTBC transmission). Full channel knowledge at the transmitter is assumed in both examples.

Figure 8.7 is a schematic of a system with antenna selection at the transmitter. The transmitter has M_T transmit antennas and P transmit RF chains. The channel between the M_T transmit and M_R receive antennas is denoted by $\mathbf{H}$. Transmission is performed through P (out of the M_T available) transmit antennas that are selected to optimize the performance criterion. There are $\binom{M_T}{P}$ distinct choices which we index using i. The channel corresponding to the ith choice is given by $\mathbf{H}_i$ $(M_R \times P)$ and comprises the P selected columns of $\mathbf{H}$.

Maximum information rate

The optimization problem can be posed as

$$C = \max_{i, \mathbf{R}_{ss}} \log_2 \det \left(\mathbf{I}_M + \frac{\rho}{P} \mathbf{H}_i \mathbf{R}_{ss} \mathbf{H}_i^H \right), \tag{8.32}$$

with $\mathrm{Tr}(\mathbf{R}_{ss}) = P$ and where $\mathbf{R}_{ss}$ is the $P \times P$ covariance matrix of the transmitted signal vector, chosen to satisfy the transmitter power constraints. We use a sub-optimal approach of equi-power allocation ($\mathbf{R}_{ss} = \mathbf{I}_P$) even though we have full channel knowledge at the transmitter. The problem now is to choose that set of P columns of $\mathbf{H}$ (selected antennas) which maximizes

$$C_i = \log_2 \det \left(\mathbf{I}_{M_R} + \frac{\rho}{P} \mathbf{H}_i \mathbf{H}_i^H \right). \tag{8.33}$$

The optimal algorithm involves an exhaustive search through all $\binom{M_T}{P}$ possible combinations. Practical algorithms that avoid serial searches and with near-optimal performance have been proposed.

Figure 8.8 is a plot of the average information rate as a function of the number of sub-set antennas, P, and the SNR for $M_T = 4$ using the optimal selection discussed

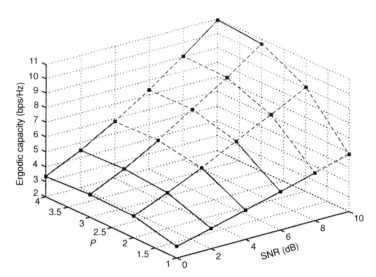

Figure 8.8: Ergodic capacity with transmit antenna selection as a function of selected antennas, P, and SNR, $M_T = 4$.

above. Notice that when $P = 3$, the capacity is almost the same as that with all four antennas utilized.

Antenna selection techniques and performance analysis can be developed for other problems including different types of channel knowledge, performance criteria and signaling-receiver choices. In most cases, assuming $\mathbf{H}$ is an $\mathbf{H}_w$ type channel, the diversity performance of the full $(M_R \times M_T)$ matrix (equal to $M_T M_R$ for diversity schemes) can be achieved through a properly selected channel $(M_R \times P)$. We now briefly discuss antenna selection based on minimizing error probability when using Alamouti coding.

Minimum SER with Alamouti encoding

Consider a MIMO system with antenna selection capability at the transmitter (as per Fig. 8.7). Assume that an OSTBC is used for transmission over the $M_R \times P$ link. The SER depends on the received SNR, which in turn depends only on the Frobenius norm of the selected channel matrix (see Chapter 6). The received SNR is given by

$$\eta = \frac{\rho}{P} \|\mathbf{H}_i\|_F^2, \tag{8.34}$$

where $\mathbf{H}_i$ $(M_R \times P)$ consists of the selected columns of $\mathbf{H}$. Therefore, the optimal antenna sub-set consists of the P columns of $\mathbf{H}$ that maximize $\|\mathbf{H}_i\|_F^2$.

For any ordered set (in decreasing order) of M_T non-negative variables, the mean of the first P variables is larger than the mean of the entire set. Hence we have

$$\frac{\|\mathbf{H}_i^{opt}\|_F^2}{P} \geq \frac{\|\mathbf{H}\|_F^2}{M_T}, \tag{8.35}$$

where

$$\mathbf{H}_i^{opt} = \arg\max_{\mathbf{H}_i} \|\mathbf{H}_i\|_F^2. \tag{8.36}$$

Further, since the squared Frobenius norm of the full $(M_R \times M_T)$ channel is greater than that of the selected $(M_R \times P)$ channel

$$\|\mathbf{H}\|_F^2 \geq \|\mathbf{H}_i^{opt}\|_F^2. \tag{8.37}$$

Combining Eqs. (8.35) and (8.37) we have

$$\frac{\rho}{P}\|\mathbf{H}\|_F^2 \geq \eta^{opt} \geq \frac{\rho}{P}\frac{P}{M_T}\|\mathbf{H}\|_F^2. \tag{8.38}$$

In other words, the upper and lower bounds on the received SNR for the optimal selection are directly dependent on the squared Frobenius norm of the full channel. For an $\mathbf{H}_w$ type channel, the upper and lower bounds in Eq. (8.38) are weighted Chi-squared variables with $2M_T M_R$ degrees of freedom and we conclude that transmit selection provides the same diversity order, $M_T M_R$.

Receive antenna selection

While we have discussed transmit selection so far, similar techniques are applicable for receive antenna selection. In all cases, the key approach is to search over the antenna sub-sets (rows or columns) to maximize the performance metric given some side information on the channel state and power constraints. For example, in the particular case of Alamouti coding, the analysis and performance for transmit selection ($P = 2$, $M_T = 3$, $M_R = 2$, i.e., choosing two out of three transmit antennas) is identical to receive selection ($M_T = 2$, $M_R = 3$, $P = 2$, i.e., choosing two out of three receive antennas). Figure 8.9 depicts the SER curves for a system with receive selection capability, assuming that Alamouti encoding is used at the transmitter.

8.6   Exploiting imperfect channel knowledge

We address the coding strategy in situations where the transmitter's estimate of the channel is in error. See Chapter 3 for transmit channel estimation and error sources. Our development closely follows [Jöngren *et al.*, 2002].

Consider a MISO channel with $M_T = 2$. Assume $\mathbf{h} = \mathbf{h}_w$. We assume the receiver has perfect (zero error) channel estimates, while the transmitter has an imperfect channel estimate, $\widehat{\mathbf{h}}$. Let the correlation between the true channel and the estimated channel at both transmitter antennas be the same:

$$\rho_{corr} = \mathcal{E}\{h_i \widehat{h}_i^*\}, \ i = 1, 2. \tag{8.39}$$

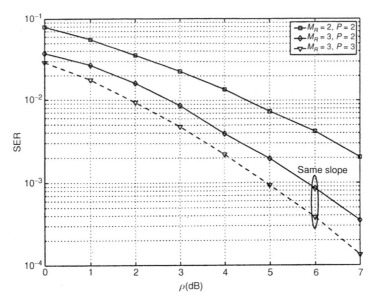

Figure 8.9: Selecting two out of three receive antennas delivers full diversity order, Alamouti encoding.

Assuming $\mathbf{h}$ and $\widehat{\mathbf{h}}$ are jointly Gaussian, the distribution of $\mathbf{h}$ conditioned on $\widehat{\mathbf{h}}$ is also Gaussian with mean $(\mathbf{m}_{\mathbf{h}|\widehat{\mathbf{h}}})$ and covariance $(\mathbf{R}_{\mathbf{hh}|\widehat{\mathbf{h}}})$ given by

$$\mathbf{m}_{\mathbf{h}|\widehat{\mathbf{h}}} = \rho_{corr}\widehat{\mathbf{h}}, \tag{8.40}$$

$$\mathbf{R}_{\mathbf{hh}|\widehat{\mathbf{h}}} = \left(1 - \left|\rho_{corr}^2\right|\right)\mathbf{I}_2. \tag{8.41}$$

When $\rho_{corr} = 1$, the transmitter has a perfect channel estimate, and when $\rho_{corr} = 0$, it has no useful estimate of the channel. Therefore ρ_{corr} is a measure of the channel error at the tranmsitter. We present a strategy that combines Alamouti coding and transmit-MRC depending on ρ_{corr}. Let the transmitter use a linear pre-filter (see Fig. 8.2). The signal model is given by

$$\mathbf{y} = \sqrt{\frac{E_s}{2}}\mathbf{hWX} + \mathbf{n}, \tag{8.42}$$

where $\mathbf{y}$ is the 1×2 received signal row vector, $\mathbf{W}$ is the 2×2 linear pre-filter and $\mathbf{n}$ is the 1×2 noise vector. The input to the pre-filter is a 2×2 codeword block $\mathbf{X}$ that is Alamouti encoded, i.e.,

$$\mathbf{X} = \begin{bmatrix} x_0 & -x_1^* \\ x_1 & x_0^* \end{bmatrix}, \tag{8.43}$$

where x_i $(i = 0, 1)$ are the constituent data symbols of the codeword. The optimization problem is to find the $\mathbf{W}$ that minimizes PEP, given the conditional statistics of the channel $(\mathbf{m}_{\mathbf{h}|\widehat{\mathbf{h}}}, \mathbf{R}_{\mathbf{hh}|\widehat{\mathbf{h}}})$. See [Jöngren *et al.*, 2002] for details. Some intuition for the solution can be developed by examining $\mathbf{W}^{opt}$ for $\rho_{corr} = 0$ and $\rho_{corr} = 1$. When $\rho_{corr} = 0$, we get $\mathbf{W}^{opt} = \mathbf{I}_2$, and the scheme reduces to standard Alamouti encoding which is of course the right solution and we get second-order diversity. When $\rho_{corr} = 1$, we get

$$\mathbf{W}^{opt} = \left[\sqrt{2}\frac{\mathbf{h}^H}{\|\mathbf{h}\|_F} \quad \mathbf{0}_{2,1} \right] \tag{8.44}$$

and the transmitted signal $\mathbf{S} = \mathbf{WX}$ becomes

$$\mathbf{S} = \left[\sqrt{2}\frac{\mathbf{h}^H}{\sqrt{\|\mathbf{h}\|_F^2}}x_0 \quad -\sqrt{2}\frac{\mathbf{h}^H}{\sqrt{\|\mathbf{h}\|_F^2}}x_1^* \right]. \tag{8.45}$$

This implies transmit-MRC (beamforming) on the symbol stream and yields an average array gain of 3 dB in addition to a second-order diversity. Clearly $\rho_{corr} = 1$ outperforms $\rho_{corr} = 0$. For any ρ_{corr}, $\mathbf{W}^{opt}$ will outperform both a pure Alamouti coding strategy (which would ignore channel knowledge and always deliver second-order diversity, but no array gain) and pure transmit-MRC on the channel estimate (which would vary in performance from no diversity and array gain to full diversity and array gain). Studies indicate significant improvement from this scheme for $\rho_{corr} > 0.8$.

The above problem for the MISO channel has also been studied by [Visotsky and Madhow, 2001]. They showed that $\mathbf{R}_{ss}^{opt}$ can be constructed by making the principal eigenvector equal to the channel mean $\widehat{\mathbf{h}}$ with the other eigenvectors being an arbitrary orthonormal set with equal associated eigenvalues. Also, similar to the optimality for covariance feedback in Section 8.3.2, [Narula et al., 1998] showed that when channel uncertainty decreases beamforming along the channel mean maximizes ergodic capacity.

Remarks

We have discussed a few ways to utilize channel knowledge at the transmitter, but many other methods are available. For example, in the 3G standard for mobile communications, the instantaneous quantized phase of the vector channel is estimated at the receiver and fed back to the transmitter for use in a transmit-MRC scheme. Transmitter optimization in the presence of such partial knowledge is a rich area of current research [Jöngren et al., 2002; Nabar et al., 2001; Narula et al., 1998].

Though most of the discussion has been for a MIMO channel, the extension to the MISO channel as a special case can be easily seen. Also, the techniques discussed in this chapter can be extended to the frequency selective case. For SC modulation we now need to consider ISI pre-equalization. Analogs of ZF, MMSE and DFE receivers for frequency selective channels can be developed including Tomlinson–Harashima pre-coding.

9 ST OFDM and spread spectrum modulation

9.1 Introduction

The preceding chapters considered MIMO signaling in conjunction with SC modulation. In this chapter, we briefly discuss how alternative modulation techniques such as OFDM and SS may be used in conjunction with multiple antenna systems. OFDM modulation offers the advantage of lower implementation complexity in systems with a large bandwidth–delay spread product. SS modulation offers the multiple access dimension as well as improved signal and interference diversity in cellular voice networks and resistance to narrowband jamming. We discuss briefly how MIMO signaling strategies developed for SC modulation may be extended to OFDM and SS modulation.

9.2 SISO-OFDM modulation

OFDM is an attractive modulation scheme used in broadband wireless systems which encounter large delay spread. The complexity of ML detection or even suboptimal equalization schemes needed for SC modulation grows exponentially with the bandwidth–delay spread product. OFDM avoids temporal equalization altogether, using a cyclic prefix (CP) technique with a small penalty in channel capacity. We briefly review OFDM modulation for a SISO channel before generalizing this to MIMO channels.

Consider a frequency selective wireless link with a single receive and transmit antenna ($M_R = M_T = 1$) and bandwidth B. We represent the baseband sampled (at $1/B$ intervals) channel impulse response (that includes effects of the physical channel and pre-/post-filtering at transmitter and receiver) by $g[l]$ ($l = 0, 1, \ldots, L - 1$), where L is the channel length. Let $s[k]$ ($k = 0, 1, 2, \ldots, N - 1$) be a sequence of N data symbols to be transmitted, each with unit average energy. We represent the data sequence as an $N \times 1$ vector $\mathbf{s} = [s[0]\ s[1] \cdots s[N - 1]]^T$. The OFDM modulation scheme follows (see schematic in Fig. 9.1).

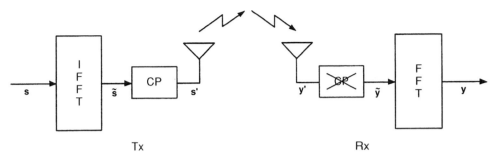

Figure 9.1: Schematic of OFDM transmission for a SISO channel.

The transmitter first performs an inverse fast Fourier transform (IFFT) operation [Brigham, 1974; Gray and Goodman, 1995] on the sequence of symbols to be transmitted. This yields the vector $\widetilde{\mathbf{s}} = [\widetilde{s}[0] \; \widetilde{s}[1] \cdots \widetilde{s}[N-1]]^T$:

$$\widetilde{\mathbf{s}} = \mathbf{D}^H \mathbf{s}, \tag{9.1}$$

where $\mathbf{D}$ is an $N \times N$ matrix whose mnth ($m, n = 1, 2, \ldots, N$) element is given by

$$[\mathbf{D}]_{m,n} = \frac{1}{\sqrt{N}} e^{-\frac{j2\pi(m-1)(n-1)}{N}}. \tag{9.2}$$

In practice, N is chosen to be a power of 2 (64–1024 are commonly used) to allow efficient implementation of the IFFT [Cooley and Tukey, 1965]. We note that since the elements of $\mathbf{s}$ are IID distributed, the central limit theorem implies that the elements of $\widetilde{\mathbf{s}}$ are Gaussian distributed for large N.

A new sequence, $\mathbf{s}'$, is now constructed by appending a CP of length $L-1$ to the vector $\widetilde{\mathbf{s}}$ and transmitted. The CP consists of the last $L-1$ symbols of the vector $\widetilde{\mathbf{s}}$. Hence, the transmitted sequence becomes $\mathbf{s}' = [\widetilde{s}[N-L+1] \cdots \widetilde{s}[N-1] \; \widetilde{s}[0] \cdots \widetilde{s}[N-1]]^T$. The elements of $\mathbf{s}'$ are transmitted serially with pulse-shaping. The vector $\mathbf{s}'$ is known as the OFDM symbol and has a duration of $T_s^{OFDM} = (N+L-1)/B$. The receiver receives a vector $\mathbf{y}'$ of length $N + 2L - 2$, that comprises the OFDM symbol convolved with the channel of length L. The receiver strips off the CP and then gathers N samples of the received signal, $\widetilde{\mathbf{y}} = [y'[0] \; y'[1] \cdots y'[N-1]]^T$ that satisfy

$$\widetilde{\mathbf{y}} = \sqrt{E_s}\widetilde{\mathbf{G}}\mathbf{s}' + \widetilde{\mathbf{n}}, \tag{9.3}$$

where E_s is the average energy available at the transmitter over time T_s, $\widetilde{\mathbf{n}}$ is the additive ZMCSCG noise vector with covariance matrix $N_0\mathbf{I}_N$ and $\widetilde{\mathbf{G}}$ is an $N \times (N + L - 1)$ Toeplitz matrix derived from the channel impulse response and is given by

$$\widetilde{\mathbf{G}} = \begin{bmatrix} g[L-1] & \cdots & g[1] & g[0] & 0 & 0 & \cdots & 0 \\ 0 & g[L-1] & \cdots & g[1] & g[0] & 0 & \cdots & 0 \\ \vdots & 0 & \ddots & \ddots & \ddots & \ddots & \ddots & \vdots \\ 0 & \vdots & 0 & g[L-1] & \cdots & g[1] & g[0] & 0 \\ 0 & 0 & 0 & 0 & g[L-1] & \cdots & g[1] & g[0] \end{bmatrix}. \tag{9.4}$$

Making use of the fact that the first $L - 1$ samples of $\mathbf{s}'$ are identical to the last $L - 1$ samples on account of the CP, it is easy to verify that Eq. (9.3) may be simplified to

$$\tilde{\mathbf{y}} = \sqrt{E_s}\mathbf{G}_c\tilde{\mathbf{s}} + \tilde{\mathbf{n}}, \tag{9.5}$$

where

$$\mathbf{G}_c = \begin{bmatrix} g[0] & 0 & \cdots & 0 & 0 & g[L-1] & \cdots & g[1] \\ g[1] & g[0] & 0 & \cdots & 0 & 0 & \ddots & \vdots \\ \vdots & g[1] & g[0] & 0 & 0 & & 0 & g[L-1] \\ g[L-1] & \vdots & g[1] & \ddots & 0 & \ddots & 0 & 0 \\ 0 & g[L-1] & \vdots & \ddots & g[0] & \ddots & \ddots & 0 \\ \vdots & 0 & g[L-1] & \ddots & g[1] & g[0] & 0 & 0 \\ \vdots & \vdots & 0 & \ddots & \vdots & \ddots & \ddots & 0 \\ 0 & 0 & \cdots & 0 & g[L-1] & \cdots & g[1] & g[0] \end{bmatrix}. \tag{9.6}$$

The CP renders the matrix $\mathbf{G}_c$ (derived from the channel impulse response) circulant, and hence the eigendecomposition of $\mathbf{G}_c$ may be expressed as [Gray, 2001]

$$\mathbf{G}_c = \mathbf{D}^H\mathbf{\Omega}\mathbf{D}, \tag{9.7}$$

where $\mathbf{\Omega} = \mathrm{diag}\{\omega[0], \omega[1], \ldots, \omega[N-1]\}$, with

$$\omega[k] = \sum_{l=0}^{L-1} g[l]e^{-\frac{j2\pi kl}{N}}, k = 0, 1, 2, \ldots, N - 1. \tag{9.8}$$

$\omega[k]$ $(k = 0, 1, 2, \ldots, N - 1)$ is the sampled frequency response of the channel, where k represents the tone index. In order to detect the received signal, the receiver performs a fast Fourier transform (FFT) on the received sequence $\tilde{\mathbf{y}}$ to yield

$$\mathbf{y} = \mathbf{D}\tilde{\mathbf{y}}, \tag{9.9}$$

where $\mathbf{y} = [y[0]\ y[1]\cdots y[N-1]]^T$. Combining Eqs. (9.1), (9.5), (9.7) and (9.9), the effective input–output relation for the channel can be expressed as

$$\begin{aligned} \mathbf{y} &= \sqrt{E_s}\mathbf{D}\mathbf{D}^H\mathbf{\Omega}\mathbf{D}\mathbf{D}^H\mathbf{s} + \mathbf{D}\tilde{\mathbf{n}} \\ &= \sqrt{E_s}\mathbf{\Omega}\mathbf{s} + \mathbf{n}, \end{aligned} \tag{9.10}$$

where we have made use of the fact that $\mathbf{D}$ is a unitary matrix, i.e., $\mathbf{D}\mathbf{D}^H = \mathbf{I}_N$, and $\mathbf{n} = \mathbf{D}\tilde{\mathbf{n}}$. Note that the elements of $\mathbf{n}$ are uncorrelated ZMCSCG with variance N_o. Equation (9.10) shows that the use of a CP in conjunction with the IFFT and FFT operations at transmitter and receiver respectively decouples the frequency selective

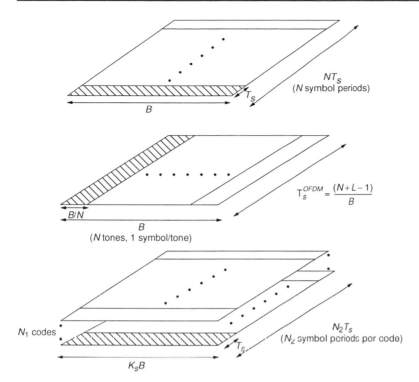

Figure 9.2: SC, OFDM and SS (multicode) modulation for SISO channels. The hashed area is one symbol.

channel into N parallel flat fading channels (each having a bandwidth B/N). The input–output relation for the kth tone in Eq. (9.10) is

$$y[k] = \sqrt{E_s}\omega[k]s[k] + n[k], \quad k = 0, 1, 2, \ldots, N - 1, \tag{9.11}$$

where $n[k]$ ($k = 0, 1, 2, \ldots, N - 1$) is the kth element of **n**. OFDM orthogonalizes the delay spread (ISI) channel into N orthogonal channels. The input–output relation is identical to that of the flat fading SISO channel. We note that in the OFDM context, E_s may be reinterpreted as the average energy available at the transmitter per tone. A schematic for OFDM modulation over SISO channels is shown in Fig. 9.1.

Figure 9.2 highlights the essential differences between SC and OFDM modulation. In SC modulation N data symbols are transmitted serially in time, each signal being transmitted over a time period T_s (the symbol period) and bandwidth B. In contrast, in OFDM modulation, N symbols are transmitted in parallel in one symbol period $T_s^{OFDM} = (N + L - 1)/B$, but are separated in frequency, each occupying a bandwidth of B/N.

Recalling from Chapter 2 that tones spaced more than the coherence bandwidth of the channel apart experience independent fading, OFDM can exploit frequency diversity via coding and interleaving across tones [Biglieri *et al.*, 1991;

Wesel and Cioffi, 1995]. This is known as coded OFDM (COFDM). While OFDM modulation avoids equalization of delay spread in wireless channels, the modulation scheme has an increased peak-to-average ratio (PAR) signal (in power) since its output waveform is approximately Gaussian [O'Neill and Lopes, 1994]. This large PAR (10–12 dB) necessitates highly linear amplifiers with large back-off and therefore higher costs. PAR reduction techniques for OFDM modulation have been developed [Muller and Huber, 1997; Tellado, 1998; Cimini and Sollenberger, 1999; Salvekar *et al.*, 2001]. Having discussed OFDM modulation for SISO channels, we now explain how this modulation scheme may be extended to MIMO channels.

9.3   MIMO-OFDM modulation

Consider a frequency selective MIMO channel with M_T transmit antennas, M_R receive antennas and bandwidth B. The channel impulse response between the ith ($i = 1, 2, \ldots, M_R$) receive antenna and the jth ($j = 1, 2, \ldots, M_T$) transmit antenna is given by $g_{i,j}[l]$ ($l = 0, 1, 2, \ldots, L - 1$), where L is the maximum channel length of all component $M_T M_R$ SISO channels. In matrix notation, the impulse response of the MIMO channel may be expressed by the sequence of matrices $\mathbf{G}[l]$ ($l = 0, 1, 2, \ldots, L - 1$), where the ijth element of the matrix $\mathbf{G}[l]$ is given by $g_{i,j}[l]$.

Consider a block of data symbols (drawn from a scalar constellation with unit average energy) of dimension $M_T \times N$ that is to be transmitted over the MIMO channel. Let the sequence to be transmitted over the jth transmit antenna be $s_j[k]$ ($k = 0, 1, 2, \ldots, N - 1$). As for the case of SISO-OFDM, the sequence to be transmitted over each transmit antenna is first subjected to an IFFT operation following which a CP is appended. At each of the receive antennas, the CP is stripped off and this is followed by an FFT operation. Analogous to OFDM modulation in SISO channels, the signal received at the ith receive antenna over the kth tone, $y_i[k]$ ($k = 0, 1, 2, \ldots, N - 1$) is given by

$$y_i[k] = \sqrt{\frac{E_s}{M_T}} \sum_{j=1}^{M_T} \omega_{i,j}[k] s_j[k] + n_i[k], \quad i = 1, 2, \ldots, M_R, \tag{9.12}$$

where E_s is the average energy allocated to the kth tone evenly divided across the transmit antennas, $n_i[k]$ represents ZMCSCG noise with variance N_o and $\omega_{i,j}[k]$ is the channel gain between the jth transmit antenna and the ith receive antenna for the kth tone. Note that $y_i[k]$ has contributions from all the transmit antennas. $\omega_{i,j}[k]$ is given by

$$\omega_{i,j}[k] = \sum_{l=0}^{L-1} g_{i,j}[l] e^{-\frac{j2\pi kl}{N}}, \quad k = 0, 1, 2, \ldots, N - 1. \tag{9.13}$$

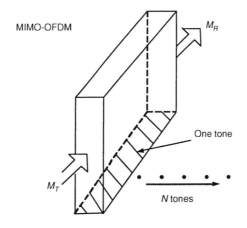

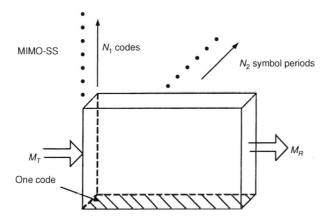

Figure 9.3: Schematic of MIMO-OFDM and MIMO-SS. Each OFDM tone or SS code admits M_T inputs and has M_R outputs.

From Eq. (9.12) it follows that the input–output relation for the MIMO system for the kth tone may be expressed as

$$\mathbf{y}[k] = \sqrt{\frac{E_s}{M_T}} \, \mathbf{H}[k]\mathbf{s}[k] + \mathbf{n}[k], \tag{9.14}$$

where $\mathbf{y}[k] = [y_1[k] \; y_2[k] \; \cdots \; y_{M_R}[k]]^T$, $\mathbf{n}[k] = [n_1[k] \; n_2[k] \; \cdots \; n_{M_R}[k]]^T$ and $\mathbf{H}[k]$ is an $M_R \times M_T$ matrix with $[\mathbf{H}[k]]_{i,j} = \omega_{i,j}[k]$. The matrix $\mathbf{H}[k]$ is the frequency response of the matrix channel corresponding to the kth tone and is related to $\mathbf{G}[l]$ via

$$\mathbf{H}[k] = \sum_{l=0}^{L-1} \mathbf{G}[l] e^{-\frac{j2\pi kl}{N}}. \tag{9.15}$$

From Eq. (9.14), just as in SISO channels, MIMO-OFDM decomposes the otherwise frequency selective channel of bandwidth B into N orthogonal flat fading MIMO channels, each with bandwidth B/N (see Fig. 9.3).

The total effective input–output relation for the MIMO-OFDM channel may be expressed as

$$\mathcal{Y} = \sqrt{\frac{E_s}{M_T}} \mathcal{H}\mathcal{S} + \mathcal{N}, \tag{9.16}$$

where $\mathcal{Y} = [\mathbf{y}[0]^T\, \mathbf{y}[1]^T\, \cdots\, \mathbf{y}[N-1]^T]^T$ is a vector of dimension $M_R N \times 1$, $\mathcal{S} = [\mathbf{s}[0]^T\, \mathbf{s}[1]^T\, \cdots\, \mathbf{s}[N-1]^T]^T$ is an $M_T N \times 1$ vector, $\mathcal{N} = [\mathbf{n}[0]^T\, \mathbf{n}[1]^T\, \cdots\, \mathbf{n}[N-1]^T]^T$ is an $M_R N \times 1$ ZMCSCG vector with $\mathcal{E}\{\mathcal{N}\mathcal{N}^H\} = N_o \mathbf{I}_{NM_R}$, and finally $\mathcal{H}$ is a block diagonal matrix of dimension $M_R N \times M_T N$ with $\mathbf{H}[k]$ ($k = 0, 1, 2, \ldots, N-1$) as the block diagonal entries. Note that Eq. (9.16) assumes that the transmit energy is divided evenly across space (transmit antennas) and frequency (tones), which is optimal when the channel is unknown to the transmitter. If the channel is known to the transmitter, transmit energy may be allocated optimally via waterpouring across space and frequency to maximize capacity as discussed in Chapter 4. However, OFDM transmission incurs on average a loss in spectral efficiency of $(L-1)/(N+L-1)$ on account of the CP. If $N \gg L$, this loss is negligible.

9.4 Signaling and receivers for MIMO-OFDM

MIMO (or MISO and SIMO) signaling can easily be overlayed on OFDM as developed in the previous section. MIMO signaling treats each OFDM tone as an independent narrowband frequency flat channel. We must take care to ensure that the modulation and demodulation parameters (carrier, phasing, FFT/IFFT, prefixes, etc.) are completely synchronized across all the transmit and receive antennas. With this precaution, every OFDM tone can be treated as a MIMO channel, and the tone index can be treated as a time index in the ST techniques developed in earlier chapters.

In this section we study how spatial diversity coding ($r_s \leq 1$) and SM ($r_s = M_T$), discussed earlier with SC modulation, may be extended to MIMO-OFDM, with an accompanying discussion on receiver architectures for these schemes. Hybrid signaling schemes such as those described in [Heath and Paulraj, 2002] or signaling for the case when the channel is perfectly known to the transmitter may be extended to MIMO-OFDM with appropriate modifications to the channel and signal model. This is not discussed here. We conclude with a brief discussion of space-frequency coded MIMO-OFDM, where the objective is to extract both space and frequency diversity.

9.4.1 Spatial diversity coding for MIMO-OFDM

Diversity techniques designed for SC modulation over flat fading channels are easily extended to OFDM modulation with the time index for SC modulation replaced by

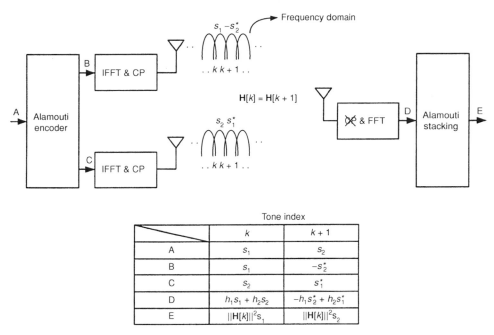

Figure 9.4: Schematic of the Alamouti transmission strategy for MIMO-OFDM. The tone index replaces the time index in SC modulation.

the tone index in OFDM. For example, consider the Alamouti scheme ($r_s = 1$), which extracts full spatial diversity in the absence of channel knowledge at the transmitter with $M_T = 2$. Recall that implementation of the Alamouti scheme requires that the channel remains constant over consecutive symbol periods. In the OFDM context, this translates to the channel remaining constant over consecutive tones, i.e., $\mathbf{H}[k] = \mathbf{H}[k + 1]$. Consider two data symbols, s_1 and s_2, to be transmitted over consecutive OFDM tones, k and $k + 1$, using the Alamouti scheme. Symbols s_1 and s_2 are transmitted over antennas 1 and 2 respectively on tone k, and $-s_2^*$ and s_1^* are transmitted over antennas 1 and 2 respectively on tone $k + 1$ within the same OFDM symbol (see Fig. 9.4).

The receiver detects the transmitted symbols from the signal received on the two tones using the Alamouti detection technique. As in SC modulation, the effective channel is orthogonalized irrespective of the channel realization and the vector detection problem collapses into scalar detection problems with the effective input–output relation for symbols s_i ($i = 1, 2$) given by

$$y_i = \sqrt{\frac{E_s}{2}} \|\mathbf{H}[k]\|_F^2 s_i + n_i, \quad i = 1, 2, \tag{9.17}$$

where n_i is ZMCSCG noise with variance $\|\mathbf{H}[k]\|_F^2 N_o$. Assuming that the $2M_R$ elements of $\mathbf{H}[k]$ undergo independent fading, the Alamouti scheme extracts $2M_R$ order diversity, just as in SC modulation. Note that the use of consecutive tones is not strictly necessary, any pair of tones can be used as long as the associated channels are equal. The technique can be generalized to extract spatial diversity over a larger number of antennas by using the OSTBC techniques developed for SC modulation. We now need a block size $T \geq M_T$ and the channel must be identical over the T tones.

An alternative technique is to use spatial diversity coding on a per tone basis across OFDM symbols in time exactly as in SC modulation. However, this requires that the channel remains constant over T OFDM symbol periods. Since the duration of an OFDM symbol $((N + L - 1)/B)$ is usually large, this may be impractical.

9.4.2   SM for MIMO-OFDM

Analogous to SM for MIMO systems with SC modulation, the objective of SM in conjunction with MIMO-OFDM is to achieve the spatial rate $r_s = M_T$ by transmitting parallel symbol streams [Bölcskei *et al.*, 2002a]. Thus, $N M_T$ scalar data symbols are transmitted over one OFDM symbol, with M_T symbols being transmitted over each tone. As for the case of SC modulation we require that $M_R \geq M_T$ in order to support the symbol streams reliably. The input–output relation for each tone may be expressed as

$$\mathbf{y}[k] = \sqrt{\frac{E_s}{M_T}} \mathbf{H}[k]\mathbf{s}[k] + \mathbf{n}[k], \quad k = 0, 1, 2, \ldots, N - 1, \tag{9.18}$$

where $\mathbf{s}[k]$ is the signal vector comprising M_T data symbols launched over the kth tone. Thus SM in MIMO-OFDM systems reduces to SM over each tone. The receiver architecture for SM is identical to that for SC modulation. Optimal performance is achieved by a vector ML receiver while sub-optimal receivers such as OSUC, MMSE and ZF may also be applied on a tone-by-tone basis. However, we can handle large delay spread in MIMO-OFDM, which creates significant complexity problems in SC modulation.

9.4.3   Space-frequency coded MIMO-OFDM

The diversity coding techniques discussed in Section 9.4.1 extract spatial diversity in a MIMO-OFDM system. However, frequency diversity may also be available if the channel is frequency selective. We recall from Chapter 2 that tones spaced greater than the coherence bandwidth of the channel experience independent fading. Let L_{eff} $(= B/B_C)$ be the number of coherence bandwidths within B. The total diversity available is then $M_T M_R L_{eff}$. In order to extract full diversity, data must be suitably spread across frequency and space [Kim *et al.*, 1998; Lang *et al.*, 1999]. Typically, the bit stream to be transmitted is first coded, then interleaved and modulated. The data symbols to be transmitted are then mapped across space and frequency by a space-frequency encoder. The receiver demodulates the received signal and estimates the transmitted space-frequency codeword, followed by deinterleaving and decoding.

As in the discussion on spatial diversity coding in Chapter 6, the main objective behind space-frequency coding [Agarwal *et al.*, 1998; Bölcskei and Paulraj, 2000b] is to design codeword matrices in such a fashion that they are able to extract all available diversity gain, making them maximally distinguishable at the receiver. In the following we assume that the channel length $L = L_{eff}$. The conditions under which this occurs have been discussed in detail in Chapter 5. Assume that the data symbols to be transmitted are encoded by the space-frequency encoder into blocks of size of $M_T \times N$. As discussed earlier, one OFDM symbol consists of N data vectors of size $M_T \times 1$. The channel is assumed to be constant (in time) over the OFDM symbol. Assuming that the channel is known to the receiver, the ML estimate of the transmitted codeword, $\mathbf{S}$, is given by

$$\widehat{\mathbf{S}} = \arg \min_{\mathbf{S}} \sum_{k=0}^{N-1} \left\| \mathbf{y}[k] - \sqrt{\frac{E_s}{M_T}} \mathbf{H}[k]\mathbf{s}[k] \right\|^2, \tag{9.19}$$

where $\mathbf{S} = [\mathbf{s}[0]\,\mathbf{s}[1] \cdots \mathbf{s}[N-1]]$ and minimization is performed over all possible codewords. Codeword design to exploit space-frequency diversity may be studied through a PEP analysis similar to that for ST codeword construction discussed in Chapter 6. Assuming individual $\mathbf{H}[k]$ are drawn from $\mathbf{H}_w$ (in general there will be correlation across tones), the probability that the receiver mistakes a transmitted codeword $\mathbf{S}^{(i)}$ for another codeword $\mathbf{S}^{(j)}$ averaged over all channel realizations, $P(\mathbf{S}^{(i)} \to \mathbf{S}^{(j)})$, is upper-bounded by [Bölcskei and Paulraj, 2000b]

$$P(\mathbf{S}^{(i)} \to \mathbf{S}^{(j)}) \le \prod_{m=1}^{r(\mathbf{G}_{i,j})} \frac{1}{1 + \lambda_m(\mathbf{G}_{i,j})\rho/4M_T}, \tag{9.20}$$

where

$$\mathbf{G}_{i,j} = \sum_{l=0}^{L-1} \left[\mathbf{Z}^l \mathbf{E}_{i,j}^T \mathbf{E}_{i,j} \mathbf{Z}^{l^H} \right] \otimes \mathbf{I}_{M_R}, \tag{9.21}$$

$$\mathbf{E}_{i,j} = \mathbf{S}^{(i)} - \mathbf{S}^{(j)}. \tag{9.22}$$

$\lambda_m(\mathbf{G}_{i,j})\,(m = 1, 2, \ldots, r(\mathbf{G}_{i,j}))$ is the mth eigenvalue of $\mathbf{G}_{i,j}$ and

$$\mathbf{Z} = \mathrm{diag}\left\{ 1, e^{-\frac{j2\pi}{N}}, \ldots, e^{-\frac{j2\pi(N-1)}{N}} \right\}. \tag{9.23}$$

We can show that the rank of $\mathbf{G}_{i,j}$ $(r(\mathbf{G}_{i,j}))$ equals M_R times the rank of the $N \times N$ matrix $\mathbf{F}_{i,j}$, where

$$\mathbf{F}_{i,j} = \mathbf{T}(\mathbf{S}^{(i)}, \mathbf{S}^{(j)})\mathbf{T}(\mathbf{S}^{(i)}, \mathbf{S}^{(j)})^H, \tag{9.24}$$

with

$$\mathbf{T}(\mathbf{S}^{(i)}, \mathbf{S}^{(j)}) = [\mathbf{E}_{i,j}^T \ \ \mathbf{Z}\mathbf{E}_{i,j}^T \ \ \cdots \ \ \mathbf{Z}^{L-1}\mathbf{E}_{i,j}^T]. \tag{9.25}$$

$\mathbf{T}(\mathbf{S}^{(i)}, \mathbf{S}^{(j)})$ is an $N \times M_T L$ matrix. Therefore, in order to achieve full space-frequency diversity of $M_R M_T L$, the design of the codeword matrices must be such that $\mathbf{T}(\mathbf{S}^{(i)}, \mathbf{S}^{(j)})$ has rank $M_T L$ for any pair of codeword matrices $\mathbf{S}^{(i)}$ and $\mathbf{S}^{(j)}$. This is achieved when $\mathbf{E}_{i,j}$ has rank M_T for all codewords $\mathbf{S}^{(i)}$ and $\mathbf{S}^{(j)}$ and each of the blocks $\mathbf{B}_l = \mathbf{Z}^l \mathbf{E}_{i,j}^T$ ($l = 0, 1, 2, \ldots, L - 1$) is linearly independent of the other blocks $\mathbf{B}_m$ ($l \neq m$) for every pair of codeword matrices $\mathbf{S}^{(i)}$ and $\mathbf{S}^{(j)}$. While a ST codeword designed to achieve full diversity in the flat fading case in Chapter 6 will have full-rank $\mathbf{E}_{i,j}$ and hence a full-rank $\mathbf{B}_l$ for all codewords $\mathbf{S}^{(i)}$ and $\mathbf{S}^{(j)}$, the linear independence of the blocks $\mathbf{B}_l$ is not guaranteed. Note that ensuring linear independence of the blocks $\mathbf{B}_l$ amounts to ensuring that the space-frequency code exploits the available frequency diversity as well.

9.5   SISO-SS modulation

In this section we briefly review the use of SS modulation [Turin, 1980; Dixon, 1994; Simon et al., 1994; Viterbi, 1995] in conjunction with SISO systems. There are two commonly used forms of SS modulation: direct sequence (DS) and frequency hopping (FH). We shall consider the more popular DS-SS technique in our discussion. Although bandwidth is an expensive commodity in wireless systems, increasing the transmission bandwidth can offer some advantages. In military applications SS modulation provides interference rejection (anti-jam) and helps "hide" the transmitted signal by lowering its power spectral density (low probability of intercept). In civilian applications, SS modulation can be used as a multiple access technique, which is the basis of a number of digital cellular standards. CDMA is a SS modulation technique in which users communicate over the same time and frequency, but are separated by codes. CDMA offers advantages of signal and interference diversity in narrowband (voice) cellular networks. Signal diversity arises from spreading the narrowband signal over a large bandwidth and capturing frequency (multipath) diversity which mitigates fading. Interference diversity arises from the presence of multiple interferers (on other codes) from inside and outside a cell, giving rise to interference averaging, a form of diversity. First, we shall introduce SS modulation for the case of flat fading and then extend the analysis to the case of frequency selective fading. We ignore the effects of multiple access interference (MAI) in the analysis.

9.5.1   Frequency flat channel

In DS-SS modulation, the data signal to be transmitted of duration T_s is spread by a binary sequence (code). Each bit of the spreading sequence is usually referred to as a chip. The code comprises K_S chips each of duration T_{chip}, where K_S is known as the spreading factor and $1/T_{chip}$ is called the chip rate. Ignoring excess bandwidth for chip pulse-shaping, the transmitted signal bandwidth $B_{chip} = 1/T_{chip} = K_S/T_s$. At the receiver, the signal is processed with a filter matched to the spreading sequence (or equivalently correlated with the spreading sequence) and then sampled to yield the decision statistic.

Single code model

Let symbol s be drawn from a constellation with unit average energy. The transmitted sequence is given by

$$\widetilde{s}[i] = s\, c[i], (i = 0, 1, 2, \ldots, K_S - 1), \tag{9.26}$$

where $c[i] = \pm 1$ represents the spreading sequence sampled at the chip interval T_{chip}. Assuming flat fading, the received signal for the channel is given by

$$\widetilde{y}[i] = \sqrt{\frac{E_s}{K_S}} g\widetilde{s}[i] + \widetilde{n}[i], \ i = 0, 1, 2, \ldots, K_S - 1, \tag{9.27}$$

where $\widetilde{y}[i]$ is the signal received over the ith chip interval, E_s/K_S is the average energy available at the transmitter over chip period T_{chip}, g is the scalar baseband channel transfer function and $n[i]$ is temporally white ZMCSCG noise with $\mathcal{E}\{|\widetilde{n}[i]|^2\} = N_o$. The receiver correlates the received sequence with the spreading sequence $c[i]$. The output of the correlator y at lag 0 is a sufficient statistic to detect s, i.e.,

$$y = \frac{1}{\sqrt{K_S}} \sum_{i=0}^{K_S-1} \widetilde{y}[i]c[i], \tag{9.28}$$

which simplifies to

$$y = \sqrt{E_s}gs + n, \tag{9.29}$$

where $g = h$ (the effective channel looks like the frequency flat channel with SC modulation), $\mathcal{E}\{n\} = 0$ and $\mathcal{E}\{|n|^2\} = N_o$. The factor $1/\sqrt{K_S}$ in Eq. (9.28) is an energy normalization factor. Equation (9.29) resembles the input–output relation for a flat fading SISO channel in conjunction with SC modulation. Detection of the transmitted symbols can be accomplished by rescaling (for g) and slicing.

Multicode model

In multicode operation, the transmitter can transmit multiple data symbols during a symbol period using different codes. Assume N_1 codes, $c_j[i]\,(j = 1, 2, \ldots, N_1, i = 0, 1, 2, \ldots, K_S - 1)$ are available to the transmitter. Let $s_j\,(j = 1, 2, \ldots, N_1)$ be the N_1 data symbols to be transmitted over one symbol period, the transmitted signal is then given by

$$\widetilde{s}[i] = \sum_{j=1}^{N_1} \sqrt{\frac{E_s}{K_S}} s_j c_j[i], \ i = 0, 1, 2, \ldots, K_S - 1, \tag{9.30}$$

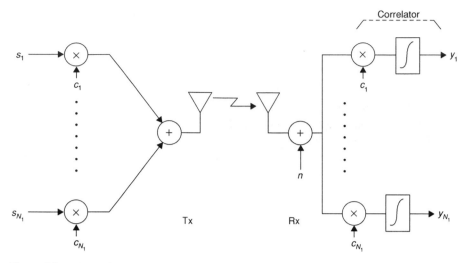

Figure 9.5: Schematic of multicode SS modulation for a SISO channel.

where E_s is the average energy available at the transmitter per code over a symbol period. The received signal $\widetilde{y}[i]$ is given by

$$\widetilde{y}[i] = \sum_{j=1}^{N_1} \sqrt{\frac{E_s}{K_S}} g s_j c_j[i] + \widetilde{n}[i], \quad i = 0, 1, 2, \ldots, K_S - 1. \tag{9.31}$$

The receiver correlates the received signal with each of the N_1 codes (see Fig. 9.5) to yield

$$y_j = \frac{1}{\sqrt{K_S}} \sum_{i=0}^{K_S-1} \widetilde{y}[i] c_j[i], \quad j = 1, 2, \ldots, N_1, \tag{9.32}$$

which may be simplified to

$$y_j = \sqrt{E_s} g s_j + \sum_{m \neq j} g \frac{\sqrt{E_s}}{K_S} R_{j,m}[0] s_m + n_j, \tag{9.33}$$

where n_j is additive ZMCSCG noise with variance N_o and

$$R_{j,m}[q] = \sum_{i=0}^{K_S-1} c_m[i] c_j[i - q] \tag{9.34}$$

represents the cross-correlation between the mth and jth codes at lag q.

The input–output relation per multicode symbol period is

$$\mathbf{y} = \sqrt{E_s} \mathbf{H} \mathbf{s} + \mathbf{n}, \tag{9.35}$$

where $\mathbf{y} = [y_1 \; y_2 \; \cdots \; y_{N_1}]^T$, $\mathbf{s} = [s_1 \; s_2 \; \cdots \; s_{N_1}]^T$, $\mathbf{n} = [n_1 \; n_2 \; \cdots \; n_{N_1}]^T$ are $N_1 \times 1$ and $\mathbf{H}$ is $N_1 \times N_1$ with

$$[\mathbf{H}]_{i,j} = \frac{g}{K_S} R_{i,j}[0]. \tag{9.36}$$

Without delay spread only $\mathbf{R}_{i,j}[0]$ contributes to the sufficient statistic for signal detection. If the codes are chosen to be orthogonal, i.e., $R_{i,j}[0] = 0$ if $i \neq j$, then $\mathbf{H}$ is diagonal and equals $g\mathbf{I}_{N_1}$. The problem is then decoupled into individual codes. When the codes are not orthogonal, we get intercode interference and must use multiuser detection techniques such as linear receivers (decorrelating or MMSE), SUC techniques and computationally expensive but optimal ML receivers [Verdu, 1998].

Now assume that a block of $N_1 N_2$ data symbols are to be transmitted with N_1 symbols (over N_1 different codes) per symbol period and N_2 symbol periods. Assume that the channel remains constant over the block. The data sequence transmitted over the jth code is $s_j[l]$ $(l = 0, 1, 2, \ldots, N_2 - 1)$ and the data sequence transmitted over the lth symbol period is the vector $\mathbf{s}[l] = [s_1[l] \; s_2[l] \; \cdots \; s_{N_1}[l]]^T$. The input–output relation per block is

$$\mathcal{Y} = \sqrt{E_s} \mathcal{H} \mathcal{S} + \mathcal{N}, \tag{9.37}$$

where $\mathcal{S} = [\mathbf{s}[0]^T \; \mathbf{s}[1]^T \; \cdots \; \mathbf{s}[N_2 - 1]^T]^T$ is $N_2 N_1 \times 1$ and $\mathcal{H} = \mathbf{I}_{N_2} \otimes \mathbf{H}$ is $N_2 N_1 \times N_2 N_1$ with $\mathbf{H}$ defined in Eq. (9.35). The vectors $\mathcal{Y}$ and $\mathcal{N}$ are $N_2 N_1 \times 1$ and contain the received signal after despreading and additive noise, respectively. The differences between multicode SS modulation and SC or OFDM modulation are highlighted in Fig. 9.2.

9.5.2 Frequency selective channel

Consider multicode transmission. The transmitted sequence is given by Eq. (9.30). Let $g[i]$ $(i = 0, 1, 2, \ldots, L - 1)$ be the sampled (at T_{chip} intervals) baseband channel impulse response with L the delay spread of the channel. The input–output relation for the channel is given by the discrete time convolution

$$\tilde{y}[i] = \sum_{k=0}^{L-1} g[i - k]\tilde{s}[k] + n[i], \; i = 0, 1, 2, \ldots, K_S + L - 1. \tag{9.38}$$

In order to detect the symbol transmitted on the jth code, the receiver correlates the received signal with the jth code at lags of $0, \ldots, L - 1$. Let $y_{j,q}$ be the correlator output (see Fig. 9.6) for the jth code at lag q, then

$$y_{j,q} = \frac{1}{\sqrt{K_S}} \sum_{i=0}^{K_S-1} \tilde{y}[i] c_j[i - q], \; q = 0, 1, 2, \ldots, L - 1, \tag{9.39}$$

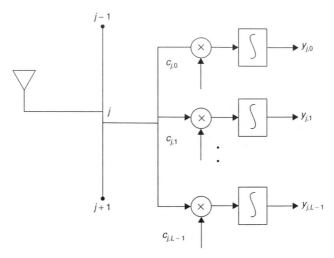

Figure 9.6: Schematic of a multilag correlator at the receiver. Only one code (c_j) is shown. $c_{j,q}$ refers to c_j code delayed by q chips.

which simplifies to

$$y_{j,q} = \frac{\sqrt{E_s}}{K_S} \sum_{i=0}^{L-1} s_j R_{j,j}[q-i]g[i] + \frac{\sqrt{E_s}}{K_S} \sum_{m=1,m\neq j}^{N_1} \sum_{i=0}^{L-1} s_m R_{j,m}[q-i]g[i] + n_{j,q},$$

(9.40)

with $\mathcal{E}\{n_{j,q}\} = 0$ and $\mathcal{E}\{|n_{j,q}|^2\} = N_o$. Ideally, the spreading codes must be designed so that the autocorrelation $R_{j,j}[q] = K_S \delta[q]$ and the cross-correlation $R_{m,j}[q] = 0$ if $m \neq j$ to avoid both self (interpath) interference and intercode interference. However, such ideal codes do not exist. In the 1960s and 1970s a lot of effort was focused on designing codes with good correlation properties [Kasami, 1966; Gold, 1967]. Maximal-length shift register (MLSR) codes were generally found to have the best correlation properties amongst all linear codes [Dixon, 1994].

The effective input–output relation for the channel is given by

$$\mathbf{y} = \sqrt{E_s}\mathbf{H}\mathbf{s} + \mathbf{n},$$

(9.41)

where $\mathbf{y} = [\mathbf{y}_1^T \ \mathbf{y}_2^T \cdots \mathbf{y}_{N_1}^T]^T$ is $LN_1 \times 1$ with $\mathbf{y}_j = [y_{j,0} \ y_{j,1} \cdots y_{j,L-1}]^T$, $\mathbf{s} = [s_1 \ s_2 \cdots s_{N_1}]^T$ is $N_1 \times 1$, $\mathbf{H}$ is $LN_1 \times N_1$, whose mnth element can be inferred from Eq. (9.40), and $\mathbf{n}$ is the noise vector. The transmitted symbols can be detected using a variety of receivers. A matched-filtering receiver leads to the well-known RAKE receiver, which is optimum only if the interpath interference is zero.

As in the flat fading case, assume that $N_1 N_2$ data symbols are to be transmitted over N_2 symbol periods with N_1 symbols being transmitted over N_1 codes over a single symbol period. The input–output relation per block is

$$\mathcal{Y} = \sqrt{E_s}\mathcal{H}\mathcal{S} + \mathcal{N},$$

(9.42)

where $\mathcal{S}$ is $N_2 N_1 \times 1$ and is defined in Eq. (9.37). The channel $\mathcal{H}$ (defined via Eq. (9.40)) is of dimension $N_1 N_2 L \times N_1 N_2$ and captures the channel, auto- and cross-interference between codes induced by the multipath and ISI at different lags. ISI can be neglected if $LT_{chip} \ll T_s$. $\mathcal{Y}$ and $\mathcal{N}$ are $N_1 N_2 L \times 1$ and represent the correlator outputs and noise respectively. There is a rich body of literature [Lupas and Verdu, 1989; Madhow and Honig, 1994; Rapajic and Vucetic, 1994; Abdulrahman et al., 1994; Verdu, 1998] covering the design of receivers for this problem. In reality, interference from other users (MAI) will also be present. The signal model with MAI is similar to multicode modulation described above with the exception that the channels of the desired user and interferers are different. If the codes assigned to various users have perfect cross-correlation properties then MAI is eliminated.

9.6 MIMO-SS modulation

Just as in MIMO-OFDM, MIMO (or MISO, SIMO) can be overlayed directly on the SISO-SS model described by Eq. (9.41) [Rooyen et al., 2000]. Let $\mathbf{G}[l]$ ($l = 0, 1, 2, \ldots, L - 1$) be the baseband sampled channel response of an $M_R \times M_T$ system. The ijth ($i = 1, 2, \ldots, M_R, j = 1, 2, \ldots, M_T$) element of $\mathbf{G}[l]$ is the response between the jth transmit antenna and ith receive antenna at delay l and L is the maximum channel length of all component $M_T M_R$ SISO links. We transmit $M_T N_1$ symbols per symbol period (T_s) using N_1 codes over M_T transmit antennas. Therefore we have $M_T N_1 N_2$ symbols per block comprising N_2 symbol periods (see Fig. 9.3). At each of the M_R receive antennas we have LN_1 outputs of correlators per symbol period corresponding to L lags and N_1 codes. We denote the symbols transmitted over code j ($j = 1, 2, \ldots, N_1$) and symbol period l ($l = 0, 1, 2, \ldots, N_2 - 1$) by the $M_T \times 1$ vector $\mathbf{s}_j[l]$. Now, setting $\mathbf{s}[l] = [\mathbf{s}_1[l]^T \; \mathbf{s}_2[l]^T \cdots \; \mathbf{s}_{N_1}[l]^T]^T$, the input–output relation for MIMO-SS modulation becomes

$$\mathcal{Y} = \sqrt{\frac{E_s}{M_T}} \mathcal{H} \mathcal{S} + \mathcal{N}. \tag{9.43}$$

Here $\mathcal{Y}$ is a vector of dimension $LM_R N_1 N_2 \times 1$ containing correlator outputs. $\mathcal{S} = [\mathbf{s}[0]^T \; \mathbf{s}[1]^T \; \cdots \; \mathbf{s}[N_2 - 1]^T]^T$ is a vector of dimension $M_T N_1 N_2 \times 1$ containing the stacked data symbols per block and $\mathcal{N}$ is the stacked noise vector, whose elements are ZMCSCG with variance N_o. Finally, $\mathcal{H}$ is the composite channel transfer matrix whose elements capture the MIMO channel $\mathbf{G}[i]$ auto- and cross-interference between codes induced by multipath and ISI at different lags. As for the case of a SISO channel, ISI can be neglected if $LT_{chip} \ll T_s$. The channel matrix $\mathcal{H}$ is typically a filled matrix and has significant block diagonal dominance when the codes have good auto- and cross-correlation properties. In the event that the MIMO channel is flat fading and the codes have perfect auto- and cross-correlation properties, $\mathcal{H}$ is in fact a block diagonal matrix with $\mathbf{G}[0]$ as the matrix along its diagonal. Equation (9.43) assumes that energy is evenly distributed across the transmit antenna array. If there is no delay spread and if the N_1 codes are orthogonal, MIMO-SS is structurally similar to MIMO-OFDM. However, in practice, with delay spread and/or quasi-orthogonal codes, the code sub-channels are not mutually orthogonal as is the case for "tone" channels in OFDM.

9.7   Signaling and receivers for MIMO-SS

Signaling schemes developed for SC modulation can be easily extended to MIMO-SS with appropriate modifications of the signal and channel model. N_1 vector symbols are transmitted over N_1 codes and a bandwidth of $K_S B$ (K_S is the spreading factor) for N_2 consecutive symbol periods. The use of spreading codes along with multiple antennas is sometimes referred to as ST spreading [Papadias, 1999; Huang *et al.* 1999]. We consider two extreme signaling schemes for MIMO-SS – spatial diversity transmission and SM. Analogous to SC modulation, intermediate schemes [Heath and Paulraj, 2002] and signaling for the case when the channel is perfectly known to the transmitter can also be designed.

9.7.1   Spatial diversity coding for MIMO-SS

Diversity transmission techniques for SC modulation can be easily adapted to MIMO-SS [Tehrani *et al.*, 1999; Hochwald *et al.*, 2001; Kim and Bhargava, 2002]. Such techniques have entered the universal mobile telecommunications system (UMTS) standard, where the base-station can support transmit diversity on forward dedicated channels. We shall highlight a few simple schemes below.

Frequency flat channel
A simple diversity scheme for channel unknown to the transmitter is as follows. The transmitter can transmit the same symbol over M_T antennas using different (M_T) orthogonal codes. Assuming M_T orthogonal codes, the output of the correlator at the receiver for the jth code (following notation in Eq. (9.32)) is

$$y_j = \sqrt{\frac{E_s}{M_T}} h_j s + n_j, \tag{9.44}$$

and a maximal ratio combiner output will be $\mathbf{h}^H \mathbf{y}$. From Section 5.4.1 for transmit diversity without channel knowledge, the received SNR becomes

$$\eta = \frac{\|\mathbf{h}\|_F^2 E_s}{M_T N_0}. \tag{9.45}$$

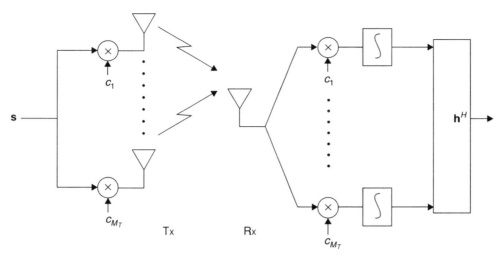

Figure 9.7: Multicode transmission will provide full M_T order diversity. We can transmit one symbol per symbol period using M_T codes.

Therefore, if $\mathbf{h} = \mathbf{h}_w$, we get M_R order diversity, but no array gain. Note that if we know the channel at the transmitter, we can get both array gain and diversity gain with the expenditure of only one code.

The receiver can separate the arrivals from different transmit antennas and then combine them using MRC (see Fig. 9.7). Performance is the same as discussed in Chapter 5.

A more efficient scheme is the Alamouti code designed for SC modulation. It can be easily extended to SS modulation (see [Hochwald *et al.*, 2001]) with the time index replaced by the code index. Assume $M_T = 2$, $M_R = 1$ and that two orthogonal codes c_1 and c_2 are available for transmission and two data symbols s_1 and s_2 are to be transmitted over a symbol period using the Alamouti scheme. Symbols s_1 and s_2 are spread by code c_1 and transmitted over antennas 1 and 2 respectively. Simultaneously $-s_2^*$ and s_1^* are spread by code c_2 and transmitted over antennas 1 and 2 respectively (see Fig. 9.8). At the receiver, the received signal is correlated with either code at each of the receive antennas. The effective channel is identical to that for SC modulation in Eq. (5.22). Further, if $\mathbf{h} = \mathbf{h}_w$, we get second-order diversity but no array gain since the channel is unknown to the transmitter. The code efficiency is one symbol per symbol per code. The above scheme can be generalized to an arbitrary number of antennas using OSTBC. Furthermore, assuming flat fading conditions, the ST codeword construction criteria for SC modulation in frequency flat channels extend directly to SS modulation with orthogonal spreading sequences, with the observation that the time index can be replaced by the code × time product index. Therefore we can derive analogous expressions for the rank and determinant criteria for MIMO-SS as discussed in Chapter 6.

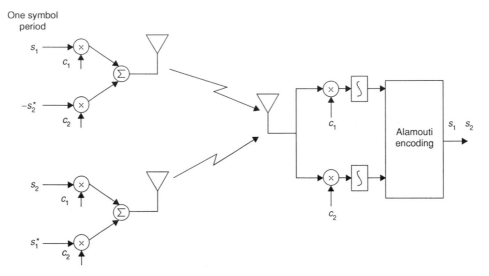

Figure 9.8: Alamouti coding with multicode SS modulation. We can transmit two symbols per symbol period using two codes.

Frequency selective channel

Consider a multiple antenna system with $M_T = 2$ and $M_R = 1$. We assume that each of the component SISO channels has two taps at lags of 0 and 1 T_{chip} (chip period) with equal average power in each of the taps. We can extract both path (frequency) and space diversity. The transmission strategy is the Alamouti scheme described above (see Fig. 9.8). At the receiver the received signal is correlated with each of the codes at lags of 0 and 1 respectively. Assume that the codes are orthogonal and have perfect auto- and cross-correlation properties. Therefore, at each lag there is no auto- or cross-code interference from the other lag (path). The output of the correlators corresponding to code c_1 at lags 0 and 1, $y_{1,0}$ and $y_{1,1}$, respectively satisfies

$$\begin{bmatrix} y_{1,0} \\ y_{1,1} \end{bmatrix} = \sqrt{\frac{E_s}{2}} \begin{bmatrix} h_1[0] & h_2[0] \\ h_1[1] & h_2[1] \end{bmatrix} \begin{bmatrix} s_1 \\ s_2 \end{bmatrix} + \mathbf{n}_1, \tag{9.46}$$

where $h_i[l]\,(l = 0, 1)$ represents the sampled (at chip rate) channel impulse response between transmit antenna i and the receive antenna, and $\mathbf{n}_1$ is the noise vector. Similarly, the output of the correlators at the receiver corresponding to code 2 satisfies

$$\begin{bmatrix} y_{2,0} \\ y_{2,1} \end{bmatrix} = \sqrt{\frac{E_s}{2}} \begin{bmatrix} h_1[0] & h_2[0] \\ h_1[1] & h_2[1] \end{bmatrix} \begin{bmatrix} -s_2^* \\ s_1^* \end{bmatrix} + \mathbf{n}_2. \tag{9.47}$$

Each corresponding row in Eqs. (9.46) and (9.47) resembles the input–output relation over consecutive symbol periods for an $M_T = 2$, $M_R = 1$ MISO system employing the Alamouti scheme in conjunction with SC modulation. Using the Alamouti decoding strategy, the effective channel can be collapsed into two scalar channels for each path

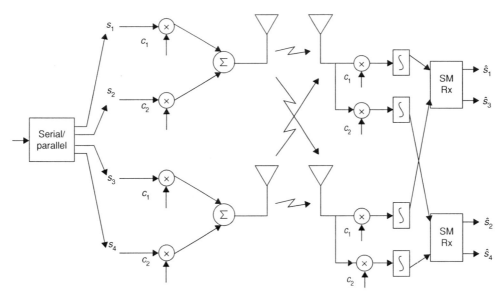

Figure 9.9: SM with multicode SS modulation, $M_T = M_R = N_1 = 2$. We get four symbols per symbol period using two codes. The presence of delay spread will require more complex receivers.

(or lag). The receiver structure is identical to the one discussed for the $M_R = 2$ case in Chapter 5 (Eqs. (5.33)–(5.40)) with the lag (path) index replacing the antenna index. This yields

$$z_i = \sqrt{\frac{E_s}{2}} \|\mathbf{H}\|_F^2 s_i + \tilde{n}_i, \quad i = 1, 2, \tag{9.48}$$

where

$$\mathbf{H} = \begin{bmatrix} h_1[0] & h_2[0] \\ h_1[1] & h_2[1] \end{bmatrix}. \tag{9.49}$$

We therefore get fourth order diversity (assuming IID fading). In more general terms, given $M_T = 2$ and L equi-power taps with IID fading across time and space, MIMO-SS in conjunction with the Alamouti scheme will extract the full $2L$ orders of diversity. In practice, the codes will not have perfect auto- and cross-correlation properties resulting in interpath and intercode interference and causing SNR loss. Also, unequal path powers or correlation will cause diversity loss.

9.7.2   SM for MIMO-SS

Similarly to MIMO-SC and MIMO-OFDM modulation, SM in conjunction with MIMO-SS delivers a spatial rate $r_s = M_T$ [Mudulodu and Paulraj, 2000; Huang *et al.*, 2002]. Therefore, a system with $M_R \geq M_T$ antennas, N_1 codes and N_2 symbol periods can transmit $M_T N_1 N_2$ independent data symbols over a period $N_2 T_s$. Figure 9.9 shows an uncoded SM system with $M_R = M_T = 2$, where four symbols are transmitted in one symbol period with two symbols (SM) over code c_1 and the other two over code c_2. At the receiver, despreading separates the combinations from the two codes. The effective input–output relation for the channel is given by Eq. (9.43). The SM receiver problem reduces to the standard case described in Chapter 7. Detection of the transmitted symbols is accomplished via optimal ML detection or sub-optimal detection schemes such as ZF, MMSE [Chen and Mitra, 2001] or OSUC. In the presence of delay spread, multilag correlation can be used. We must expect interpath and intercode interference. If these are ignored, we can build RAKE combiners after SM processing (see Fig. 9.9). More optimal techniques would combine interference mitigation for interstream (SM), intercode and interpath dimensions.

A number of schemes can be developed using muticodes to span SM and diversity. In terms of code efficiency, naive transmit diversity (see Fig. 9.7) offers $1/M_T$ symbols per period per code, OSTBC schemes (see Fig. 9.8) offer one symbol per period per code and pure SM (see Fig. 9.9) offers M_T symbols per period per code.

10 MIMO-multiuser

10.1 Introduction

Another approach in ST wireless is to deploy multiple antennas at the base to support multiple users with one or more antennas per user terminal (see Fig. 10.1). Assuming a single antenna per user, the forward link from a base to the users is a vector broadcast channel and the reverse link is a vector multiple access channel. With multiple antenna terminals, we get the corresponding matrix channels. We refer to this class of channels as MIMO-MU. The single user case studied in earlier chapters is referred to as MIMO-SU (single user) to differentiate it from the multiuser case. The base-station communicates with the multiple users simultaneously in the same frequency channel by exploiting differences in spatial signatures at the base-antenna array induced by spatially dispersed users. This technique is also known as SDMA [Gerlach, 1995; Ottersten, 1996; Roy, 1997; Lotter and van Rooyen, 1998; Vandenameele, 2001]. The value of SDMA in wireless is not so much because of its multiple access capability, but rather that it allows channel reuse within a cell to increase spectral efficiency.

Another view of MIMO-MU is that it extends the usual scalar (SISO) multiuser channel to vector (SIMO, MISO) or matrix (MIMO) multiuser channels.

MIMO-MU has been used in satellite communication, where a frequency channel is reused in angle at the satellite using beamforming. The users located on the ground have LOS paths to the satellite that are completely free of scattering and hence have zero angle spread. The satellite communicates with two users at well-separated angles by using co-channel pencil beams that have controlled cross interference through good sidelobe control. The satellite employs beams instead of cells in a cluster format with frequency reuse between clusters. Cluster size and sidelobe levels determine co-channel interference.

In terrestrial cellular systems MIMO-MU implies reuse within the sector (or cell). This is more complicated due to scattering in such environments. Wavefronts may have large angle spreads and therefore random channels or signatures. Therefore even well-separated users may have potentially overlapping channels. Also, users may have identical spatial channels at the base-station if their signals are scattered by the same dominant scatterer making separability hard to guarantee. With SS modulation (CDMA),

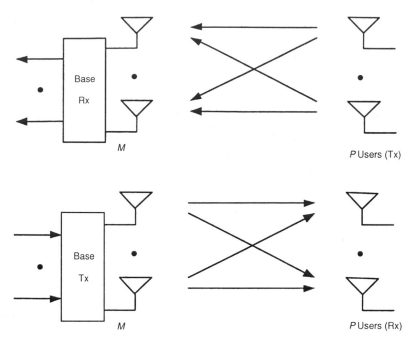

Figure 10.1: MIMO-MU reverse link (multiple access) channel and forward link (broadcast) channels shown for P terminals and M antennas at the base-station.

the problem of guaranteeing spatial channel separability is greatly mitigated since users have quasi or fully orthogonal temporal spreading codes.

MIMO-SU and MIMO-MU pose different problems from several viewpoints.

(a) MIMO-SU is a point-to-point link with a defined link capacity. In MIMO-MU, the link is a multiple access channel on the reverse link and a broadcast channel on the forward link, where the link rates (to or from different users) are characterized in terms of a capacity region [Cover, 1972, 1998; Cover and Thomas, 1991].

(b) In MIMO-MU, each user link has a target data rate and SER performance that are typically equal for all users (fairness). In contrast, in MIMO-SU, only the sum rate performance of the overall link matters since all the streams or sub-channels are delivered to the same user. If one stream has a poor SNR, MIMO-MU will experience an outage. MIMO-SU on the other hand has the advantage of stream diversity.

(c) In MIMO-MU, where users have near–far distribution in range [Pickholtz *et al.*, 1982], there may be significant differences in path loss from each user to the base. Not all of these differences can be compensated for by power control due to peak power constraints and power control errors. As a result, there can be large differences in SINR in the links. This benefits the stronger user, but hurts the weaker user. In MIMO-SU there is no near–far problem.

(d) In MIMO-SU, coding at the transmitter and decoding at the receiver can be done with co-operation between the co-located transmitting antennas. In MIMO-MU, users can co-operate in encoding on the forward link at the base and in decoding on the reverse link. The users, however, cannot co-operate in decoding on the forward link or during encoding on the reverse link. Limited co-operation inbetween may be possible in terms of power and rate.

(e) In MIMO-SU the capacity on the forward link is identical to that on the reverse link (for the same transmit power) if the channel is known at both the transmitter and receiver. In MIMO-MU, the relationship between the multiple access and broadcast capacity regions is still a subject of research.

(f) MIMO-SU suffers only a small penalty in capacity without channel knowledge at the transmitter. MIMO-MU has a much larger penalty on the forward link if the channel is not known to the transmitter.

In this chapter we assume only single antenna terminals and that the full channel is known to the base and to all users. Because there are distributed users, obtaining complete channel knowledge at the transmitters is even more complicated than that described for the single user case in Chapter 3. We refer to the forward link MIMO-MU channel as the MIMO broadcast channel (MIMO-BC) and the reverse link MIMO-MU channel as the MIMO multiple access channel (MIMO-MAC).

10.2 MIMO-MAC

10.2.1 Signal model

Consider a system with M antennas at the base-station and P users, each equipped with one antenna. The model can be extended to multiple antennas at each user, but that is not attempted here. Assuming a frequency flat channel, the channel between the ith ($i = 1, 2, \ldots, P$) user and the base-station is given by a complex Gaussian $M \times 1$ vector, $\mathbf{h}_i$. Assume s_i is a complex data symbol transmitted by the ith user with average energy $\mathcal{E}\{|s_i|^2\} = E_{s,i}$ ($i = 1, 2, \ldots, P$). Note that $E_{s,i}$ in general will not be the same for each user since each user will employ power control to compensate for differences in path loss.

The signal received at the base-station is an $M \times 1$ vector, $\mathbf{y}$, given by (see Section 3.9)

$$\mathbf{y} = \sum_{i=1}^{P} \mathbf{h}_i s_i + \mathbf{n}$$
$$= \mathbf{Hs} + \mathbf{n}, \tag{10.1}$$

where $\mathbf{s} = [s_1 \ s_2 \ \cdots \ s_P]^T$ is a $P \times 1$ vector, $\mathbf{H} = [\mathbf{h}_1 \ \mathbf{h}_2 \ \cdots \ \mathbf{h}_P]$ is an $M \times P$ matrix and $\mathbf{n}$ is the $M \times 1$ ZMCSCG spatially white noise vector with covariance matrix $N_o \mathbf{I}_m$. Note that the elements of $\mathbf{H}$ also reflect path loss differences between the various users necessitating a departure from the normalization that every element of $\mathbf{H}$ has unit power. Further, we observe that M must be equal to or greater than P to obtain acceptable spatial separability of the users. Assuming that the user signals are uncorrelated, the covariance matrix of the vector **s**, $\mathbf{R}_{ss} = \mathcal{E}\{\mathbf{ss}^H\}$, becomes

$$\mathbf{R}_{ss} = \text{diag}\{E_{s,1}, E_{s,2}, \ldots, E_{s,P}\}. \tag{10.2}$$

The reverse link channel is a multiple access channel. The data rate that can be reliably maintained by all users simultaneously is characterized by a capacity region. In the following, we consider a deterministic problem and consider a sample realization of the channel **H** and assume that it is perfectly known to the receiver (i.e., at the base-station). What makes this problem different from that in MIMO-SU is that co-ordinated encoding is not allowed at the transmitters which are geographically dispersed.

10.2.2   Capacity region

We study the capacity region for two different receiver decoding strategies at the base-station – joint decoding and independent decoding. Joint decoding implies that the signals are decoded in a co-operative fashion, while independent decoding assumes that the signals are decoded independently in parallel.

Joint decoding
Joint decoding implies that the signals are detected optimally at the receiver via ML detection. Let $\mathcal{T}$ be a sub-set of the set $\{1, 2, \ldots, P\}$ and $\mathcal{T}'$ represent its complement. We denote the covariance matrix of the signals transmitted from the terminals indexed by $\mathcal{T}$ by $\mathbf{R}_{ss,\mathcal{T}}$ and the corresponding $M \times c(\mathcal{T})$ channel matrix by $\mathbf{H}_{\mathcal{T}}$ ($c(\mathcal{T})$ is the cardinality of the set $\mathcal{T}$). Representing the rate that can be reliably (error free) maintained for the ith user by R_i ($i = 1, 2, \ldots, P$) (in bps/Hz) and assuming Gaussian signaling for each user, the capacity region has been shown to satisfy [Suard *et al.*, 1998]

$$\sum_{k \in \mathcal{T}} R_k \leq \log_2 \det \left(\mathbf{I}_M + \frac{1}{N_o} \mathbf{H}_{\mathcal{T}} \mathbf{R}_{ss,\mathcal{T}} \mathbf{H}_{\mathcal{T}}^H \right) \quad \text{bps/Hz} \tag{10.3}$$

for all $2^P - 1$ possible non-empty sub-sets $\mathcal{T}$ of the set $\{1, 2, \ldots, P\}$. For example, the capacity region for a two-user system ($P = 2$) satisfies the following inequalities:

$$R_1 \leq \log_2 \left(1 + \frac{E_{s,1}}{N_o} \|\mathbf{h}_1\|_F^2 \right), \tag{10.4}$$

$$R_2 \leq \log_2 \left(1 + \frac{E_{s,2}}{N_o} \|\mathbf{h}_2\|_F^2 \right), \tag{10.5}$$

$$R_1 + R_2 \leq \log_2 \det \left(\mathbf{I}_2 + \frac{E_{s,1}}{N_o} \mathbf{h}_1 \mathbf{h}_1^H + \frac{E_{s,2}}{N_o} \mathbf{h}_2 \mathbf{h}_2^H \right). \tag{10.6}$$

The rate region is shown in Fig. 10.2. Along the bold line the sum-rate $R_1 + R_2$ is constant and is the maximum achievable sum-rate, C_{MC}. Every point along this

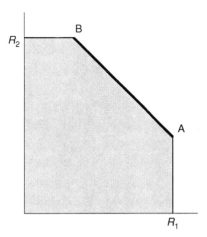

Figure 10.2: Capacity region for MIMO-MAC with joint decoding at the receiver. The bold line indicates the maximum achievable sum-rate on the reverse link.

line is achieved by each user transmitting at the maximum available power. To achieve the lower corner point A, user 1 builds Gaussian codewords at full rate $R_1 = \log_2\left(1 + (E_{s,1}/N_o)\|\mathbf{h}_1\|_F^2\right)$, thus assuming no interference. User 2 builds codewords assuming that the signal from user 1 is additional noise. The upper corner point B may be achieved in a similar fashion, with user 1 designing codewords treating user 2 as additional noise and user 2 designing codewords with full rate $R_2 = \log_2\left(1 + (E_{s,2}/N_o)\|\mathbf{h}_2\|_F^2\right)$. All other points along the bold line can be achieved by time-sharing between the two schemes (an alternative technique is rate-splitting proposed by [Rimoldi and Urbanke, 1996]). The capacity region for more than two users will, in general, be polyhedral.

We note that while ML decoding is optimal, the multiuser reverse link sum-rate capacity, C_{MC}, given by Eq. (10.3) can also be achieved via a MMSE receiver with successive cancellation as reported in [Varanasi and Guess, 1997]. We summarize the proof below.

For clarity assume that all users have the same average energy at the transmitter. Then the sum rate may be expressed as

$$\sum_{k=1}^{P} R_k = \log_2 \det\left(\mathbf{I}_M + \frac{E_{s,i}}{N_o}\mathbf{H}\mathbf{H}^H\right)$$

$$= \sum_{k=1}^{P} \log_2\left(1 + \frac{E_{s,i}}{N_o}\mathbf{h}_k^H\left(\mathbf{I}_M + \frac{E_{s,i}}{N_o}\mathbf{H}_{(k)}\mathbf{H}_{(k)}^H\right)^{-1}\mathbf{h}_k\right). \tag{10.7}$$

$\mathbf{H}_{(k)}$ is the channel matrix obtained by removing users (columns) with indices $k, k+1, \ldots, P$. A more detailed proof follows in Chapter 12 (see Eq. (12.3)). Consider the term corresponding to $k = P$. It is easily verified that this term corresponds to a capacity obtained by extracting the Pth user through MMSE filtering. If the user signals at a rate, R_P, less than or equal to this capacity, the signal can be decoded without any error and then subtracted from the received signal. The effective channel reduces to an $M \times (P - 1)$ matrix, $\mathbf{H}_{(P)}$, corresponding to the remaining $P - 1$ users. The $(P - 1)$th user (with a capacity corresponding to the term indexed by $k = P - 1$) may now be similarly decoded and subtracted from the received signal. This procedure is repeated until all users' signals are extracted without error. Note that for this procedure to work, first a decoding order has to be decided upon and then the users must signal with the correct rate assignments specific to that order. Each such ordering corresponds to one maximum sum-rate achieving vertex of the polyhedral capacity region. Since there are $P!$ possible orderings of the users, there are a corresponding number of vertices. Other rate assignments may be obtained through a convex combination (time-sharing) of these vertices.

Independent decoding

Recall that independent decoding attempts to recover each user's signal treating all other signals as interfering noise. The received signal covariance matrix, $\mathbf{R_{yy}} = \mathcal{E}\{\mathbf{yy}^H\}$, is given by

$$\mathbf{R_{yy}} = \mathbf{HR_{ss}H}^H + N_o\mathbf{I}_M. \tag{10.8}$$

The capacity region for independent decoding is the set of all rates satisfying [Suard *et al.*, 1998]

$$R_i \leq \log_2\left(\frac{\det\left(\mathbf{R_{yy}}\right)}{\det\left(\mathbf{R_{yy}} - E_{s,i}\mathbf{h}_i\mathbf{h}_i^H\right)}\right), \quad i = 1, 2, \ldots, P. \tag{10.9}$$

The maximum rate for each user is achieved via MMSE reception for each user. The capacity region for independent decoding relative to the joint decoding region for two users is shown in Fig. 10.3. Note that the MMSE decoding maximum rates are found by projecting the corner points A and B back on to the axes. For more than two users the capacity region will, in general, be a cuboid.

Discussion

Focusing on the two-user scenario, we see that the capacity region depends strongly on the geometry of $\mathbf{h}_1$ relative to $\mathbf{h}_2$ and the power available to the individual users. Note that when $\mathbf{h}_1$ is orthogonal to $\mathbf{h}_2$, the two decoding schemes (joint and independent) have equal and square capacity regions (see Fig. 10.4). This is so since the signals transmitted by the users can be perfectly separated and do not appear as interference to each other. At the other extreme, the smallest capacity region occurs when $\mathbf{h}_1$ is parallel to $\mathbf{h}_2$, as the two users cannot be spatially separated. Typically, with random channels the spatial separability of the users will improve as the number of base-station antennas M increases.

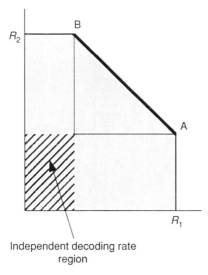

Independent decoding rate
region

Figure 10.3: Capacity region for MIMO-MAC with independent decoding at the receiver. The maximum sum-rate achieved through independent decoding will, in general, be less than that for joint decoding.

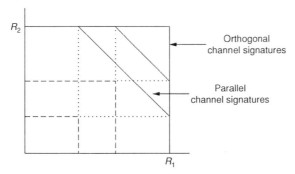

Figure 10.4: Influence of the relative geometry of channel signatures on the capacity region for MIMO-MAC. Rectangular regions correspond to independent decoding for arbitrary channels. Pentagonal (polyhedral) regions correspond to joint decoding. Regions overlap for orthogonal signatures (optimal).

In a random fading channel, the capacity region is also random and a given sum-rate can be sustained only with a certain level of reliability. Figure 10.5 plots the CDF of the maximum sum-rate, C_{MC}, for a two-user MIMO-MAC system with $M = 2$ and $M = 10$, $\mathbf{H} = \mathbf{H}_w$ and $E_{s,i}/N_o = 10$ dB for each user. As expected, joint decoding outperforms independent decoding at all outage levels. Also, the maximum sum-rate achieved at any outage level increases with an increase in the number of base-station antennas. Further, we note that the difference in maximum sum-rate achieved by the two schemes decreases with an increase in the number of base-station antennas. This

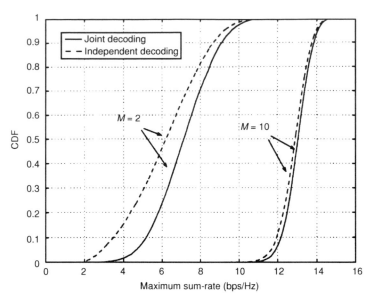

Figure 10.5: CDFs of maximum sum-rate for MIMO-MAC with joint and independent decoding at receiver. The difference between the decoding schemes decreases with increasing M.

can be attributed to better separability (orthogonality) of the spatial signatures with increasing M.

So far, we have discussed the sum-rate that can be achieved when the channel is known perfectly to the transmitter. In order to achieve a certain point in the capacity region, co-ordination between the users is necessary. This can be accomplished if all users are aware of the channel $\mathbf{H}$ and transmit at rates according to a pre-determined strategy. Alternatively, the base-station can determine and notify each user of the correct transmission rate. See also discussion in Chapter 12 on different co-ordination strategies.

The capacity region of MIMO-MAC for the case when the number of antennas at the users is greater than 1 is a convex hull of the pentagonal regions obtained by choosing different transmit signal covariance matrices $\mathbf{R}_{s_i s_i}$ with power constraints $\mathrm{Tr}[\mathbf{R}_{s_i s_i}] = E_{s,i}$. The boundary of the region is generally curved, except at the sum point where it is a straight line. Each point on the boundary is achievable with a different set of optimal covariance matrices $\mathbf{R}_{s_i s_i}^{opt}$ and is the corner point of the corresponding pentagonal region.

Like in the single antenna discussion, we need to decode users in an ordered manner corresponding to the operating point on the rate region. The optimal covariance $\mathbf{R}_{s_i s_i}^{opt}$ for the sum rate is obtained by solving a convex optimization problem. A more efficient numerical solution was developed by [Yu *et al.*, 2001b], using an iterative waterfilling procedure on the modes of each user's MIMO channel. This leads to a solution in which each user waterfills to his own channel with an effective noise equal to the additive noise and the interference from the other $P - 1$ users.

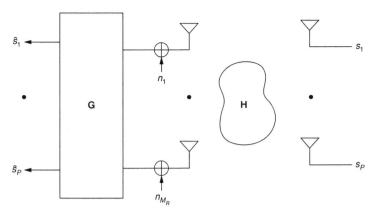

Figure 10.6: Schematic of linear processing at the receiver for MIMO-MAC. In principle the design of **G** is similar to that for MIMO-SU with HE.

10.2.3   Signaling and receiver design

Coding and power optimization at the user terminals for random fading multiple access channels has not been adequately studied. Given the target rate (or power) at each user, scalar coding techniques similar to horizontal encoding (SM-HE) studied in Chapter 6 are applicable with different user rates. We assume the channel is known to the users and the base.

Assuming single antenna terminals and, of course, independent encoding, all receiver architectures such as ZF, MMSE, OSUC and ML are applicable to the multiple access channel with the same tradeoffs between complexity and diversity/SNR performance. Of course, ML receivers are optimum. The performance of the different receivers described in Fig. 7.12 still applies with $M_T = M$ and $M_R = P$.

If power co-ordination between users is possible, the sub-optimum MMSE (independent decoding) receiver in Section 6.4 can be recast as a SINR balancing receiver (see Fig. 10.6). Let **G** of dimension $P \times M$ be a linear receiver at the base-station, and $\mathbf{g}_i$ ($i = 1, 2, \ldots, P$) be its ith row. **G** operates on the received signal **y** to output the $P \times 1$ vector $\widehat{\mathbf{s}}$, whose elements are estimates of the user signals. The SINR per user at the output of **G** is given by

$$\text{SINR}_i = \frac{|\mathbf{g}_i \mathbf{h}_i|^2 E_{s,i}}{\left(\sum_{j=1, j \neq i}^{P} |\mathbf{g}_i \mathbf{h}_j|^2 E_{s,j}\right) + N_o \|\mathbf{g}_i\|_F^2}. \tag{10.10}$$

The problem is to find the user powers $E_{s,i}$ ($i = 1, 2, \ldots, P$) and the filter **G** such that $\text{SINR}_i \geq \text{SINR}_{T,i}$ ($i = 1, 2, \ldots, P$), the target SINR for the ith user, subject to $E_{s,i} < E_{peak,i}$, where $E_{peak,i}$ is the peak power per constraint per user. Depending on the channel, peak power constraints and the target SINR, there may or may not be a feasible solution to the problem in Eq. (10.10). SINR balancing has been studied in

[Schubert and Boche, 2002], while the problem of power control from an information theoretic perspective has been studied in [Hanly and Tse, 1998].

10.3 MIMO-BC

10.3.1 Signal model

Once again, we assume a frequency flat channel. Let signals s_i ($i = 1, 2, \ldots, P$) with average energy $E_{s,i}$ be the signals transmitted from the base-station to the P users. The total average power (energy per symbol period) is constrained by $\sum_{i=1}^{P} E_{s,i} = E_s$. We assume the base-station has perfect knowledge of the forward link channel **H**. The signals travel over different vector channels to each of the P users. This is clearly a broadcast channel. Denoting the signal received at the ith terminal by y_i, the forward link signal model (see also Section 3.9) is given by

$$\mathbf{y} = \mathbf{H}\mathbf{s} + \mathbf{n}, \tag{10.11}$$

where $\mathbf{y} = [y_1 \, y_2 \, \cdots \, y_P]^T$ is a $P \times 1$ vector, **H** is the $P \times P$ forward link channel (including prefiltering) matrix, $\mathbf{s} = [s_1 \, s_2 \, \cdots \, s_P]^T$ is a $P \times 1$ vector containing the signals transmitted at the antennas and $\mathbf{n} = [n_1 \, n_2 \, \cdots \, n_P]^T$ is the $P \times 1$ additive ZMCSCG noise vector with variance N_o in each dimension. Again, the elements of **H** are not normalized to unit average power since they include a different path loss and shadow loss for each user. What differentiates this problem from MIMO-SU is that co-ordinated decoding is not allowed at the receivers which are geographically dispersed.

10.3.2 Forward link capacity

The MIMO-BC channel in general belongs to the class of non-degraded Gaussian broadcast channels [Caire and Shamai, 2000]. The capacity region for such a channel remains an unsolved problem. An achievable region was first derived in [Caire and Shamai, 2000] using the so-called "writing on dirty paper" result [Costa, 1983] in which the authors also demonstrate that the sum-rate broadcast capacity, C_{BC}, equals the maximum sum-rate of this achievable (dirty paper coded) region for a MIMO-BC channel with two users and two antennas at the base-station (see Fig. 10.7 for a schematic).

Furthermore, a duality between the achievable rate region and the multiple access capacity region that simplifies computation of the achievable rate region has been demonstrated [Jindal *et al.*, 2001; Vishwanath *et al.*, 2002]. An extension of the work in [Caire and Shamai, 2000] to a channel with more than two users, each having multiple receive antennas, was done in [Yu and Cioffi, 2001b], and the sum-rate capacity of a

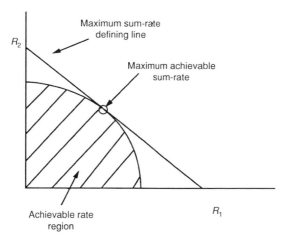

Figure 10.7: Schematic of the achievable rate region for a two-user MIMO-BC. The maximum sum-rate of the achievable region equals the sum-rate capacity of MIMO-BC.

Gaussian broadcast channel has been shown to satisfy [Yu and Cioffi, 2001a]

$$C_{BC} = \min_{\mathbf{R}_{nn} > 0.[\mathbf{R}_{nn}]_{k,k} = N_o} \max_{\mathrm{Tr}(\mathbf{R}_{ss}) = E_s} \log_2 \frac{\det\left(\mathbf{H}\mathbf{R}_{ss}\mathbf{H}^H + \mathbf{R}_{nn}\right)}{\det\left(\mathbf{R}_{nn}\right)}, \tag{10.12}$$

where $\mathbf{R}_{ss} = \mathcal{E}\{\mathbf{s}\mathbf{s}^H\}$, $\mathbf{R}_{nn} = \mathcal{E}\{\mathbf{n}\mathbf{n}^H\}$ and the optimization in Eq. (10.12) allows for coloration of the noise $\mathbf{n}$. Equation (10.12) may be interpreted as a Gaussian mutual information game in which a signal player chooses a transmit covariance matrix to maximize the mutual information and a noise player chooses a fictitious noise correlation to minimize the mutual information. A detailed discussion of these results is beyond the scope of this book.

10.3.3 Signaling and receiver design

The transmit coding problem at the base for the random fading multiple access channel is also largely an open problem. The base broadcasts the user signals after joint encoding and spatial weighting such that each user gets its desired SINR and can decode the intended signal. Clearly, users at the base-station can co-operate with each other in the encoding process. Co-operation between the users can also include choosing rates or power ($E_{s,i}$ ($i = 1, 2, \ldots, P$)), usually subject to a sum power constraint. The power optimization is a min–max problem that results in a least favorable interference correlation. Power co-ordination is again an open problem particularly in the context of IID fading wireless channels. We now discuss alternative transmit schemes. We assume that the channel is known to the transmitting base and to the receiving terminals.

Our general approach is to use a linear or a non-linear pre-filter as the spatial encoding scheme. The individual user signals are temporally encoded as in SM-HE with an appropriate rate per user. See Fig. 10.8.

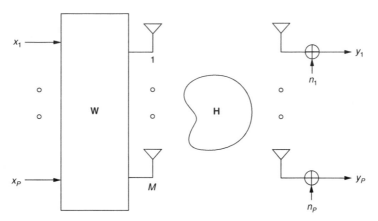

Figure 10.8: Schematic of linear pre-filtering at the base-station in MIMO-BC.

ZF Interference cancellation

A ZF linear pre-filter $\mathbf{W}_{ZF}$ (see Fig. 10.8) transmits user signals towards the intended user with nulls steered in the "direction" of the other users. This is the transmitter analog of the ZF receivers studied in Chapter 7 with $M \geq P$ and $\mathbf{H}$ is full row rank. The users will receive no interference, because of perfect nulling. The ith ($i = 1, 2, \ldots, P$) column of the ZF pre-filtering matrix, $\mathbf{w}_{ZF,i}$, is given by

$$\mathbf{w}_{ZF,i} = \frac{\mathbf{h}_i^{(\dagger)}}{\sqrt{\|\mathbf{h}_i^{(\dagger)}\|_F^2}}, \tag{10.13}$$

where $\mathbf{h}_i^{(\dagger)}$ is the ith column of $\mathbf{H}^\dagger$ and $E_{s,i}$ may now be chosen subject to the power constraint $\sum_{i=1}^{P} E_{s,i} = E_s$ to meet the target SNR for each user. If two or more user channels, $\mathbf{h}_i$, are close to each other, $\|\mathbf{h}_i^{(\dagger)}\|_F^2$ will become large and some users will receive very little power (power reduction), and we get the transmitter dual of the noise enhancement problem observed in ZF receivers. Figure 10.9 illustrates the power efficiency problem. The unit power weight vector $\mathbf{w}_{ZF,1}$ is orthogonal to $\mathbf{h}_2$, but delivers very little power along $\mathbf{h}_1$, the desired user direction. The problem worsens as $\mathbf{h}_1$ and $\mathbf{h}_2$ become closer.

MMSE with co-ordination

The naive ZF solution, as we saw, is power inefficient. As with the MMSE receiver, a better pre-filter can trade interference reduction for signal power inefficiency. Let $\{\mathbf{w}_i\}$ be the weight vectors for each user. Then, the SINR at the ith user is given by

$$\text{SINR}_i = \frac{|\mathbf{h}_i \mathbf{w}_i|^2 E_{s,i}}{\left(\sum_{j=1, j \neq i}^{P} |\mathbf{h}_i \mathbf{w}_j|^2 E_{s,j}\right) + N_o}, \tag{10.14}$$

where $\mathbf{h}_i$ ($i = 1, 2, \ldots, P$) is the ith row of $\mathbf{H}$. The design variables, $\mathbf{w}_i$ and $E_{s,i}$, are chosen to meet the target SINR set for each user, $\text{SINR}_{T,i}$, subject to the total

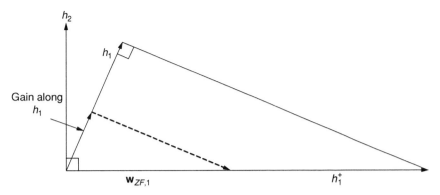

Figure 10.9: Schematic illustrating the power penalty problem. $\mathbf{w}_{ZF,1}$ has gain $\ll 1$ along $\mathbf{h}_1$.

power constraint $\sum_{i=1}^{P} E_{s,i} = E_s$ and norm constraint $\|\mathbf{w}_i\|_F^2 = 1$ $(i = 1, 2, \ldots, P)$. Depending on the channel and the power constraint there may or may not be a feasible solution to the problem in Eq. (10.14). SINR balancing for MIMO-BC has been studied in [Farsakh and Nossek, 1995; Gerlach and Paulraj, 1996; Rashid-Farrokhi *et al.*, 1998; Madhow *et al.*, 1999; Schubert and Boche, 2002]. Again the problem is not convex and is computationally non-trivial. Unlike interference pre-subtraction and ZF pre-filtering, SINR balanced pre-filtering may leave some residual interference to be handled by the receiver. The signal received at the ith user is given by

$$y_i = \mathbf{h}_i \mathbf{w}_i s_i + \sum_{j=1, j \neq i}^{P} \mathbf{h}_i \mathbf{w}_j s_j + n_i. \tag{10.15}$$

There are two possible receiver schemes for dealing with this interference. In the first, the user can ignore the interference and treat it as noise. In which case, the receiver needs to know only the signal channel which is $\mathbf{h}_i \mathbf{w}_i$. A better approach is to use multiuser detection and jointly detect all user signals individually at each user. In this case, the ith user needs to know the full pre-filtering matrix $\mathbf{W}$ and $\mathbf{h}_i$.

Matched filtering
Another pre-filtering method is to match each signal's pre-filter to its channel. The ith column $\mathbf{w}_{MRC,i}$ of the pre-filtering matrix $\mathbf{W}_{MRC}$ is given by

$$\mathbf{w}_{MRC,i} = \frac{\mathbf{h}_i^H}{\sqrt{\|\mathbf{h}_i\|_F^2}}. \tag{10.16}$$

This pre-filter does not attempt to eliminate interference between users. This occurs only if the rows of $\mathbf{H}$ are mutually orthogonal, a zero-probability event in practice. As before, $E_{s,i}$ must be chosen subject to the total transmit power constraint to meet the target SINR for each user. If $\mathbf{H} = \mathbf{H}_w$, the orthogonality of $\mathbf{H}$ and hence link performance will improve with increasing M.

Dirty paper coding (DPC) or interference pre-subtraction

ZF interference cancellation has a significant signal power penalty as we saw above. Pre-coding can be used to mitigate this problem. This technique is motivated by the surprising "writing on dirty paper" result [Costa, 1983] that showed that if the transmitter has perfect side information about an additive interference at the receiver (about which the receiver has no information), the transmitter can use optimal coding along with "layered" interference pre-subtraction and transmit at the same rate as if there were no interference. In MIMO-BC, DPC can be applied at the transmitter when choosing the codewords for different receivers. The transmitter first picks a codeword for receiver 1 and then chooses the codeword for receiver 2 with full (non-causal) knowledge of the codeword intended for receiver 1. Therefore the codeword for receiver two can be pre-subtracted such that user 2 will not see the signal to user 1 as interference. Similarly, the codeword for user 3 is chosen such that it does not see signals to users 1 and 2 as interference. Codeword choices with sum power constraints define the dirty paper region. When the users have a single antenna, it has been shown that DPC reduces to scalar coding with transmit beamforming.

We use a QR decomposition [Golub and Van Loan, 1989] variation of the ZF interference cancellation discussed above to outline this approach. Let $\mathbf{H} = \mathbf{RQ}$ be the QR decomposition of $\mathbf{H}$, where $\mathbf{R}$ is a $P \times P$ lower triangular matrix and $\mathbf{Q}$ is a $P \times M$ matrix such that $\mathbf{QQ}^H = \mathbf{I}_P$. Now, letting

$$\mathbf{W} = \mathbf{Q}^H, \tag{10.17}$$

where $\mathbf{W}$ has been defined in Chapter 8 (see Fig. 8.2), the effective input–output relation for the channel is

$$\mathbf{y} = \mathbf{Rs} + \mathbf{n}, \tag{10.18}$$

or equivalently

$$y_i = \sum_{j=1}^{i} [\mathbf{R}]_{i,j} s_j + n_i, \ i = 1, 2, \ldots, P. \tag{10.19}$$

Therefore the broadcast channel is decomposed into P channels with interference between channels. Because $\mathbf{R}$ is lower triangular, the channel to the first user is the usual SISO channel, the second channel has interference from the first channel, the third channel has interference from the first and second channels and so on. An interference pre-subtraction scheme for each parallel sub-channel can eliminate the effect of mutual interference completely. A naive interference pre-subtraction scheme would

Figure 10.10: Modulo operation to reduce the power penalty in interference pre-subtraction.

be to replace the transmitted symbols s_i $(i = 1, 2, \ldots, P)$ by s_i', where

$$s_i' = s_i - \frac{1}{[\mathbf{R}]_{i,i}} \sum_{j=1}^{i-1} [\mathbf{R}]_{i,j} s_j'. \tag{10.20}$$

With interference pre-subtraction, Eq. (10.19) reduces to

$$y_i = [\mathbf{R}]_{i,i} s_i + n_i, \quad i = 1, 2, \ldots, P, \tag{10.21}$$

thus completely eliminating interference between users.

One approach to reducing transmit power is to extend the ideas of modulo arithmetic in Tomlinson–Harashima pre-coding for ISI pre-subtraction [Tomlinson, 1971; Harashima and Miyakawa, 1972]. This can be done using a vector quantizer generalization of Tomlinson–Harashima pre-coding (see Fig. 10.10) with modulo operation on Voronoi regions replacing the modulo operation on the real line. See [Ginis and Cioffi, 2000; Erez *et al.*, 2000; Yu and Cioffi, 2001b] for more details. This technique can also be seen as a transmit variant of OSUC receivers studied in Chapter 7. The practicality of pre-subtraction ideas needs to be more thoroughly explored, particularly since perfect non-causal channel knowledge at the transmitter is required.

10.4   Outage performance of MIMO-MU

MIMO-MU is in theory an attractive approach to increasing spectral efficiency in wireless links, particularly since the users need to have only one antenna each. However, MIMO-MU with non-spread modulation has a number of problems that can make its implementation difficult. First, in random fading channels, user separability cannot be guaranteed since two user channels may become close and this will become a source of outage or link failure beyond the usual problems in wireless systems caused by fading and interference. Another problem is the near–far problem mentioned earlier in the chapter. Perhaps the most difficult issue is the need for accurate channel knowledge at the transmitter in the forward link. This fundamental requirement is hard to satisfy, as pointed out in Chapter 3. The above problems have so far blocked the deployment of MIMO-MU in practical systems for non-spread modulation. See Section 10.6 for comments on MIMO-MU with spread modulation. Further insights into these issues are demonstrated by the simulation studies described below.

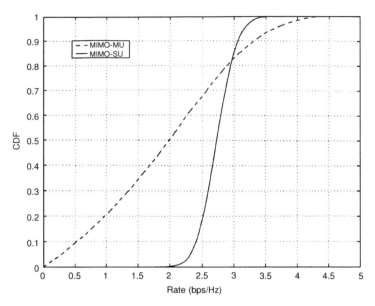

Figure 10.11: Forward link capacity CDFs of MIMO-SU and MIMO-BC with ZF pre-filtering. MIMO-SU outperforms MIMO-BC by a factor of 5 at the 10% outage level.

10.4.1 MU vs SU – single cell

Consider the alternative scenarios of supporting a forward link using MIMO-SU with $\mathbf{H} = \mathbf{H}_w$ and $M_T = M_R = 5$ and MIMO-BC with $M_T = P = 5$ and a single antenna at the terminal. In the MIMO-SU case, the five users are time multiplexed and each user receives 20% of the information rate on the forward link. Figure 10.11 plots the CDF of the information rate received by a random user for two cases: (a) MIMO-BC transmission, where the base has perfect channel knowledge and uses ZF transmit beamforming (not the more optimal DPC); (b) MIMO-SU transmission, where the base has no channel knowledge but time multiplexes the five users equally. The transmit power is normalized to deliver an average of 10 dB SNR to each user. We see from Fig. 10.11 that at 10% outage, 20% of the information rate in MIMO-SU is still five times greater than the single user information rate in MIMO-BC. This is despite the MIMO-SU transmitter not knowing the channel. Thus, MIMO-BC is at a big disadvantage. The comparison can be even more unfavorable for MIMO-BC if we add near–far power differences, channel coupling due to dominant scatterers and errors in the forward channel estimation. The source of this disadvantage is that in MIMO-SU all the spatial sub-channels support the link adding diversity, whereas in MIMO-BC only one sub-channel is available. However, if $M > P$, we can add diversity and the MU-SU gap will reduce.

Influence of channel estimation errors

We now examine the effect of the channel estimation error on the outage performance of a MIMO-BC system. Figure 10.12 plots the CDF of the SINR for one user for varying

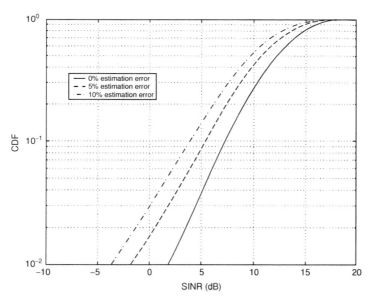

Figure 10.12: SINR CDF with varying degrees of channel estimation error for MIMO-BC with ZF pre-filtering. SINR degrades rapidly with an increasing degree of channel estimation error.

degrees of channel estimation error. We assume ZF pre-filtering with $M = 3$ and $P = 2$ and $E_{s,i}/N_o = 10$ dB for both users. The graph shows that the penalty in SINR at the 1% outage level is approximately 4 dB at 5% channel estimation error and 7 dB at 10% estimation error. Therefore, the performance of MIMO-BC degrades rapidly with the channel estimation error.

10.4.2 MU single cell vs SU multicell

Since the role of multiple antennas is to improve spectral efficiency, it also logical to compare MIMO-MU (channel reuse within the cell) with another common technique for improving spectral efficiency in wireless networks – channel reuse between cells. See also Chapter 11 for a discussion on channel reuse. Consider two alternative scenarios for supporting a forward link: a single user MISO configuration with $M_T = 3$, $M_R = 1$ and a MIMO-BC configuration with $M = 3$ and $P = 2$. We assume $\mathbf{H} = \mathbf{H}_w$ with $E_s/N_o = 13$ dB in both cases. The transmit power is equally divided between the two users for MIMO-MU. Perfect channel knowledge at the transmitter is assumed in both cases. MISO-SU uses transmit-MRC, while MIMO-BC uses ZF pre-filtering. In a multicell environment, the spectral efficiency can be improved either by increasing reuse within the cell (MIMO-BC) or by reducing the reuse factor (increasing reuse between cells). Figure 10.13 shows the CDF of SINR for the MIMO-BC and MISO-SU systems. We notice an 8 dB penalty at 1% outage for MIMO-BC compared with MISO-SU. In a multicell environment, we can halve the reuse factor for about a 6 dB

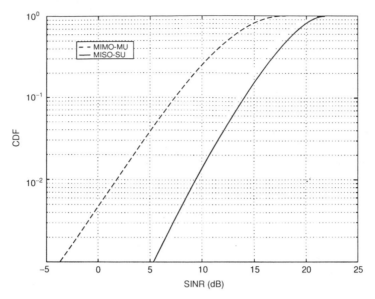

Figure 10.13: Forward link SINR CDFs of MISO-SU and MIMO-BC with ZF pre-filtering. Halving the reuse factor with MISO-SU is an attractive alternative to using MIMO-BC.

increase in SINR. Therefore, it would appear that MIMO-SU transmission using half the reuse factor scores over MIMO-MU with the same reuse factor. Note that these results do not include the problems associated with MIMO-BC, which are likely to hurt MIMO-BC even further. Again, if $M >> P$, the performance penalties will reduce.

10.5　MIMO-MU with OFDM

Since OFDM decomposes the ISI channel into orthogonal tone channels, the general theory of MIMO-MU for frequency flat channels applies on a tone-by-tone basis. The key area of interest is now the per tone power allocation problem in the reverse and forward links. This is the waterpouring problem extended to multiuser channels. It has been shown that for the multiple access problem, a co-ordinated approach leads to a simultaneous waterpouring solution and in the forward link it is a min–max formulation that converges to a waterpouring solution based on a Nash equilibrium.

10.6　CDMA and multiple antennas

The theory of multiuser methods using SS modulation (CDMA) for scalar (SISO) antenna channels has been an area of research and practice for many decades [Cover and Thomas, 1991; Viterbi, 1995; Verdu, 1998]. The extensions to vector (MISO,

SIMO) or matrix (MIMO) channels has so far been largely addressed from coding and receiver viewpoints. We discuss some examples of ST coding and ST receivers for CDMA in Chapters 9 and 12. If the multiuser codes are guaranteed to be perfectly orthogonal at the transmitter and at the receiver, then the CDMA channel decomposes into independent code channels and the multiuser channel decouples into single user channels and the full ST theory for SC modulation described in this book carries over. In practice, true orthogonality cannot be achieved, complicating the picture. Nevertheless, with a reasonably large spreading factor ($K_S \approx 100$), quasi-orthogonality is available and we can approximate the multiple access interference as additional additive noise. This approximation again converts the problem to a single user case and this is indeed a pragmatic (though sub-optimal) approach to extending ST techniques to CDMA.

The use of multiple antennas in CDMA adds a further (or bonus) degree of separability based on differences in spatial channels in addition to that provided by spreading codes. Reverse link and forward link schemes can exploit this spatial dimension as well as the code dimension to enhance overall multiuser performance. Multiuser separation by spatial dimension alone is no longer needed, and this dramatically reduces the difficulties described previously in meeting the outage performance of space only multiuser methods. Indeed simple beamforming or just directional antennas can be used to reduce multiuser interference and increase multiuser capacity [Naguib, 1996; Adachi et al., 1998]. CDMA standards such as UMTS and CDMA 2000 reuse the same frequency channel within a cell through simple sectoring antennas.

The advantages of CDMA for exploiting the spatial dimension come at the price of bandwidth expansion and therefore do not necessarily imply better spectral efficiency. One could argue, however, that ST techniques for multiuser operation are more robust with highly spread modulation compared with non-spread modulation. However, highly spread CDMA is not practical for broadband links due to the large post-spread bandwidth requirement.

11 ST co-channel interference mitigation

11.1 Introduction

When ST links are employed in a multi-cell environment with frequency reuse, the links suffer from CCI (co-channel interference). CCI mitigation has been studied for many years and has been used in a very limited form in wireless networks. The use of directional antennas and antenna arrays has long been recognized as an effective technique for reducing CCI.

The abstract CCI (or simply interference) channel model consists of a number of co-channel but independent transmit–receive links that interfere with each other. Since the transmitters or receivers are not co-located, there is no possibility of co-ordination between the transmitters acting together or the receivers acting together. In other words, on the reverse link, the reference base cannot easily co-operate with the other co-channel bases to jointly decode the signals. Likewise, on the forward link, the reference base cannot easily co-operate with the other transmitting bases to jointly encode the signals. In this sense, interference channels are further removed from the MIMO-MU channels where joint decoding and joint encoding are possible in multiple access and broadcast channels respectively.

The value of CCI mitigation in wireless networks is that it enables better frequency reuse and hence improves network spectrum efficiency. In non-spread modulation, a reuse factor of typically 3 or greater is required to achieve adequate signal to interference ratio (SIR). A switch to a lower reuse factor (usually a rhombic number) will need a corresponding step increase in CCI reduction and therefore has poor (large) granularity. In SS modulation, the cellular reuse factor is typically 1 with further reuse within the cell, and we can show that any CCI mitigation translates to improved capacity with good (low) granularity.

The capacity of interference channels is largely an unsolved problem. Our approach in this chapter is to ignore capacity issues and focus on exploiting multiple antennas for CCI cancellation at receive and CCI avoidance at transmit.

CCI mitigation has been addressed in a number of books [Verdu, 1993; Lee, 1995; Rappaport, 1996; Liberti and Rappaport, 1999; Giannakis *et al.*, 2000]. In this chapter we very briefly overview CCI mitigation using antennas and address techniques for forward and reverse links. Our development assumes SC modulation. We begin with SIMO signal models for reverse link CCI and discuss receiver options and their performance. We then extend these to MIMO signal models. Next, we discuss briefly the forward link problem where CCI mitigation is an avoidance (or pre-cancellation) problem. We then discuss how these receivers can be extended to SS and OFDM modulation. Finally, we discuss interference diversity techniques to mitigate CCI.

11.2   CCI characteristics

CCI characterization (power and SINR statistics) has been widely studied [Lee, 1995; Stüber, 1996]. CDF curves for CCI averaged over the entire cell or in an annular cell region have been characterized. The ST channel characterization of CCI has received much less attention. Since CCI originates from a distant cell, it will, in general, have "large range" characteristics such as higher delay and angle spreads, a lower or zero K-factor, smaller XPD and, in general be closer to IID channels.

CCI characteristics for TDMA
Figure 11.1 shows typical reverse link and the forward link CCI sources in TDMA. Typically one or two strong interferers are present. Due to power control differences, the reverse link and the forward link have different interference characteristics. Also inter- and intra-cell propagation delays, along with delay spread can result in the signal and interference having misaligned TDMA slots. This can make interference cancellation very difficult.

CCI characteristics for CDMA
Figure 11.2 shows typical reverse link and forward link CCI sources in CDMA (IS-95 type systems). The spatial reuse factor is 1. Quasi-orthogonal codes give rise to CCI, which in this case is also known as MAI. Typically there are about 3–5 interferers (base-stations) in the forward link and 15–25 interferers (terminals) in the reverse link. Approximately 40% of interference is from outside the cell [Viterbi, 1995].

11.3   Signal models

Throughout our discussion we assume that the desired user and co-channel interferers have synchronized time slots and the same frequency and modulation scheme. We assume SC modulation.

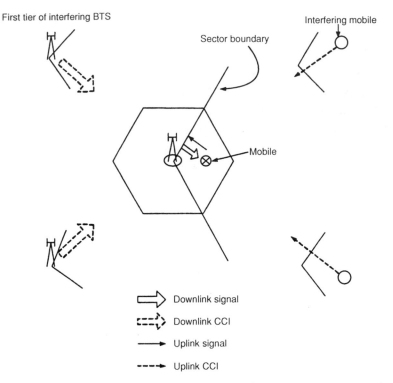

First tier of interfering BTS

Interfering mobile

Sector boundary

Mobile

⇨ Downlink signal

⊏⊏⊅ Downlink CCI

⟶ Uplink signal

⟶ Uplink CCI

Figure 11.1: Typical TDMA CCI model. Typically there are one or two strong interferers in the reverse and forward links (SINR $\approx$ 6–14 dB in the Global System for Mobile Communications (GSM)).

11.3.1 SIMO interference model (reverse link)

Consider a reverse scenario with one desired signal and N co-channel interfering signals arriving at an antenna array after being scattered by the medium (see Fig. 11.3). Both the desired and interfering signals emanate from terminals with one transmit antenna each.

Frequency flat channel

The signal model at the base-station developed in Chapter 3 for a single user in a frequency flat SIMO channel can be extended to the signal and interference model given by

$$\mathbf{y} = \mathbf{h}_0 s_0 + \sum_{i=1}^{N} \mathbf{h}_i s_i + \mathbf{n}, \tag{11.1}$$

where $\mathbf{y}$ ($M_R \times 1$) is the received signal vector, s_0 is the desired user's signal with $\mathcal{E}\{|s_0|^2\} = E_{s,0}$, s_i is the signal from the ith ($i = 1, 2, \ldots, N$) interferer with $\mathcal{E}\{|s_i|^2\} = E_{s,i}$, $\mathbf{h}_i$ ($M_R \times 1$) is the channel from the ith interferer to the base-station and $\mathbf{n}$ is the $M_R \times 1$ ZMCSCG noise with covariance matrix $N_o \mathbf{I}_{M_R}$. Note that just as in the

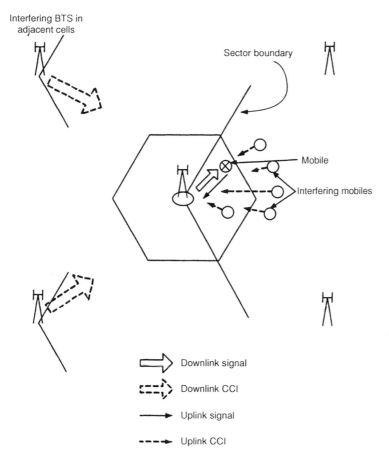

Figure 11.2: Typical CDMA CCI model. SINR ≈ -15 to -8 dB. A spreading (processing) gain of 20 dB makes the signal detectable.

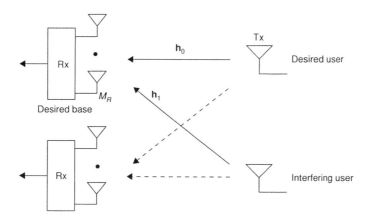

Figure 11.3: SIMO interference channel (reverse link). Only one interfering user is shown.

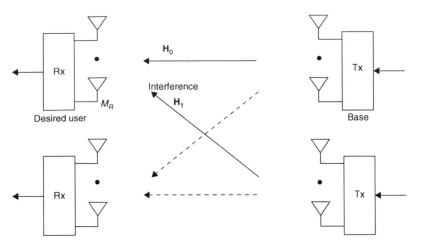

Figure 11.4: MIMO interference channel. Only one interfering user is shown.

MIMO-MAC (Chapter 10), the elements of $\mathbf{h}_0$ and $\mathbf{h}_i$ need not be normalized to reflect differences in path loss for the different users.

Frequency selective channel
The block signal model developed in Chapter 3 for a single user in a frequency selective environment is extended to multiple users as follows:

$$\mathbf{Y} = \mathbf{H}_0 \mathcal{S}_0 + \sum_{i=1}^{N} \mathbf{H}_i \mathcal{S}_i + \mathcal{N}, \tag{11.2}$$

where

$$\mathbf{H}_i = \begin{bmatrix} h_{1,i}[L-1] & \dots & h_{1,i}[0] \\ \vdots & \vdots & \vdots \\ h_{M_R,i}[L-1] & \dots & h_{M_R,i}[0] \end{bmatrix}, \quad \mathcal{S}_i = \begin{bmatrix} s_i[k-L+1] & \dots & s_i[k+T-L] \\ \vdots & \vdots & \vdots \\ s_i[k] & \vdots & s_i[k+T-1] \end{bmatrix},$$

$$\tag{11.3}$$

with $\mathcal{E}\{|s_i[k]|^2\} = E_{s,i}$.

11.3.2 MIMO interference channel (any link)

A similar scenario applies to MIMO except that the desired and interfering signals each emanate from multiple (M_T) transmit antennas (see Fig. 11.4). The model is similar for both links.

Frequency flat channel
The signal model developed in Chapter 3 for a single user in a frequency flat MIMO channel can be extended to signal plus interference as follows:

$$\mathbf{y} = \mathbf{H}_0 \mathbf{s}_0 + \sum_{i=1}^{N} \mathbf{H}_i \mathbf{s}_i + \mathbf{n}, \tag{11.4}$$

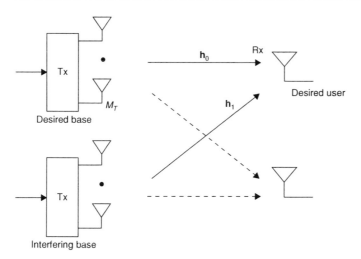

Figure 11.5: MISO interference channel (forward link). Only one interfering user is shown.

where $\mathbf{s}_0$ with $\mathcal{E}\{\mathbf{s}_0\mathbf{s}_0^H\} = (E_{s,0}/M_T)\mathbf{I}_{M_T}$ is the signal from the desired user, $\mathbf{H}_0$ is the $M_R \times M_T$ desired user's channel, $\mathbf{H}_i$ is a matrix of dimension $M_R \times M_T$ corresponding to the channel of the ith interferer and $\mathbf{s}_i$ with $\mathcal{E}\{\mathbf{s}_i\mathbf{s}_i^H\} = (E_{s,i}/M_T)\mathbf{I}_{M_T}$ is the $M_T \times 1$ vector corresponding to the transmit signal of the ith interferer. The frequency selective signal model follows from Chapter 3 and is not presented here.

11.3.3 MISO interference channel (forward link)

Consider a scenario with $N + 1$ interfering base-stations, each serving one intended user per cell. See Fig. 11.5. Assume a frequency flat channel. Each base-station is equipped with multiple transmit antennas and each mobile user has a single antenna. We assume linear pre-filtering (beamforming) at the transmitting base. The signal model developed for the frequency flat MISO channel in Chapter 3 may be extended for signal and interference as follows:

$$y_0 = \mathbf{h}_0\mathbf{w}_0 s_0 + \sum_{i=1}^{N} \mathbf{h}_i \mathbf{w}_i s_i + n_0, \qquad (11.5)$$

where y_0 is the scalar received signal at the zeroth user (within the reference cell), $\mathbf{w}_0$ is the $M_T \times 1$ pre-coding weight vector used to steer energy towards the desired user, $\mathbf{h}_i$ is the $1 \times M_T$ channel between the zeroth user and the ith base-station, s_i is the signal transmitted to the ith user with $\mathcal{E}\{|s_i|^2\} = E_{s,i}$ and n_0 is ZMCSCG noise at the reference user. Once again, the elements of $\mathbf{h}_i$ are not normalized to reflect differences in path loss.

Note that only the transmit channel $\mathbf{h}_0$ is observable at the reference base-station and can be determined using the techniques explained in Chapter 3. The other channels are much more difficult to determine as they correspond to users outside the reference cell.

Equation (11.5), however, contains the channels $\mathbf{h}_i$ that are not easily observable. We assume that $\|\mathbf{w}_i\|_F^2 = 1$ to maintain the transmit power constraint at each base-station. The channel model can easily be extended to the frequency selective case. Likewise, a signal and interference model with multiple receive antennas can be developed.

The forward link problem is related to the broadcast channel in Chapter 10. The capacity and/or rate regions will depend on the type/degree of channel information and the co-operation model between the base-stations.

11.4 CCI mitigation on receive for SIMO

We now generalize the SIMO receivers discussed in Chapter 7 to cancel CCI. The approach is similar to that used in MIMO receivers for SM-HE with MSI replacing CCI. The differences in the spatial channels between the desired user and co-channel interferer are leveraged to mitigate CCI. Note that the differences in the temporal channel alone (without the spatial dimension) offer little value for CCI reduction and what little leverage is available is related to excess bandwidth and pulse shaping [Paulraj and Papadias, 1997].

11.4.1 Frequency flat channel

Space-ML (S-ML) receiver
From Eq. (11.1) the desired signal is obtained by solving the multiuser (joint decoding) detection problem

$$\hat{s}_0 = \arg \min_{s_0, s_1, s_2, \cdots, s_N} \left\| \mathbf{y} - \mathbf{h}_0 s_0 - \sum_{i=1}^{N} \mathbf{h}_i s_i \right\|_F^2 . \tag{11.6}$$

The receiver needs to know the signal channel $\mathbf{h}_0$ and the interfering channels $\mathbf{h}_i$ ($i = 1, 2, \ldots, N$). The performance of this receiver is similar to the performance of the ML decoder for SM. Assuming that $\mathbf{h}_i$ ($i = 0, 1, 2, \ldots, N$) are $\mathbf{h}_w$ type channels, the ML receiver extracts M_R order signal diversity. The two main disadvantages of the ML receiver are that it requires channel knowledge of the interferers (which is hard to obtain) and the high decoding complexity (which is exponential in the number of interferers).

Space-MMSE (S-MMSE) receiver
The MMSE filter is computationally attractive. It is easy to verify (see Chapter 7) that the MMSE filter, $\mathbf{g}_{MMSE}$ ($1 \times M_R$), is given by

$$\mathbf{g}_{MMSE} = \arg \min_{\mathbf{g}} \mathcal{E}\{|\mathbf{g}\mathbf{y} - s_0|^2\}$$

$$= \mathbf{R}_{s_0 \mathbf{y}} \mathbf{R}_{\mathbf{yy}}^{-1}$$

$$= E_{s,0} \mathbf{h}_0^H \mathbf{R}_{\mathbf{yy}}^{-1}, \tag{11.7}$$

where $\mathbf{R}_{s_0\mathbf{y}} = \mathcal{E}\{s_0\mathbf{y}^H\} = E_{s,0}\mathbf{h}_0^H$ and $\mathbf{R}_{\mathbf{yy}} = \mathcal{E}\{\mathbf{yy}^H\}$ is the covariance matrix of the received signal estimated at the receiver. The output SINR can be shown to be

$$\text{SINR}_{MMSE} = \frac{\det(\mathbf{R}_{\mathbf{yy}})}{\det(\mathbf{R}_{\mathbf{yy}} - E_{s,0}\mathbf{h}_0\mathbf{h}_0^H)} - 1. \tag{11.8}$$

A more detailed analysis is presented in [Winters *et al.*, 1994].

The S-MMSE receiver performance when $\mathbf{h}_i$ ($i = 0, 1, 2, \ldots, N$) are all $\mathbf{h}_w$ channels is identical to that of the MIMO MMSE receiver with SM-HE. Recall from Chapter 7 that, at high SNR, the MMSE and ZF receivers converge in performance and the diversity order is equal to $M_R - M_T + 1$. This implies that we get $M_R - (N + 1) + 1 = M_R - N$ order signal diversity through use of the S-MMSE (see [Winters *et al.*, 1994]). The above comments, of course, only apply when the interference power is comparable with or larger than the signal power (medium or low SIR). If the SIR is large, the MMSE receiver can ignore the interference and will converge in performance with a MRC receiver extracting full M_R order diversity.

Space-MRC (S-MRC) receiver

The S-MRC receiver ($\mathbf{g}_{MRC} = \mathbf{h}_0^H$) has been covered in Chapter 5. The S-MRC ignores the CCI but has an inherent interference suppression property since the signal and interference channels have different signatures. The SINR after MRC processing for a given $\mathbf{h}_0$ becomes

$$\text{SINR}_{MRC} = \frac{E_{s,0}(\mathbf{h}_0^H\mathbf{h}_0)^2}{\sum_{i=1}^{N} E_{s,i}|\mathbf{h}_0^H\mathbf{h}_i|^2 + \mathbf{h}_0^H\mathbf{h}_0 N_o}. \tag{11.9}$$

If noise is dominant ($N_o \gg E_{s,i}$), we have

$$\text{SINR}_{MRC} \approx \rho\,\mathbf{h}_0^H\mathbf{h}_0 = \rho\|\mathbf{h}_0\|_F^2. \tag{11.10}$$

Further, if $\mathbf{h}_0 = \mathbf{h}_w$, from Section 5.3 the S-MRC receiver offers M_R order diversity. Also,

$$\mathcal{E}\{\text{SINR}_{MRC}\} = \mathcal{E}\{\|\mathbf{h}\|_F^2\}\rho = M_R\,\rho, \tag{11.11}$$

yielding an array gain of M_R.

Applying the Cauchy–Schwartz inequality, $|\mathbf{h}_0^H\mathbf{h}_i|^2 \leq \|\mathbf{h}_0\|_F^2\|\mathbf{h}_i\|_F^2$, to Eq. (11.9) we can show that

$$\frac{E_{s,0}|\mathbf{h}_0^H\mathbf{h}_0|^2}{\sum_{i=1}^{N} E_{s,i}|\mathbf{h}_0^H\mathbf{h}_i|^2 + \mathbf{h}_0^H\mathbf{h}_0 N_o} \geq \frac{E_{s,0}\mathbf{h}_0^H\mathbf{h}_0}{\sum_{i=1}^{N} E_{s,i}\mathbf{h}_i^H\mathbf{h}_i + N_o}, \tag{11.12}$$

which implies that the SINR after MRC processing is greater than the input SINR (RHS of Eq. (11.12)), which is the ratio of the total signal power to the interference plus noise power at the array output. The exact SINR gain depends on the signal to interference power ratio and the relative geometry of the signal and the interfering channels ($\mathbf{h}_0^H\mathbf{h}_i$).

Table 11.1. *Receivers for CCI cancellation – frequency flat channels*

	Diversity order	Noise enhancement loss	Channel knowledge
S-ML	M_R	Nil	$\mathbf{h}_i$ $(i = 0, 1, 2, \ldots, N)$
S-MMSE	$M_R - N$	Medium	$\mathbf{R}_{yy}, \mathbf{h}$
S-MRC	M_R	Nil	$\mathbf{h}$

S-ML, S-MMSE and S-MRC tradeoffs

- S-ML extracts M_R order diversity. This receiver is computationally hard and needs CCI channel knowledge ($\mathbf{h}_i$).
- S-MMSE extracts approximately $M_R - N$ diversity at high SNR and low–medium SIR. It is simpler computationally and does not need the CCI channel. At high SIR, it converges in performance with S-MRC with M_R order diversity.
- S-MRC does not directly attempt to cancel CCI. It provides only array gain and M_R order signal diversity with an incidental reduction in interference that depends on the relative geometry of interference and signal channels.

Table 11.1 summarizes the performance of the three receivers.

11.4.2 Frequency selective channel

ST-ML receiver

In a delay spread channel, the ML receiver is a multiuser detection problem given by

$$\hat{\mathcal{S}}_0, \hat{\mathcal{S}}_1, \hat{\mathcal{S}}_2, \ldots, \hat{\mathcal{S}}_N = \arg \min_{\mathcal{S}_0, \mathcal{S}_1, \mathcal{S}_2, \ldots, \mathcal{S}_N} \left\| \mathbf{Y} - \mathbf{H}_0 \mathcal{S}_0 - \sum_{i=1}^{N} \mathbf{H}_i \mathcal{S}_i \right\|_F^2 . \tag{11.13}$$

As in the frequency flat channel, knowledge of the desired user's channel as well as the interfering channel is necessary. The ST-ML receiver offers space (M_R) and path diversity ($\approx L_{eff}$) and mitigates CCI and ISI. This receiver suffers from very high decoding complexity and requires knowledge of the interfering channels.

ST-MMSE receiver

In presence of delay spread, a ST-MMSE receiver has to cancel ISI as well as CCI. The ST-MMSE receiver for ST equalization of SIMO frequency selective channels has been discussed in Chapter 7. The filter weight $\mathbf{g}_{MMSE}$ $(1 \times M_R T)$ vector is given by

$$\mathbf{g}_{MMSE} = \arg \min_{\mathbf{g}} \mathcal{E}\{|\mathbf{g}\mathcal{Y}[k] - s_0[k - L + \Delta_D]|^2\}$$

$$= E_{s,0} \mathbf{1}_{\Delta_D, T+L-1} \overline{\mathcal{H}}_0^H \mathbf{R}_{yy}^{-1}, \tag{11.14}$$

where $\mathcal{Y} = \mathrm{vec}(\mathbf{Y}^T)$ ($\mathbf{Y}$ is defined in Eq. (11.2)) and $\mathbf{R}_{y,y} = \mathcal{E}\{\mathcal{Y}\mathcal{Y}^H\}$ now also incorporates the CCI contribution and $\overline{\mathcal{H}}_0$ is the signal channel as defined in Eq. (7.24).

Figure 11.6 plots the SER curves for an ST-MMSE receiver with BPSK modulation, root raised-cosine pulse-shaping (40% excess bandwidth) and one interferer. The

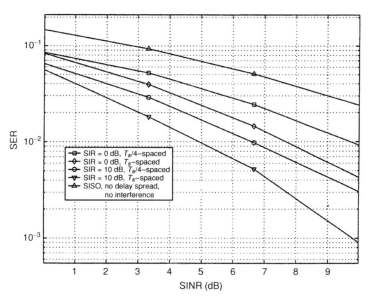

Figure 11.6: Performance of the ST-MMSE receiver for one user and a single interferer with one transmit antenna each. The base-station has two receive antennas. The performance degrades with decreasing delay spread and decreasing SIR.

physical channel taps for both the interferer and the desired signal are independent and equi-powered with spacings of $T_s/4$ or T_s. For a fixed SIR, the performance degrades with decreasing delay spread. For a fixed delay spread the performance degrades with decreasing SIR. At high SIR, the performance is equivalent to a SIMO link with a ST-MMSE receiver.

Discussion

The temporal taps in the ST-MMSE mitigate ISI but are of little use for mitigating CCI. The ability of the temporal taps to cancel CCI depends on the excess bandwidth and the channel delay spread. In practice this offers negligible value. Spatial taps (antennas) are effective for CCI mitigation but not necessarily useful for mitigating ISI. Therefore, in ST-MMSE the spatial dimension primarily mitigates CCI and the temporal dimension primarily mitigates ISI. Oversampling is typically used in practice to reduce the effect of jitter and enable robust equalization. The role of oversampling in improving diversity is marginal, and its role in improving CCI mitigation is also extremely limited [Paulraj and Papadias, 1997].

ST-MMSE-ML receiver

As seen earlier, the ST-ML receiver mitigates CCI and ISI and has the best performance but is very difficult to implement. The ST-MMSE receiver on the other hand is easier to implement but loses frequency diversity in the presence of CCI. The ST-MMSE-ML (see Fig. 11.7) offers a compromise two-stage receiver with ST-MMSE in the first stage and MLSE in the second stage. The ST-MMSE cancels CCI alone, capturing $\approx M_R - N$ spatial diversity, and the scalar MLSE handles ISI, capturing $\approx L_{eff}$ order path diversity.

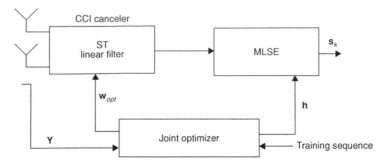

Figure 11.7: ST-MMSE-ML receiver. The first stage eliminates CCI while passing through the ISI to the second stage ML receiver.

The first stage uses the training sequence ($\mathcal{F}$) convolved with the channel as a reference signal thereby canceling only CCI and not ISI. The equivalent channel at the output of the first stage is now a scalar channel which contains ISI passed through the ST-MMSE stage. The MLSE receiver (with metrics suitably modified to take into account noise coloring from the first stage) equalizes the ISI. For a more detailed treatment refer to [Liang and Paulraj, 1996; Liang, 1998].

ST-ML, ST-MMSE and ST-MMSE-ML tradeoffs

- ST-MLSE captures $M_R L_{eff}$ spatial and temporal diversity. ST-MLSE will outperform ST-MMSE especially in presence of significant ISI. However, the receiver is, in general, complex to implement and requires knowledge of the interferer channels $\mathbf{H}_i$. Furthermore, the ST-MLSE implementation is complicated if the CCI also has delay spread.
- ST-MMSE extracts $M_R L_{eff} - N L_{eff}$ spatial and temporal diversity ($N L_{eff}$ loss due to the CCI). It is easier to implement since it requires knowledge of only $\mathbf{R}_{yy}$ and $\mathbf{H}_0$. For high SIR, the ST-MMSE receiver reduces to ST-MRC with $\approx M_R L_{eff}$ diversity.
- ST-MMSE-ML extracts spatial diversity in the first stage ($\approx M_R - N$) and path diversity ($\approx L_{eff}$) in the second stage. This receiver therefore combines the strengths of ST-MMSE and ST-MLSE and appears to be a good practical compromise.

Table 11.2 summarizes the performance of the three receivers.

Finally, we should state that in a cellular network the goal should be to maximize the minimum SINR for each user. CCI cancelling is perhaps the most difficult way to improve SINR. Simpler techniques implemented in wireless networks include dynamic channel assignment, user grouping and power control.

11.5   CCI mitigating receivers for MIMO

We now consider only frequency flat channels for the MIMO case. The analysis presented above where each CCI source has one antenna is extended to cover the case in

Table 11.2. *Receivers for CCI mitigation – frequency selective channels*

	Diversity order	Noise enhancement	Channel knowledge
ST-MLSE	$M_R L_{eff}$	Nil	$\mathbf{H}_0, \mathbf{H}_i$
ST-MMSE	$< (M_R - N)L_{eff}$	Medium	$\mathbf{H}_0, \mathbf{R}_{\mathcal{YY}}$
ST-MMSE-ML	$\approx M_R L_{eff} - N$	Low	$\mathbf{H}_0, \mathbf{R}_{\mathcal{YY}}$

which each CCI source has multiple antennas (M_T) with MIMO encoding. The M_T antennas at each co-channel source should be treated as M_T independent interferers. This implies that for each CCI source, we need M_T antennas at the receiver to cancel it and in general the receiver will need $M_R = M_T N + M$ antennas at the receiver ($M_T N$ for CCI and M for desired diversity for the user signal). However, in cases in which the co-channel users employ space-time diversity coding, this requirement can be reduced. We discuss one example below.

11.5.1 Alamouti coded signal and interference ($M_T = 2$)

Consider the case in which there is one signal and one co-channel source with two transmit antennas each, both using Alamouti coding. A simplistic approach would require $M_R = 3$ receive antennas, two to cancel the interferer and one for signal reception. Intelligent processing [Naguib *et al.*, 1998a] can reduce this requirement to $M_R = 2$ antennas.

Consider an $M_T = 2$ (desired signal, CCI) and $M_R = 2$ model (see Fig. 11.8). We expand the model in Eq. (5.37) to include an interfering source to get

$$\mathbf{y} = \sqrt{\frac{E_{s,0}}{2}}\mathbf{H}_{0,eff}\mathbf{s}_0 + \sqrt{\frac{E_{s,1}}{2}}\mathbf{H}_{1,eff}\mathbf{s}_1 + \mathbf{n}, \qquad (11.15)$$

where $\mathbf{y}$ and $\mathbf{n}$ are 4×1 received signal and noise vectors (the vector index is a mixture of the receive antenna index and the Alamouti symbol index), $\mathbf{H}_{0,eff}$ and $\mathbf{H}_{1,eff}$ are orthogonal (i.e., $\mathbf{H}_{0,eff}^H\mathbf{H}_{0,eff} = (1/2)\|\mathbf{H}_{0,eff}\|_F^2\mathbf{I}_2$, $\mathbf{H}_{1,eff}^H\mathbf{H}_{1,eff} = (1/2)\|\mathbf{H}_{1,eff}\|_F^2\mathbf{I}_2$) matrices with dimension 4×2, $\mathbf{s}_0$ and $\mathbf{s}_1$ are the 2×1 signal and interference vectors and $E_{s,0}$ and $E_{s,1}$ are the signal and interference powers.

The MMSE CCI canceling receiver can be formulated as

$$\mathbf{G} = \arg\min_{\mathbf{G}} \|\mathbf{G}\mathbf{y} - \mathbf{s}_0\|_F^2, \qquad (11.16)$$

where $\mathbf{G}$ is a 2×4 filter matrix. In the absence of interference, clearly we should choose $\mathbf{G} = \mathbf{H}_{0,eff}^H$ which delivers the component symbols of $\mathbf{s}_0$ with gain $\|\mathbf{H}_{0,eff}\|_F^2$ and fourth-order diversity as discussed in Chapter 5. In presence of interference, we can show that the MMSE solution is

$$\mathbf{G} = \sqrt{\frac{E_{s,0}}{2}}\mathbf{H}_{0,eff}^H\mathbf{R}_{yy}^{-1}. \qquad (11.17)$$

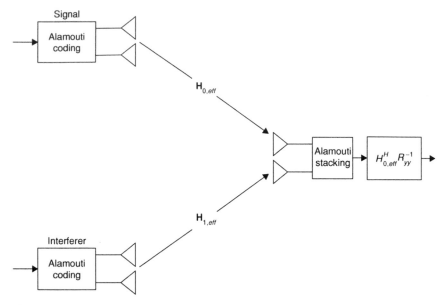

Figure 11.8: MIMO interference cancellation for Alamouti coded interference.

This receiver cancels the Alamouti coded ($M_T = 2$) interferer with just one receive antenna leaving the other antenna to support Alamouti coded signal reception. At high SIR, the performance of the receiver in Eq. (11.16) converges to the standard Alamouti scheme with fourth-order diversity. At low–medium SIR, the receiver will provide second-order diversity (see [Naguib *et al.*, 1998a] for more details).

In general, we can show that the number of receive antennas required per interferer is not the number of transmit antennas per interferer, but rather the effective spatial rate (see Chapter 6) of the signaling scheme employed in the system. Therefore, if the system uses SM ($r_s = M_T$), we need M_T antennas at the receiver to cancel each CCI source whereas with OSTBC ($r_s = 1$), we need one receive antenna per CCI source.

11.6 CCI mitigation on transmit for MISO

As pointed out in the introduction to this chapter, CCI mitigation in the forward link is a technique mostly used by the base-station to avoid generating CCI for the vulnerable user. The general approach is pre-filtering or beamforming at transmit to eliminate or reduce CCI generation at the interferee while maximizing signal power at the desired user.

11.6.1 Transmit-MRC or matched beamforming

If the reference base-station has knowledge of the channel $\mathbf{h}_0$ to its own user, then the transmit-MRC technique is possible. The beamforming weight vector $\mathbf{w}_0$ is

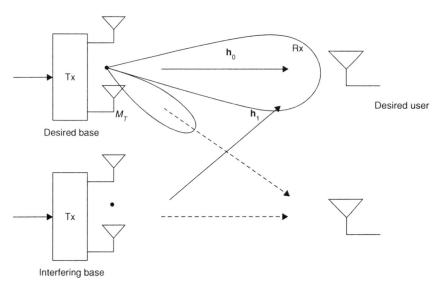

Figure 11.9: Transmit beamforming may give rise to intercell interference.

given by

$$\mathbf{w}_{MRC,0} = \mathbf{h}_0^H / \sqrt{\|\mathbf{h}_0\|_F^2}, \qquad (11.18)$$

and the SINR to the reference user becomes

$$SINR_0 = \frac{E_{s,0}\|\mathbf{h}_0\|_F^2}{\sum_{i=1}^{N} E_{s,i} \left\|\mathbf{h}_i \mathbf{w}_{MRC,i}\right\|_F^2 + N_o}. \qquad (11.19)$$

We assume that all other interfering base-stations also perform transmit-MRC beamforming to their respective users. Clearly no attempt is made to reduce interference (see Fig. 11.9). However, with an increasing number of antennas M_T, the beampattern will become spatially selective. Several options are possible: (a) regular beams, which need compact arrays and no scattering, (b) irregular amoeba-shaped beams due to regular beams reshaped by scattering, and (c) beams with heaving grating lobes due to the large antenna spacing. In all cases, the average SINR will improve as M_T increases. The diversity in the forward link is maintained at order M_T assuming channel $\mathbf{h}_0 = \mathbf{h}_w$, where $\mathbf{h}_i$ is the channel from the ith base to the desired user.

11.6.2 Transmit ZF or nulling beamformer

A transmit ZF beamformer places nulls in the direction of the interfered users thereby ensuring no CCI is delivered to these users. Let $\mathbf{h}_0$ be the usual $1 \times M_T$ channel from base 0 to the desired user. Adopting a different notation from Section 11.3.3, let $\mathbf{h}_i$ be the channel from base 0 to ith, interfering user (cell). Then, following Eq. (10.13) with $M_T \geq N + 1$, where $\mathbf{H}$, a matrix with rows $\mathbf{h}_0, \mathbf{h}_1, \ldots, \mathbf{h}_N$, is full row rank, the

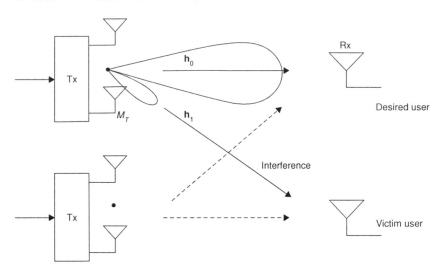

Figure 11.10: Schematic of a nulling beamformer. Nulls are formed in the direction of the victim users by exploiting differences in spatial signatures.

$M_T \times 1$ prefiltering vector $\mathbf{w}_{ZF,0}$ is given by

$$\mathbf{w}_{ZF,0} = \frac{\mathbf{h}_0^{(\dagger)}}{\sqrt{\|\mathbf{h}_0^{(\dagger)}\|_F^2}}, \qquad (11.20)$$

where $\mathbf{h}_0^{(\dagger)}$ is the first column of $\mathbf{H}^\dagger$. If another user's channel (say $\mathbf{h}_1$) comes close to the desired user's channel, $\|\mathbf{h}_0^{(\dagger)}\|_F^2$ will become large and the desired user will receive very little power (power reduction) as we saw in Chapter 10. Figure 11.10 depicts a ZF beamformer shown only at the desired base. Note that pre-coding such as a vector extension of the Tomlinson–Harashima technique is not applicable (at least in the low SINR case) since the interference generating base and the serving base are different. The diversity offered by the ZF beamformer will be $M_T - N$.

11.6.3 Max SINR beamforming with coordination

A max SINR transmit beamformer trades the signal power delivered to the desired user for interference generated to other users. If power co-ordination is possible, the overall performance can be further enhanced by adjusting the transmit powers to minimize mutual interference.

Let $\mathbf{w}_i$, $E_{s,i}$ be the pre-filter weight vector and transmit power at the ith user. Let $\mathbf{h}_i$ be the channel from the ith base to the reference user (index 0). Therefore, $\mathbf{h}_0$ is the channel from the reference (index 0) base. If all the bases exchange weight vectors, powers and the channels from themselves to other users, a solution similar to Eq. (10.14) can be formulated. The SINR at the reference user is given by

$$\text{SINR}_0 = \frac{E_{s,0} \|\mathbf{h}_0 \mathbf{w}_0\|_F^2}{\sum_{i=1}^{N} E_{s,i} \|\mathbf{h}_i \mathbf{w}_i\|_F^2 + N_o}. \tag{11.21}$$

A centralized processor will compute the SINR for each base and then search for an optimum $\mathbf{w}_i$ and $E_{s,i}$ such that these SINRs equal or exceed a target SINR, subject to the constraint $E_{s,j} \leq E_{peak,j}$. Again, this is a non-convex problem and a solution may or may not exist. The diversity offered by the SINR balancing beamformer will be $M_T - N$. The problem can also be formulated when the transmit power is kept constant (no power control) and only the beamforming weights $\mathbf{w}_i$ are adaptable.

In a non-co-operative situation, each user will transmit full power with a transmit filter matched to the channel. For OFDM modulation, there is a further degree of freedom of power allocation across tones, and a non-co-operative strategy leads to a Nash equilibrium type solution [Saraydar et al., 2002].

11.7   Joint encoding and decoding

If there is full (bits or signals level) co-ordination between base-stations, i.e., the different base-stations work as a single base with centralized processing, the problem is no longer that of an interference channel and will reduce to that of a broadcast or multiple access channel for the forward and reverse links respectively. The base-stations can now use joint encoding (interference pre-subtraction with pre-coding) for the forward link and joint decoding (ML) reception for the reverse link. However, the transmit power per base on the forward link has to be individually constrained and a sum constraint would not be logical.

Note that the MIMO channel between all users and base stations needs to be known, which is a difficult task. Also note that the channel cannot be modeled as standard IID, due to significant differences of signal powers at different stations, with the highest power available on the serving link (reference base station to reference user). The reverse link problem reduces to a standard MIMO-MAC whose capacity region is well understood. Some results for the down link proposed [Shamai and Zaidel, 2001] DPC. However, the per base station power constraint invalidates the duality results between MIMO-MAC and MIMO-BC discussed in Chapter 10. Also, the relationship between the dirty paper region and the sum rate capacity is not yet fully established.

11.8   SS modulation

In SS modulation (CDMA), interference from other users can be either MAI (from within the same cell) or CCI interference from other cells. All interferers have distinct spreading codes (time signatures) and this can be exploited to reduce interference.

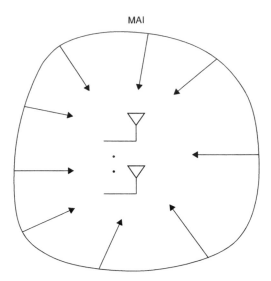

Figure 11.11: Quasi-isotropic interference field caused by large number of interferers.

Designing receivers that mitigate MAI has been a major research area [Lupas and Verdu, 1989; Madhow and Honig, 1994; Rapajic and Vucetic, 1994; Abdulrahman *et al.*, 1994; Verdu, 1998]. The techniques parallel those used in Chapter 7 for handling MSI with different spatial signatures. Typically, there are a large number of co-channel users $(0.1K_S–0.25K_S)$. CCI interference is best treated as white noise.

With the introduction of the spatial dimension, the problem of dealing with MAI becomes one of working with ST signatures as opposed to time-only signatures in CDMA. Also, interferers are dispersed in angle as seen from the base-station and therefore in general have different spatial signatures. However, CCI which was temporally white can be spatially colored and adaptive spatial processing can be useful.

If we assume perfect autocorrelation of spreading codes, we can ignore interchip interference (ICI). Furthermore, if we model MAI as white noise, the problem of ST CDMA processing becomes one of dealing with temporally white and spatially colored noise. In this case MAI and CCI become indistinguishable. We discuss a ST receiver and transmitter structure for this special case below. Since there are a large number of co-channel users (due to multiple code channels), these usually far exceed the number of antennas and therefore appear to arrive from all directions to generate a smooth noise field (see Fig. 11.11).

11.8.1 ST-RAKE

We formulate a simple ST receiver for an idealized interference model, where we assume zero ICI and model MAI and CCI as temporally additive white noise at the receiver. We follow the notation in Section 11.4 and first consider the case of frequency flat fading. If $\mathbf{R_{yy}}$ is the spatial covariance matrix of signal plus interference plus noise at the receiver, and $\mathbf{h}_0$ is the channel for the desired signal (assuming no multipath for now), the MMSE receiver is given by

$$\mathbf{g}_{MMSE} = E_{s,0}\mathbf{h}_0^H \mathbf{R_{yy}}^{-1}. \tag{11.22}$$

In the presence of multipath, assume that the channel has two paths separated by one chip delay with the spatial channels corresponding to delays 0 and 1 given by $\mathbf{h}_0[0]$ and $\mathbf{h}_0[1]$ respectively. Since the spreading sequence is assumed to have perfect autocorrelation the two paths can be perfectly resolved at the receiver via a correlator with no ICI. We assume $\mathbf{R_{yy}}$ is the same for both lags and that the received signal and noise for each path are uncorrelated. The output of the ST-RAKE is given by

$$z = \mathbf{g}[0]\mathbf{y}[0] + \mathbf{g}[1]\mathbf{y}[1], \tag{11.23}$$

where $\mathbf{y}[0]$, $\mathbf{y}[1]$ are the vector outputs $(M_R \times 1)$ of the correlators at lags 0 and 1 respectively and $\mathbf{g}[0]$, $\mathbf{g}[1]$ are the weight vectors $(1 \times M_R)$ for lags 0 and 1 respectively and are given by

$$\mathbf{g}[j] = E_{s,0}\mathbf{h}_0[j]^H \mathbf{R_{yy}}^{-1}, \ \ j = 1, 2. \tag{11.24}$$

We assume that $\mathbf{R_{yy}}$ is the same at both lags. The ST-RAKE offers an array gain of M_R and a diversity gain of $2M_R$ with two paths. The channels $\mathbf{h}_0[i]$ can be estimated using pilot symbols or codes. Blind methods have also been developed [Naguib, 1996].

The ST-RAKE is a SIMO generalization of the standard RAKE receiver. The ST-RAKE can be improved by explicitly modeling the ICI which has been assumed to be zero. This can be done by expanding the signal model (see Section 9.5) appropriately [Ramos *et al.*, 2000]. The next step in generalizing the model is to allow for intercode interference (i.e., MAI). This leads to the ST linear receivers studied by [Papadias *et al.*, 1998; Huang *et al.*, 1999; Huaiyu and Poor, 2001].

11.8.2 ST pre-RAKE

The transmit analog of a ST-RAKE is a ST pre-RAKE [Esmailzadeh and Nakagawa, 1993; Montalbano *et al.*, 1998], which is a matched transmit beamformer. Consider a channel with M_T transmit antennas and a single receive antenna. Assuming two paths with one chip delay and no ICI, let $\mathbf{h}[0]$ and $\mathbf{h}[1]$ be the MISO channels associated with the two paths. The pre-RAKE output at the transmit ($\tilde{\mathbf{s}}[i]$) antennas (see Eq. (9.26)) after spreading is given by

$$\tilde{\mathbf{s}}[i] = \frac{s}{\sqrt{2}}\left(c[i]\frac{\mathbf{h}_0[0]^H}{\|\mathbf{h}_0[0]\|_F} + c[i+1]\frac{\mathbf{h}_0[1]^H}{\|\mathbf{h}_0[1]\|_F}\right), \tag{11.25}$$

where s is the data symbol and $c[i]$ is the spreading code. In other words we transmit with beamforming onto each path with appropriate delay pre-correction and phasing such that the paths combine at the user antenna at the same delay in phase to provide ST maximum ratio combining. This also reduces (but does not eliminate) the delay spread at the receiver. In general it is very hard to estimate path phases at the transmitter (i.e., $\mathbf{h}_0[i]$ will be known to only within a phase ambiguity, see Chapter 3). In this case delay pre-correction will perform very poorly since the arriving paths may combine destructively robbing diversity. In this case we should employ transmit beamforming but without delay pre-correction, so that the physical channel paths remain distinct and can be used by the receiver's RAKE for diversity. This scheme still provides the advantage of transmit beamforming for each path. The ST pre-RAKE offers an array gain of M_T and a diversity gain of $2M_T$ at the receive antenna with delay pre-correction and co-phasing.

Discussion

The spatial dimension offers significant advantages in CDMA and we discuss some of these below. They arise from differences in properties between the codes and the spatial dimension for mitigating CCI and MAI. The leverage of the spatial dimension offers a rich source of diversity $(M_T M_R)$. The spreading code offers minimal (B/B_c) frequency (path) diversity. The potential leverage of MAI reduction using complex receivers (linear, ML, etc.) over a simple MF (or RAKE) receiver is limited in CDMA networks. This is because about 30–40% of the total interference (CCI) comes from outside the user's cell and can realistically be treated only as noise. Also, the interference to noise ratio has to be kept between 0 and 3 dB, to maintain power control stability in CDMA networks. Therefore, even if we eliminate all intracell MAI via advanced receivers, the total SINR can improve only by 3–4 dB. Note that temporal processing cannot provide gain against CCI or thermal noise. In the spatial dimension, however, the leverage of antennas is not limited, is proportional to the number of antennas and is applicable to all interfering sources, i.e., MAI, CCI and noise. In addition when using long codes in CDMA, the interference signature changes from symbol to symbol, making the receiver filter design change from symbol to symbol and very complex. On the other hand the signature in the spatial dimension is stable at least over the span of a coherence time T_C. Finally, since there are only a few antennas in the array, calculating the optimum interference canceling receiver is computationally much simpler. The corresponding dimension in the temporal domain is K_S.

There are also significant differences in the robustness of ST techniques between CDMA and non-spread modulation. Consider the reverse link. In CDMA, MAI and CCI will both appear as a large number of relatively weak (post-despreading) interferers dispersed in angle (spatially white). Therefore, in CDMA, a simple matched beamformer will offer MAI suppression proportional to the number of antennas. In non-spread modulation (TDMA), there is usually one or, at maximum, two (but strong)

CCI sources and the receiver now needs to null or at least strongly suppress such sources. Nulling is very error sensitive while matched-filtering is not. This leads to more robust ST receivers for CDMA. Similar comments apply to the forward link, where channel estimates are often poor, making the robustness advantage of CDMA even more valuable.

11.9   OFDM modulation

The ST CCI techniques strongly parallel those used in SC modulation discussed earlier in this chapter. OFDM is used with a time division multiplexing (TDM)/time division multiple access (TDMA) system and therefore we can expect only one or two interferers as opposed to many in CDMA. In frequency selective channels, we should expect the signal and CCI channels to vary across the OFDM tones and the CCI mitigation principles developed for the frequency flat channel have to be applied to each tone separately. The waterpouring solution will vary depending on the type of pre-filter used and whether we adopt a co-operative or competitive strategy. The interested reader is referred to [Li and Sollenberger, 1999; Kapoor *et al.*, 1999; Li *et al.*, 2001; Stamoulis *et al.*, 2002] for a discussion.

11.10   Interference diversity and multiple antennas

Canceling CCI using multiple antennas is a powerful way of mitigating CCI. However, it does come at the cost of complexity and also, as seen later (see Section 12.4.1), signal diversity. Another complementary approach to reducing the effect of CCI is known as interference diversity.

In a cellular system, there is wide variation of SIR from user to user depending on user and interferer locations, power levels and fade amplitudes. Clearly any SIR above the target value (say 9 dB in GSM) is of little value to that user; but a SIR below the target can result in a poor or even unusable link. Therefore, the goal of the designer should be to ensure that the SIRs are the same for all users. While signal diversity and power control reduce signal variability, these by themselves, are often not adequate to reduce SIR variability. We briefly discuss interference diversity in TDMA and CDMA which reduces interference variability. In TDMA networks, a number of techniques can be used to reduce the interferer's variability and bring the SIR as close as possible to its mean value. Cell-to-cell randomized frequency hopping and power control based on SIR are commonly used techniques. Use of multiple antennas at the receiver is yet another technique. If the interference exhibits independent space selective fading, the antennas become a source of interference diversity. We discuss a simple example below to demonstrate the value of interference diversity for SC modulation.

12

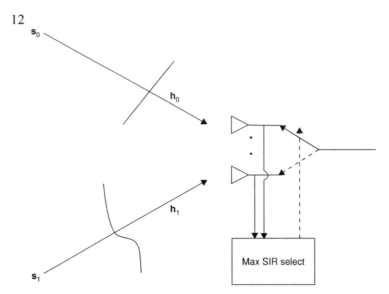

Figure 11.12: Signal amplitude is constant across the array. Interference amplitude has IID fading across the array.

We assume a frequency flat Rayleigh channel for both signal and interference (single interferer). The SISO signal model is given by

$$y = \sqrt{E_{s,0}}h_0 s_0 + \sqrt{E_{s,1}}h_1 s_1 + n. \tag{11.26}$$

The receiver is equipped with M_R antennas and has antenna selection capability (see Fig. 11.12), where only one antenna is chosen out of M_R antennas. We assume the signal amplitude is constant across the receive antennas and that the interference varies across the antennas. Given that all receive antennas have the same signal amplitude, the antenna with the minimum interference power is selected for reception. Figure 11.13 plots four different scenarios: (a) $M_T = M_R = 1$, no interference (which is replaced by noise) – the SER curve corresponds to the standard Rayleigh curve, (b) $M_T = M_R = 1$, interference only – the SER curve coincides with the Rayleigh curve, (c) $M_T = M_R = 2$, interference only, receive selection, we get ≈ 3 dB gain, (d) $M_T = M_R = 4$, interference only, receive selection, we get ≈ 5 dB gain.

Observe, that though we use the term "interference diversity", the slope of the curves does not change with multiple antennas. Performance improvement is visible in the form of a SIR gain alone. Increasing the number of receive antennas increases interference diversity and improves effective SIR.

It is realistic to assume that the signal also exhibits space selective fading. This will result in signal diversity gain (increased slope of the SER curve) as well as SIR gain due to interference diversity.

The above remarks also carry over to CDMA. However, CDMA enjoys some inherent advantages. On the reverse link, we have already noted that there are a large number

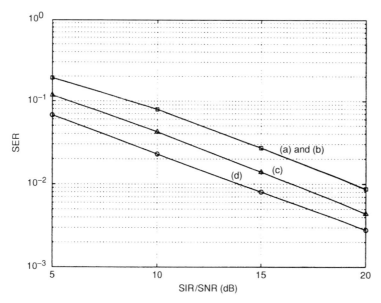

Figure 11.13: Interference diversity through receive antenna selection offers SIR gain: (a) no interferer; (b) one interferer, no diversity; (c) one interferer, selection with $M_R = 2$; (d) one interferer, selection with $M_R = 4$.

(15–20) of interferers. Therefore, this provides an implicit averaging effect and controls interference variability. On the forward link, once again, the interferers are the base-stations, emitting a pilot and a large number of user signals, each with a different power level. This also provides a degree of interference averaging.

12 Performance limits and tradeoffs in MIMO channels

12.1 Introduction

The previous chapters motivated and discussed various ST signaling schemes with the emphasis on improving error rate performance. In this chapter we examine the performance limits of ST wireless links and study the performance tradeoffs of specific signaling and receiver schemes, when the channel is unknown to the transmitter. We consider a block fading channel mode where the channel is drawn from a random (fading) distribution but remains constant over the entire transmitted codeword (packet). For such channels we establish a link between the packet error rate (PER) and outage capacity and begin by examining the PER vs signaling rate vs SNR tradeoffs in fading channels. We show how this leads to fundamental limits on error performance and answers questions about the diversity vs spatial multiplexing tradeoff that arises in MIMO systems. We then examine the spectral efficiency of signaling schemes such as OSTBC and SM with horizontal encoding, and with receivers such as MMSE or OSUC. Finally we discuss practical issues in system design in light of the performance metrics explored in this chapter.

12.2 Error performance in fading channels

We begin by discussing why a transmitter in the absence of instantaneous channel knowledge cannot achieve zero packet error performance for any non-zero signaling rate in a block fading channel. By block fading we imply that the channel is randomly drawn from a given distribution and is then held constant for every channel use across the length of the transmitted codeword. Note that there is a significant difference between the concept of capacity for fading and non-fading channels. In a non-fading (say AWGN) channel error-free transmission is guaranteed if we signal (through optimal coding) at a rate equal to or below the channel capacity. In a block fading MIMO channel, we recall that for any non-zero signaling rate there is always a finite probability that the channel is unable to support it. We recall that the $x\%$ outage capacity is the rate that the channel

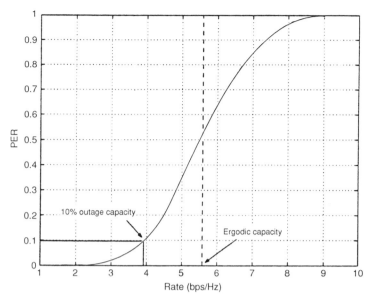

Figure 12.1: PER (outage probability) vs rate, SNR = 10 dB, $M_T = M_R = 2$, $\mathbf{H} = \mathbf{H}_w$. 10% PER corresponds to a signaling rate of approximately 3.9 bps/Hz.

can support with a $(100 - x)\%$ probability. If we use very large block (packet) size, and optimal coding, the PER performance will be binary – the packet is always decoded successfully if the channel supports the rate and is always in error otherwise. Therefore, if the transmitter does not know the channel, the PER will equal the outage probability for that signaling rate (outage capacity). We can infer the PER from the information rate CDF curve computed for the given average SNR. Note that PER is the actual error probability of a codebook and differs from the PEP discussed in Chapter 6, which is the probability of error between two codewords constituting the code.

12.3   Signaling rate vs PER vs SNR

We now examine the fundamental limits on signaling rate vs PER vs SNR and study tradeoffs between these parameters. We illustrate this with the following example. Consider a MIMO channel, $\mathbf{H} = \mathbf{H}_w$, $M_T = M_R = 2$, and the case in which the transmitter has no knowledge of the channel except for the SNR, ρ. A reasonable strategy for the transmitter is to compute the information rate CDF and choose the signaling rate for which the PER (i.e., outage probability) is at the desired level. The pairwise relationship between these parameters (signaling rate, PER and SNR) is discussed next.

Figure 12.1 plots the PER vs rate for a fixed average SNR, $\rho = 10$ dB. This is the familiar information rate CDF curve from Chapter 4. Figure 12.2 plots the PER vs SNR

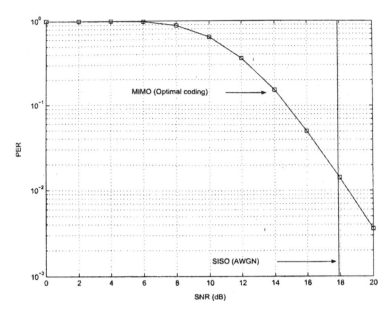

Figure 12.2: PER (outage probability) vs SNR, rate = 6 bps/Hz, $M_T = M_R = 2$, $\mathbf{H} = \mathbf{H}_w$. We get fourth-order diversity at high SNR.

for a fixed rate of 6 bps/Hz. Notice that this curve implies that the PER cannot be zero and that it depends on the average SNR much like PER or BER in uncoded (or sub-optimally coded) AWGN channels. The magnitude of the slope of the PER curve plotted on a log–log scale has been shown to be $M_T M_R$ [Zheng and Tse, 2001]. This indicates that for fixed rate transmission, optimal coding delivers full $M_T M_R$ order spatial diversity. In comparison, the PER curve for a SISO AWGN channel with a signaling rate of 6 bps/Hz is a vertical line at $\rho = 18$ dB, i.e., an error is always made (assuming asymptotic block size) if we attempt to transmit at 6 bps/Hz on the SISO AWGN channel when $\rho < 18$ dB. In other words, below 18 dB, a fading channel has better PER than an AWGN channel.

Figure 12.3 plots the outage capacity vs average SNR for a fixed PER of 10%. We notice that at high SNR the outage capacity increases by $M_T = M_R = 2$ bps/Hz for every 3 dB increase in SNR. The magnitude of the outage capacity vs SNR curve is $\min(M_R, M_T)$ bps/Hz/3 dB. This is the multiplexing gain offered by the channel. We can conclude that with optimal coding and a fixed PER, an increase in SNR may be leveraged to increase the transmission rate at $\min(M_T, M_R)$ bps/Hz/3 db slope.

Figure 12.4 plots a three-dimensional surface of the PER vs rate vs average SNR. The surface represents the fundamental limit of fading channels, assuming optimal coding and a large enough block size. An optimal coding scheme achieves this surface. The region to the right of this surface is the achievable region, where practical signaling and receivers lie. We have seen that with optimal coding, for a given transmission rate,

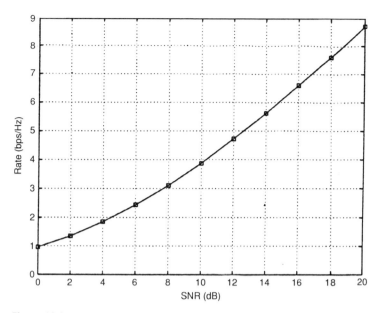

Figure 12.3: Rate vs SNR, PER = 10%, $M_T = M_R = 2$, $\mathbf{H} = \mathbf{H}_w$. The capacity increase is linear with second-order diversity.

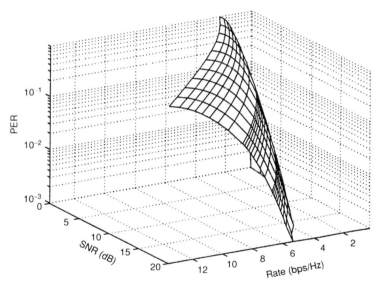

Figure 12.4: Optimal signaling limit surface (PER vs rate vs SNR), $M_T = 2$, $M_R = 2$, $\mathbf{H} = \mathbf{H}_w$. The achievable region is to the right of the surface.

we can trade SNR for PER at $M_T M_R$ slope (full diversity gain), and conversely for a fixed error rate, we can trade SNR for transmission rate at min (M_T, M_R) bps/Hz/3 dB slope (full multiplexing gain). Sub-optimal coding schemes, as we shall see later in this chapter, do not achieve the same tradeoffs.

12.4   Spectral efficiency of ST coding/receiver techniques

As is evident from previous discussions, the specific ST coding or ST receivers discussed in Chapters 6 and 7 respectively may not be optimal in the sense that they may limit the maximum achievable information rate. We now discuss the maximum achievable information rate (spectral efficiency[1]) of different ST coding and receiver schemes. Note that since we are using the information rate as a metric, we are implying that the limiting rates are achieved through optimal channel coding. First we show that DE with MMSE based layer peeling at the receiver (see Chapters 6 and 7 respectively), also referred to as D-BLAST [Foschini and Gans, 1998], is an optimal scheme.

12.4.1   D-BLAST

The MIMO channel capacity for a deterministic channel unknown to the transmitter is

$$C = \log_2 \det \left(\mathbf{I}_{M_R} + \frac{\rho}{M_T} \mathbf{H}\mathbf{H}^H \right), \tag{12.1}$$

which can be expanded as

$$
\begin{aligned}
C &= \log_2 \det \left(\mathbf{I}_{M_R} + \frac{\rho}{M_T} \mathbf{h}_{M_T} \mathbf{h}_{M_T}^H + \frac{\rho}{M_T} \mathbf{H}_{(M_T)} \mathbf{H}_{(M_T)}^H \right) \\
&= \log_2 \det \left(\mathbf{I}_{M_R} + \frac{\rho}{M_T} \mathbf{H}_{(M_T)} \mathbf{H}_{(M_T)}^H \right) \\
&\quad + \log_2 \left(1 + \frac{\rho}{M_T} \mathbf{h}_{M_T}^H \left(\mathbf{I}_{M_R} + \frac{\rho}{M_T} \mathbf{H}_{(M_T)} \mathbf{H}_{(M_T)}^H \right)^{-1} \mathbf{h}_{M_T} \right),
\end{aligned}
\tag{12.2}
$$

where $\mathbf{H}_{(M_T)}$ is the channel matrix with the M_Tth antenna (column) removed. Repeating this procedure, the capacity may be expressed as [Varanasi and Guess, 1997]

$$C = \sum_{i=1}^{M_T} \log_2 \left(1 + \frac{\rho}{M_T} \mathbf{h}_i^H \left(\mathbf{I}_{M_R} + \frac{\rho}{M_T} \mathbf{H}_{(i)} \mathbf{H}_{(i)}^H \right)^{-1} \mathbf{h}_i \right), \tag{12.3}$$

where $\mathbf{h}_i$ is the ith column of $\mathbf{H}$ and $\mathbf{H}_{(i)}$ is the matrix obtained by removing columns with indices $i, i + 1, \ldots, M_T$ from $\mathbf{H}$ ($\mathbf{H}_{(1)}$ is an empty matrix).

Recall that in D-BLAST transmission and reception discussed in Chapters 6 and 7 (see Fig. 6.10), each layer, consisting of several frames/time slots, is transmitted over all M_T transmit antennas through a stream rotator. The frames corresponding to any one layer (given that the previous layers are already decoded) are extracted one after the

---

[1] For ease of presentation we shall continue to abuse the terminology and sometimes refer to this as "capacity".

other using an MMSE receiver. The frames are recombined into one composite frame (corresponding to the entire layer) and optimally decoded.

The SINR for the ith frame with MMSE reception is given by

$$\frac{\rho}{M_T}\mathbf{h}_i^H \left(\mathbf{I}_{M_R} + \frac{\rho}{M_T}\mathbf{H}_{(i)}\mathbf{H}_{(i)}^H\right)^{-1} \mathbf{h}_i, \tag{12.4}$$

so that the information rate associated with that frame, C_i, is given by

$$C_i = \log_2 \left(1 + \frac{\rho}{M_T}\mathbf{h}_i^H \left(\mathbf{I}_{M_R} + \frac{\rho}{M_T}\mathbf{H}_{(i)}\mathbf{H}^H(i)\right)^{-1} \mathbf{h}_i\right). \tag{12.5}$$

Further, the noise is independent across the frames, so that the information rate of any layer/stream (assuming coding across frames) is

$$C_{layer} = \frac{1}{M_T}\sum_{i=1}^{M_T} C_i. \tag{12.6}$$

With M_T such layers in parallel, the capacity of D-BLAST is given by

$$C_{D\text{-}BLAST} = \sum_{i-1}^{M_T} C_i = C. \tag{12.7}$$

Therefore, in a fading channel, the CDF of the capacity of the D-BLAST signaling scheme matches the CDF of the information rate of the channel. In other words, the information rate guaranteed $(100 - x)\%$ of the time for the D-BLAST scheme is equal to the $x\%$ outage capacity of the channel.

We note here that the "wasted ST" (the triangular block in Fig. 6.10) has been ignored in this discussion. The ST wastage can be significant if the frame size is not appropriately chosen.

12.4.2   OSTBC

OSTBC is a class of STBC that guarantees full $(M_T M_R)$ spatial diversity (see Chapter 6). The SNR of the decoded stream is $(\rho/M_T)\|\mathbf{H}\|_F^2$. The capacity for a given channel realization $\mathbf{H}$ is [Hassibi and Hochwald, 2001; Papadias and Foschini, 2002]

$$C_{OSTBC} = r_s \log_2 \left(1 + \frac{\rho}{M_T}\|\mathbf{H}\|_F^2\right), \tag{12.8}$$

where r_s is the spatial code rate defined in Chapter 6. The rate loss is now easily verifiable. The capacity of the MIMO channel may alternatively be represented as

$$C = \log_2 \det \left(\mathbf{I}_{M_R} + \frac{\rho}{M_T} \mathbf{H} \mathbf{H}^H \right)$$

$$= \log_2 \prod_{k=1}^{r} \left(1 + \frac{\rho}{M_T} \lambda_k \right)$$

$$= \log_2 \left(1 + \frac{\rho}{M_T} \|\mathbf{H}\|_F^2 + \frac{\rho}{M_T}^2 (.) + \dots \right)$$

$$\geq C_{OSTBC}, \tag{12.9}$$

where λ_k are the eigenvalues of $\mathbf{H}\mathbf{H}^H$. Equation (12.9) follows from $\lambda_k \geq 0$, ($k = 1, 2, \dots, r$) and $\sum_{k=1}^{r} \lambda_k = \text{Tr}(\mathbf{H}\mathbf{H}^H) = \|\mathbf{H}\|_F^2$.

We can conclude that the CDF characteristic of the channel capacity with OSTBC is poorer than that with optimal coding. By poorer we mean that the transmission rate supported by OSTBC for any given outage will be less than that for the optimal scheme. Alternatively, the outage in OSTBC for a given transmission rate will be higher than that for optimal coding.

12.4.3 ST receivers for SM

We now study how the specific receiver design affects the capacity (spectral efficiency) of MIMO links, just as we saw how ST coding schemes affect capacity. Our focus is on non-ML receivers that are sub-optimal. From Chapter 7, we know that sub-optimal receivers include MMSE and OSUC receivers. For the remainder of the section, we assume HE with optimal channel encoding for each stream.

MMSE and OSUC for SM using HE
In the MMSE receiver, the M_T streams (packets) are first separated and then decoded independently. In the OSUC receiver, the symbol streams are separated layer by layer with ordering using the MMSE receiver and optimal decoding. In both cases, the M_T decoded streams are then recombined into a single composite packet. Assume that the SINR for the kth packet with MMSE or OSUC reception is denoted by η_k. The composite packet is guaranteed to be decoded correctly only when the packet with the worst SINR is decoded correctly. Given that the channel is unknown at the transmitter, the governing statistics for the system are the statistics of the worst SINR. The capacity of the system may be written as [Papadias and Foschini, 2002]

$$C_{recv} = M_T \log_2(1 + \min(\eta_1, \eta_2, \dots, \eta_{M_T})). \tag{12.10}$$

The performances of the two receivers are compared in Figs. 12.5 and 12.6 with OSUC significantly outperforming MMSE. A reasonable explanation is as follows. At higher SNRs the weakest stream (and correspondingly also the one that limits performance) in OSUC is generally the first decoded stream. Due to the ordering

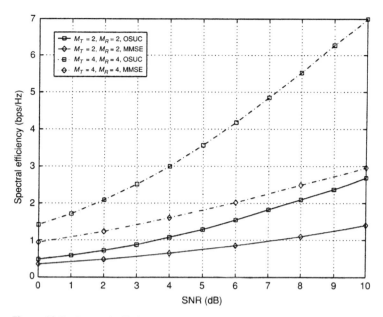

Figure 12.5: Spectral efficiency at 10% outage vs SNR for MMSE and OSUC receivers with horizontal encoding. OSUC clearly outperforms MMSE.

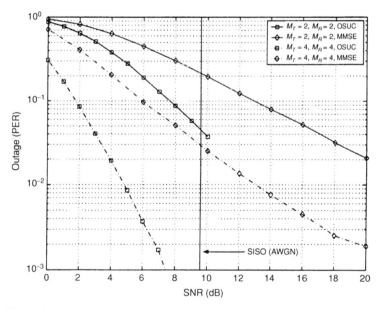

Figure 12.6: PER vs SNR for MMSE and OSUC receivers, rate = 2 bps/Hz. OSUC has higher slope (diversity) than MMSE.

process (selecting the stream with the highest SINR), the SNR statistics of the weakest stream are the statistics of the maximum SINR out of M_T different SINRs at the first layer. On the other hand, the statistics of the weakest stream for the MMSE receiver are the statistics of the minimum SINR at the first layer. The performance differential

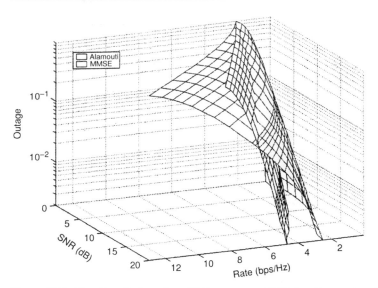

Figure 12.7: Signaling limit surface (PER vs rate vs SNR) for Alamouti coding and SM-HE with a MMSE receiver, $M_T = M_R = 2$, $\mathbf{H} = \mathbf{H}_w$. Crossover in the surfaces motivates the diversity vs multiplexing problem.

can be substantial especially for a large number of streams (transmit antennas) and is illustrated in Fig. 12.5, where the performance gap is seen to be superlinear in the number of antennas.

We end this section with a note of caution. The performance of the receivers is particularly sensitive to the encoding scheme employed at the transmitter. The development above assumes SM-HE, i.e., the transmitted streams are horizontally encoded at the same rate. This need not be the case in general. For example, the capacity of the MMSE receiver for low and medium values of SNR with vertical encoding/decoding is higher than the capacity of the OSUC receiver with horizontal encoding. See [Gore *et al.*, 2002a] for more details. Further improvements may be possible with some kind of feedback information on the channel state.

Rate vs SNR vs PER for OSTBC and SM-HE
We now consider performance limits for the schemes studied above: (a) OSTBC with optimal outer encoding and ML reception and (b) SM with optimal HE and MMSE reception. The Alamouti coding scheme is used as the OSTBC. Figure 12.7 plots the limit surfaces for the schemes. Notice that these curves lie in the achievable region of Fig. 12.4.

Figure 12.8 plots a PER vs SNR slice of Fig. 12.7 with the signaling rate kept fixed at 6 bps/Hz. The optimal curve has also been shown for comparison. Notice that the slope of the PER curve for SM is 2 and for the Alamouti curve it is 4. Further, for high PER (30%), SM outperforms Alamouti coding. However, due to the higher slope of the

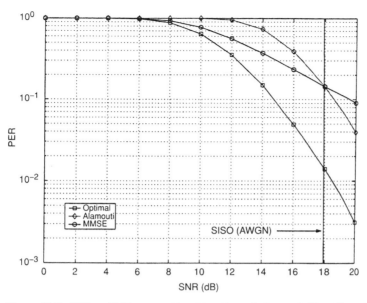

Figure 12.8: PER vs SNR, rate = 6 bps/Hz, $M_T = 2$, $M_R = 2$, $\mathbf{H} = \mathbf{H}_w$. Alamouti coding achieves fourth order diversity (optimal). SM-HE with MMSE reception has a lower slope (diversity).

Alamouti scheme, at low PER (1%) the situation reverses. Therefore for sub-optimal schemes, the question of diversity vs multiplexing is of practical relevance and the answer depends on the target PER.

12.4.4   Receiver comparison: Varying M_T/M_R

In the previous discussions, a tacit assumption was made that $M_T \leq M_R$ if SM is used (this assumption is not needed for diversity coding). In this section we explore the effect of the M_T/M_R ratio on the performance of SM with HE and compare the spectral efficiency of different receiver schemes with the channel capacity (see [Oyman *et al.*, 2002a]). Figure 12.9 plots the spectral efficiency guaranteed at 90% reliability for varying M_T/M_R for ZF and MMSE receivers with $r_s = M_T$ and $\mathbf{H} = \mathbf{H}_w$. We note that the outage capacity increases almost linearly with M_T, until we reach $M_T = M_R$, after which it improves logarithmically with M_T.

The capacity of the ZF receiver shows a maximum at $M_T/M_R = 0.4$, and its performance falls off rapidly with increasing M_T/M_R. The capacity of the MMSE receiver has a maximum close to $M_T/M_R = 0.5$ and falls off more gently at higher ratios. The spectral efficiency with ZF and MMSE receivers lies close to the optimal curve till $M_T/M_R = 0.2$. Thereafter, the performance of these receivers falls off rapidly. The performance trend of the ZF receiver can be understood by considering the interplay of the number of streams and the diversity of each stream. At low M_T/M_R the diversity order of each stream is high although the SM order is low. At higher M_T/M_R, the

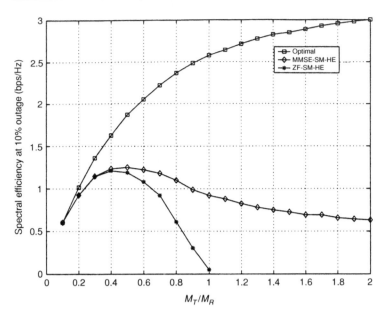

Figure 12.9: Spectral efficiency at 90% reliability vs M_T/M_R ($M_R = 10$) for various receivers with SM-HE. The optimal curve increases first linearly and then logarithmically.

diversity order of each stream is low even though the SM order is high. Note that these curves are somewhat similar to those that arise in CDMA with variable user loading [Verdu, 1998].

12.5   System design

System design for fading channels fundamentally involves a tradeoff between three system parameters – PER, rate and SNR. The space dimension in MIMO channels considerably enriches the problem by allowing a multiplicity of spatial rate schemes. Further, one scheme may outperform another for different target parameters. Given a multiplicity of options to choose from, a useful design tool would be a performance comparison of various schemes with all parameter dependencies clearly visible. One such tool could be the framework developed in this chapter. However we have assumed optimal channel coding. Sub-optimal coding and finite block lengths will change the tradeoffs discussed here and the limit curves may have to be reinterpreted as a lower bound on system performance. Figure 12.10 provides some insights into system performance when either no outer coding or sub-optimal outer coding is used. We assume a block fading channel with $M_T = M_R = 2$ and no channel knowledge at the transmitter. The three curves depict the achievable spectral density (bps/Hz) vs SNR at a PER (outage) of 10% for: (a) optimal channel coding, (b) Alamouti encoding with a

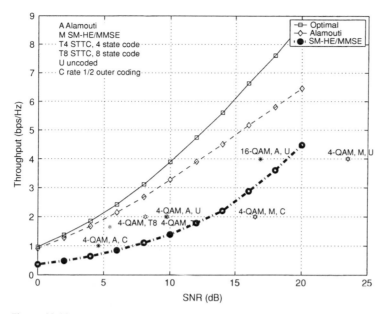

Figure 12.10: Throughput vs SNR at PER of 10%. Sub-optimal signaling causes performance loss.

ML receiver and (c) SM-HE with a MMSE receiver. The optimal curve outperforms the Alamouti scheme, which outperforms SM-HE (in the SNR region plotted). Observe that the Alamouti curve has a lower rate vs SNR slope, while the SM-HE has a better slope, indicating that it will outperform Alamouti at high enough SNR.

Figure 12.10 also plots sample performance points when either no outer coding or sub-optimal outer coding is used. We assume 100 symbols per block with target PER at 10%. Observe that coded (rate 1/2 code with hard decoding) 4-QAM/Alamouti has a 6 dB advantage over the uncoded case which has only 1 bps/Hz rate advantage. Similarly, the rate 1/2 coded 4-QAM/SM-HE with MMSE has an 8 dB gain over the uncoded system. We also plot the performance of STTC with four and eight states (presented earlier in Chapter 6). The eight-state code has an SNR advantage over the four-state encoder. Note that the performance of the four-state STTC is close to the 4-QAM Alamouti coded case. The STTC results are for a block size of 130 symbols.

Clearly there are a number of dimensions to be explored including different PER set points, SNR regions, ST coding, receiver schemes etc.

12.6   Comments on capacity

MIMO capacity depends on the degree of co-ordination allowed between antennas/users (see Fig. 12.11). If the antennas can co-ordinate both encoding and decoding, we get

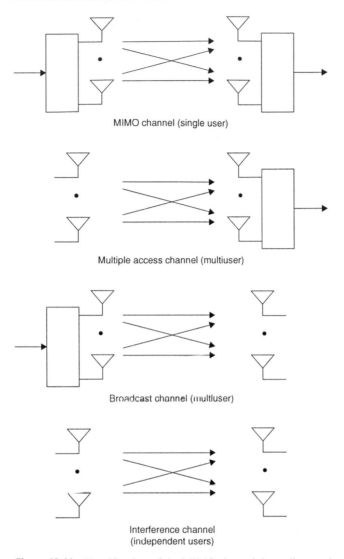

MIMO channel (single user)

Multiple access channel (multiuser)

Broadcast channel (multiuser)

Interference channel
(independent users)

Figure 12.11: Classification of the MIMO channel depending on the degree of co-ordination between antennas at transmitter and receiver.

the standard MIMO channel discussed in Chapter 4. On the other hand if co-ordination is possible in encoding but not in decoding, we get the broadcast channel discussed in Chapter 10. If co-ordination is possible in decoding but not in encoding, we get the multiple access channel also studied in Chapter 10. If no co-ordinated encoding or decoding is allowed we get the interference channel studied in Chapter 11.

Factors affecting capacity include: (a) the type of channel knowledge available at the transmitter (exact, statistical or no knowledge) which was studied in Chapters 4 and 8; (b) the power constraint at the transmitter (sum power or peak power), studied in Chapter 8 (c) fading channels; and (d) co-operative or non-co-operative power co-ordination. With some abuse of the concept of capacity (strictly speaking the spectral efficiency), the capacity also depends on the specific coding (OSTBC, SM-HE) at the transmitter and the decoding scheme (ML, OSUC) at the receiver examined earlier in this chapter.

There are a number of open problems in the capacity of ST channels. These include interference channels, broadcast channels, and situations with different assumptions on channel knowledge at the receiver/transmitter and power constraints.

References[1]

[Abdulrahman et al., 1994] M. Abdulrahman, A. Sheikh and D. Falconer. Decision feedback equalization for CDMA in indoor wireless communications. *IEEE J. Sel. Areas Comm.*, **12(4)**, 698–706, May 1994.

[Abhayapala et al., 1999] T. Abhayapala, R. Kennedy and R. Williamson. Isotropic noise modelling for nearfield array processing. *Proc. IEEE Workshop on App. of Sig. Proc. to Audio and Acoustics*, 11–14, New Paltz, NY, October 1999.

[Adachi et al., 1986] F. Adachi, M. Feeney, A. Willianson and J. Parsons. Crosscorrelation between the envelopes of 900 MHZ signals received at a mobile radio base station. *Proc. IEE*, **133(6)**, 506–512, October 1986.

[Adachi et al., 1998] F. Adachi, M. Sawahashi and H. Suda. Wideband ds-cdma for next-generation mobile communication systems. *IEEE Comm. Mag.*, **36(9)**, 56–69, September 1998.

[Agarwal et al., 1998] D. Agarwal, V. Tarokh, A. Naguib and N. Seshadri. Space-time coded OFDM for high data rate wireless communication over wideband channels. *Proc. IEEE VTC*, **3**, 2232–2236, May 1998.

[Al-Dhahir and Sayed, 2000] N. Al-Dhahir and A. Sayed. The finite-length multi-input multi-output MMSE DFE. *IEEE Trans. Sig. Proc.*, **48(10)**, 2921–2936, October 2000.

[Al-Dhahir et al., 2001] N. Al-Dhahir, A. Naguib and A. Calderbank. Finite-length MIMO decision feedback equalization for space-time block-coded signals over multipath-fading channels. *IEEE Trans. Veh. T.*, **50(4)**, 1176–1182, July 2001.

[Alamouti, 1998] S. Alamouti. A simple transmit diversity technique for wireless communications. *IEEE J. Sel. Areas Comm.*, **16(8)**, 1451–1458, October 1998.

[Andersen, 2001] J. Andersen. Constraints and possibilities of adaptive antennas for wireless broadband", *Proc. International Conference on Antennas and Propagation, ICAP*, **1**, 220–225, Manchester, UK, April 2001.

[Andrews et al., 2001] M. Andrews, P. Mitra and R. Carvalho. Tripling the capacity of wireless communication using electromagnetic polarization. *Nature*, **409**, 316–318, January 2001.

[Ariyavisitakul, 2000] S. Ariyavisitakul. Turbo space-time processing to improve wireless channel capacity. *IEEE Trans. Comm.*, **48(1)**, 1347–1359, August 2000.

[Ashikhmin et al., 2002a] A. Ashikhmin, G. Kramer and S. ten Brink. Design of LDPC codes, multi-antenna modulation and detection. *IEEE Trans. Comm.*, Submitted.

---

[1] The publisher has used its best endeavors to ensure URLs for external websites referred to in this book are correct and active at the time of going to press. However, the publisher has no responsibility for the websites and can make no guarantee that a site will remain live or that the content is or will remain appropriate.

[Ashikhmin *et al.*, 2002b] A. Ashikhmin, G. Kramer and S. ten Brink. Extrinsic information transfer functions: a model and two properties. *Proc. Conf. on Inf. Sciences and Systems, Princeton University*, NJ, March 2002.

[Austin, 1967] M. Austin. Decision-feedback equalization for digital communication over dispersive channels. *MIT Lincoln Laboratory Tech. Report No. 437*, August 1967.

[Bahl *et al.*, 1974] L. Bahl, J. Cocke, F. Jelinek and J. Raviv. Optimal decoding of linear codes for minimizing symbol error rate. *IEEE Trans. IT*, **20**, 284–287, March 1974.

[Balaban and Salz, 1992] P. Balaban and J. Salz. Optimum diversity combining and equalization in digital data transmission with applications to cellular mobile radio – part ii: Numerical results. *IEEE Trans. Comm.*, **40(5)**, 895–907, May 1992.

[Baum, 2001] D. Baum. Simulating the SUI channel models. *IEEE 802.16 BWA working group*, November 2001.

[Baum *et al.*, 2000] D. Baum, D. Gore, R. Nabar, S. Panchanathan, K. Hari, V. Erceg and A. Paulraj. Measurement and characterization of broadband MIMO fixed wireless channels at 2.5 GHz. *Proc. IEEE ICPWC*, 203–206, Hyderabad, India, December 2000.

[Bello, 1963] P. Bello. Characterization of randomly time-variant linear channels. *IEEE Trans. Comm. Syst.*, **11**, 360–393, 1963.

[Benedetto *et al.*, 1998] S. Benedetto, D. Divsalar, G. Montorsi and E. Pollara. Serial concatenation of interleaved codes: performance analysis, design and iterative decoding. *IEEE Trans. Inf. Theory*, **44(3)**, 909–926, May 1998.

[Berrou *et al.*, 1993] C. Berrou, A. Glavieux and P. Thitimajshima. Near Shannon limit error-correcting coding and decoding: Turbo-codes. *Proc. IEEE ICC*, 1064–1070, Geneva, Switzerland, May 1993.

[Bertoni, 1999] H. Bertoni. *Radio Propagation for Modern Wireless Systems*. Prentice Hall, Upper Saddle River, NJ, 1999.

[Biglieri, 2002] E. Biglieri. Performance of space-time codes for a large number of antennas. *IEEE Trans. Inf. Theory*, **48(7)**, 1794–1803, July 2002.

[Biglieri *et al.*, 1991] E. Biglieri, D. Divsalar, P. McLane and M. Simon. *Introduction to Trellis-Coded Modulation with Applications*. Macmillan, New York, NY, 1991.

[Biglieri *et al.*, 1998] E. Biglieri, J. Proakis and S. Shamai. Fading channels: Information-theoretic and communications aspects. *IEEE Trans. Inf. Theory*, **44(6)**, 2619–2692, October 1998.

[Blum and Winters, 2002] R. Blum and J. Winters. On optimum MIMO with antenna selection. *IEEE Comm. Letters*, **6(8)**, 322–324, August 2002.

[Blum *et al.*, 2001] R. Blum, L. Geoffrey, J. Winters and Q. Yan. Improved space-time coding for MIMO-OFDM wireless communications. *IEEE Trans. Comm.*, **49(11)**, 1873–1878, November 2001.

[Bölcskei and Paulraj, 2000a] H. Bölcskei and A. Paulraj. Performance of space-time codes in the presence of spatial fading correlation. *Proc. Asilomar Conf. on Signals, Systems and Computers*, **1**, 687–693, Pacific Grove, CA, November 2000.

[Bölcskei and Paulraj, 2000b] H. Bölcskei and A. Paulraj. Space-frequency coded broadband OFDM systems. *Proc. IEEE WCNC*, **1**, 1–6, Chicago, IL, September 2000.

[Bölcskei *et al.*, 2001] H. Bölcskei, A. Paulraj, K. Hari, R. Nabar and W. Lu. Fixed broadband wireless access: state of the art, challenges and future directions. *IEEE Comm. Mag.*, **39(1)**, 100–108, January 2001.

[Bölcskei *et al.*, 2002a] H. Bölcskei, D. Gesbert and A. Paulraj. On the capacity of OFDM based spatial multiplexing systems. *IEEE Trans. Comm.*, **50(2)**, 225–234, February 2002.

[Bölcskei *et al.*, 2002b] H. Bölcskei, R. Heath and A. Paulraj. Blind channel identification and equalization in OFDM based multiantenna systems. *IEEE Trans. Signal Proc.*, **50(1)**, 96–109, January
2002.

[Bouzekri and Miller, 2001] H. Bouzekri and S. Miller. Analytical tools for space-time codes over
quasi-static fading channels. *Proc. IEEE GLOBECOM*, **2**, 1118–1121, November 2001.

[Braun and Dersch, 1991] W. Braun and U. Dersch. A physical mobile radio channel model. *IEEE
Trans. Veh. Tech.*, **40(2)**, 472–482, May 1991.

[Brigham, 1974] E. Brigham. *The Fast Fourier Transform*. Prentice Hall, Englewood Cliffs, NJ, 1974.

[Brutel and Boutros, 1999] C. Brutel and J. Boutros. Euclidean space lattice decoding for joint detection in CDMA systems. *Proc. IEEE Inf. Theory and Comm. Workshop*, 129, 1999.

[Caire and Shamai, 2000] G. Caire and S. Shamai. On achievable rates in a multi-antenna broadcast downlink. *Proc. Allerton Conf. on Communication, Control and Computing*, Monticello, IL,
October 2000.

[Capon *et al.*, 1967] J. Capon, R. Greenfield and R. Kolker. Multidimensional maximum likelihood
processing of a large aperture seismic array. *Proc. IEEE*, **55(2)**, 192–211, February 1967.

[Chen and Mitra, 2001] W. Chen and U. Mitra. An improved blind adaptive MMSE receiver for fast
fading DS-CDMA channels. *IEEE Trans. Comm.*, **19(8)**, 1531–1543, August 2001.

[Chizhik *et al.*, 2000] D. Chizhik, G. Foschini and R. Valenzuela. Capacities of multi-element transmit
and receive antennas: Correlations and keyholes. *Electronic Letters*, **36(22)**, 1099–1100, June 2000.

[Chizhik *et al.*, 2002] D. Chizhik, G. Foschini, M. Gans and R. Valenzuela. Keyholes, correlations,
and capacities of multielement transmit and receive antennas. *IEEE Trans. Wireless Comm.*, **1(2)**,
361–368, April 2002.

[Chuah *et al.*, 1998] C. Chuah, J. Kahn and D. Tse. Capacity of multi-antenna array systems in indoor
wireless environment. *Proc. IEEE GLOBECOM*, **4**, 1894–1899, Sydney, Australia, November
1998.

[Chuah *et al.*, 2002] C. Chuah, D. Tse, J. Kahn and R. Valenzuela. Capacity scaling in MIMO wireless
systems under correlated fading. *IEEE Trans. Inf. Theory*, **48(3)**, 637–650, March 2002.

[Cimini and Sollenberger, 1999] L. Cimini and N. Sollenberger. Peak-to-average power ratio reduction of an OFDM signal using partial transmit sequences. *Proc. IEEE ICC*, **1**, 511–515, Vancouver,
Canada, June 1999.

[Cioffi, 2002] J. Cioffi. *Class Reader for EE379a – Digital Communication: Signal Processing*.
Stanford University, Stanford, CA. Available online at http://www.stanford.edu/class/ee379a,
2002.

[Cioffi and Forney, 1997] J. Cioffi and G. D. Forney. Generalized decision-feedback equalization
for packet transmission with ISI and Gaussian noise: *in Communication, Computation, Control
and Signal Processing*. Kluwer, Boston, MA, 1997. Edited by A. Paulraj, V. Roychowdhury and
C. Schaper.

[Cooley and Tukey, 1965] J. Cooley and J. Tukey. An algorithm for the machine calculation of the
complex Fourier series. *Math. Comp.*, **19**, 297–301, 1965.

[COST 231 TD(973) 119-REV 2 (WG2), 1991] COST 231 TD(973) 119-REV 2 (WG2). Urban transmission loss models for mobile radio in the 900- and 1,800-MHz bands. September 1991.

[Costa, 1983] M. Costa. Writing on dirty paper. *IEEE Trans. Inf. Theory*, **29(3)**, 439–441, May 1983.

[Courant and Robbins, 1996] R. Courant and H. Robbins. *What is Mathematics? An Elementary
Approach to Ideas and Methods*. Oxford University Press, Oxford, England, 2nd edition, 1996.

[Cover, 1972] T. Cover. Broadcast channels. *IEEE Trans. Inf. Theory*, **18(1)**, 2–14, January 1972.

[Cover, 1998] T. Cover. Comments on broadcast channels. *IEEE Trans. Inf. Theory*, **44(6)**, 2524–2530,
October 1998.

[Cover and Thomas, 1991] T. Cover and J. Thomas. *Elements of Information Theory*. Wiley, New York, 1991.

[Damen et al., 2000] O. Damen, A. Chkeif and J. Belfiore. Lattice code decoder for space-time codes. *IEEE Comm. Letters*, **4(5)**, 161–163, May 2000.

[Demmel, 1988] J. Demmel. The probability that a numerical analysis problem is difficult. *Math. Comp.*, **50(182)**, 449–480, April 1988.

[Ding and Li, 1994] Z. Ding and Y. Li. On channel identification based on second-order cyclic spectra. *IEEE Trans. Sig. Proc.*, **42(5)**, 1260–1264, May 1994.

[Dixon, 1994] R. C. Dixon. *Spread Spectrum Systems with Commercial Applications*. Wiley, New York, 3rd edition, 1994.

[Durgin, 2000] G. Durgin. Theory of stochastic local area channel modeling for wireless communications. PhD thesis, Virginia Polytechnic Institute and State University, December 2000.

[Edelman, 1989] A. Edelman. Eigenvalue and condition numbers of random matrices. PhD thesis, MIT, May 1989.

[El Gamal and Hammons, 2001] H. El Gamal and R. Hammons. A new approach to layered space-time coding and signal processing. *IEEE Trans. Inf. Theory*, **47(6)**, 2321–2334, September 2001.

[Eggers et al., 1993] P. Eggers, J. Tøftgård and A. Oprea. Antenna systems for base station diversity in urban small and micro cells. *IEEE J. Sel. Areas Comm.*, **11(7)**, 1046–1057, September 1993.

[Eng et al., 1996] T. Eng, N. Kong and L. Milstein. Comparison of diversity combining techniques for Rayleigh fading channels. *IEEE Trans. Comm.*, **44(9)**, 1117–1129, September 1996.

[Erceg et al., 1992] V. Erceg, S. Ghassemzadeh, M. Taylor, D. Li and D. Schilling. Urban/suburban out-of-sight propagation modeling. *IEEE Comm. Mag.*, **30(6)**, 56–61, June 1992.

[Erceg et al., 1997] V. Erceg, S. Fortune, J. Ling, A. Rustako and R. Valenzuela. Comparison of the WISE propagation tool prediction with experimental data collected in urban microcellular environments. *IEEE J. Sel. Areas Comm.*, **15(4)**, 677–684, May 1997.

[Erceg et al., 1999a] V. Erceg, L. Greenstein, S. Tjandra, S. Parkoff, A. Gupta, B. Kulic, A. Julius and R. Bianchi. An empirically based path loss model for wireless channels in suburban environments. *IEEE J. Sel. Areas Comm.*, **17(7)**, 1205–1211, July 1999.

[Erceg et al., 1999b] V. Erceg, D. Michelson, S. Ghassemsadeh, L. Greenstein, A. Rustako, P. Guerlain, M. Dennison, R. Roman, D. Barnickel, S. Wang and R. Miller. A model for the multipath delay profile of fixed wireless channels. *IEEE J. Sel. Areas Comm.*, **17(3)**, 399–410, March 1999.

[Erceg et al., 2002] V. Erceg, P. Soma, D. Baum and A. Paulraj. Capacity obtained from multiple-input multiple-output channel measurements in fixed wireless environments at 2.5 GHz. *Proc. IEEE ICC*, **1**, 396–400, New York, NY, April/May 2002.

[Erez et al., 2000] U. Erez, S. Shamai and R. Zamir. Capacity and lattice strategies for cancelling known interference. *Proc. Int. Symp. on Inf. Theory and its Applications*, Honolulu, HW, November 2000.

[Ertel et al., 1998] R. Ertel, P. Cardieri, K. Sowerby, T. Rappaport and J. Reed. Overview of spatial channel models for antenna array communication systems. *IEEE Personal Comm.*, **5(1)**, 10–22, February 1998.

[Esmailzadeh and Nakagawa, 1993] R. Esmailzadeh and M. Nakagawa. Prerake diversity combination for direct sequence spread spectrum mobile communication systems. *IEICE Trans. Comm.*, **E76-B(8)**, 1008–1015, August 1993.

[Farsakh and Nossek, 1995] C. Farsakh and J. Nossek. Channel allocation and downlink beamforming in an SDMA mobile radio system. *Proc. IEEE Int. Symp. on PIMRC*, **2**, 687–691, Toronto, Canada, September 1995.

[Feher, 1995] K. Feher. *Wireless Digital Communications*. Feher/Prentice Hall Digital and Personal Wireless Communication Series, Upper Saddle River, NJ, 1995.

[Feng and Leung, 2001] X. Feng and C. Leung. A new optimal transmit and receive diversity scheme. *Proc. IEEE PACRIM*, **2**, 538–541, Victoria, Canada, August 2001.

[Fincke and Pohst, 1985] U. Fincke and M. Pohst. Improved methods for calculating vectors of short length in a lattice, including a complexity analysis. *Mathematics of Computation*, **44**, 463–471, April 1985.

[Fitz *et al.*, 1999] M. Fitz, J. Grimm and S. Siwamogsatham. A new view of performance analysis techniques in correlated Rayleigh fading. *Proc. IEEE WCNC*, **1**, 139–144, New Orleans, LA, September 1999.

[Forney, 1972] G. D. Forney. Maximum-likelihood sequence estimation of digital sequences in the presence of intersymbol interference. *IEEE Trans. Inf. Theory*, **IT-18**, 363–378, May 1972.

[Foschini, 1996] G. Foschini. Layered space-time architecture for wireless communication in a fading environment when using multi-element antennas. *Bell Labs Tech. J.*, 41–59, 1996.

[Foschini and Gans, 1998] G. Foschini and M. Gans. On limits of wireless communications in a fading environment when using multiple antennas. *Wireless Pers. Comm.*, **6(3)**, 311–335, March 1998.

[Freedman *et al.*, 2001] A. Freedman, A. Sadri, A. Sarajedini, D. Trinkwon, E. Verbin, O. Kelman and J. Shen. Channel models for 30 km and 50 km range. *IEEE 802.16 BWA Working Group*, June 2001.

[Ganesan and Stoica, 2001] G. Ganesan and P. Stoica. Space-time block codes: a maximum SNR approach. *IEEE Trans. Inf. Theory*, **47(4)**, 1650–1656, May 2001.

[Gauthier *et al.*, 2000] E. Gauthier, A. Yongacoglu and J.-Y. Chouinard. Capacity of multiple antenna systems in Rayleigh fading channels. *Canadian J. Electr. Comp. Eng.*, **25(3)**, 105–108, July 2000.

[Gerlach, 1995] D. Gerlach. Adaptive transmitting antenna arrays at the base station in mobile radio networks. PhD dissertation, Department of Electrical Engineering, Stanford University, 1995.

[Gerlach and Paulraj, 1996] D. Gerlach and A. Paulraj. Base station transmitting arrays for multipath environments. *Signal Processing* (Elsevier Science), **54**, 59–73, 1996.

[Gesbert *et al.*, 2000] D. Gesbert, H. Bölcskei, D. Gore and A. Paulraj. MIMO wireless channels: capacity and performance prediction. *Proc. IEEE GLOBECOM*, **2**, 1083–1088, San Francisco, CA, November/December 2000.

[Giannakis *et al.*, 2000] G. Giannakis, P. Stoica and Y. Hua, editors. *Signal Processing Advances in Wireless and Mobile Communications*, Volume 2: *Trends in Single- and Multi-User Systems*. Prentice Hall, Upper Saddle River, NJ, 2000.

[Ginis and Cioffi, 2000] G. Ginis and J. Cioffi. A multi-user precoding scheme achieving crosstalk cancellation with application to DSL systems. *Proc. Asilomar Conf. on Signals, Systems and Computers*, **2**, 1627–1631, Pacific Grove, CA, October/November 2000.

[Ginis and Cioffi, 2001] G. Ginis and J. Cioffi. On the relation between V-BLAST and the GDFE. *IEEE Comm. Letters*, **5(9)**, 364–366, September 2001.

[Godavarti *et al.*, 2001a] M. Godavarti, A. Hero and T. Marzetta. Min-capacity of a multiple antenna wireless channel in a static Rician fading environment. *Proc. IEEE ISIT*, 57, Washington, DC, June 2001.

[Godavarti *et al.*, 2001b] M. Godavarti, T. Marzetta and S. Shamai. Capacity of a mobile multiple-antenna wireless link with isotropically random Rician fading. *Proc. IEEE ISIT*, 323, Washington, DC, June 2001.

[Gold, 1967] R. Gold. Optimum binary sequences for spread spectrum multiplexing. *IEEE Trans. Inf. Theory*, **14**, 154–156, 1967.

[Golden *et al.*, 1999] G. Golden, G. Foschini, R. Valenzuela and P. Wolniansky. Detection algorithm and initial laboratory results using V-BLAST space-time communication architecture. *Electron. Lett.*, **35**(1), 14–16, January 1999.

[Golub and Van Loan, 1989] G. Golub and C. Van Loan. *Matrix Computations*. Johns Hopkins University Press, Baltimore, 2nd edition, 1989.

[Gong and Letaief, 2000] Y. Gong and K. Letaief. Performance evaluation and analysis of space-time coding in unequalized multipath fading links. *IEEE Trans. Comm.*, **48**(11), 1778–1782, November 2000.

[Gore and Paulraj, 2001] D. Gore and A. Paulraj. Space-time block coding with optimal antenna selection. *Proc. IEEE ICASSP*, **4**, 2441–2444, Salt Lake City, May 2001.

[Gore and Paulraj, 2002] D. Gore and A. Paulraj. MIMO antenna sub-set selection for space-time coding. *IEEE Trans. Sig. Proc.*, **50**(10), 2580–2588, October 2002.

[Gore *et al.*, 2002a] D. Gore, A. Gorokhov and A. Paulraj. Joint MMSE vs V-BLAST and receive selection. *Proc. Asilomar Conf. on Signals, Systems and Computers*, November 2002.

[Gore *et al.*, 2002b] D. Gore, R. Heath and A. Paulraj. On performance of the zero forcing receiver in presence of transmit correlation. *Proc. IEEE ISIT*, 159, Lausanne, Switzerland, July 2002.

[Gore *et al.*, 2002c] D. Gore, S. Sandhu and A. Paulraj. Delay diversity codes for frequency selective channels. *Proc. IEEE ICC*, **3**, 1949–1953, New York, NY, April/May 2002.

[Gorokhov, 2000] A. Gorokhov. Capacity of multi-antenna Rayleigh channel with a limited transmit diversity. *Proc. IEEE ISIT*, 411, Sorrento, Italy, June 2000.

[Gorokhov, 2001] A. Gorokhov. Outage error probability for space-time codes over Rayleigh fading. *Proc. IEEE ISIT*, 242, Washington, DC, June 2001.

[Gorokhov, 2002] A. Gorokhov. Antenna selection algorithms for MEA transmision systems. *Proc. IEEE ICASSP*, **3**, 2857–2860, Orlando, FL, May 2002.

[Grant, 2002] A. Grant. Rayleigh fading multiple-antenna channels. *EURASIP J. Appl. Signal Processing*, **2002**(3), 316–329, March 2002.

[Gray, 2001] R. Gray. *Toeplitz and Circulant Matrices: A Review*. Available online at http://www.isl.stanford.edu/-gray/toeplitz.pdf, 2001.

[Gray and Goodman, 1995] R. Gray and J. Goodman. *Fourier Transforms*. Kluwer, Norwell, MA, 1995.

[Guey *et al.*, 1996] J. Guey, M. Fitz, M. Bell and W. Kuo. Signal design for transmitter diversity wireless communication systems over Rayleigh fading channels. *Proc. IEEE VTC*, **1**, 136–140, Atlanta, GA, 1996.

[Hammons and El Gamal, 2000] R. Hammons and H. El Gamal. On the theory of space-time codes for psk modulation. *IEEE Trans. Inf. Theory*, **46**(2), 524–542, March 2000.

[Hanly and Tse, 1998] S. Hanly and D. Tse. Multiaccess fading channels. I. Polymatroid structure, optimal resource allocation and throughput capacities. *IEEE Trans. Inf. Theory*, **44**(7), 2796–2815, November 1998.

[Harashima and Miyakawa, 1972] H. Harashima and H. Miyakawa. Matched-transmission technique for channels with intersymbol interference. *IEEE Trans. Comm.*, **COM-20**(4), 774–780, August 1972.

[Hassibi, 1999] B. Hassibi. An efficient square-root algorithm for BLAST. Technical report, Bell Labs, 1999. Available online at http://cm.bell-labs.com/who/hochwald/papers/squareroot/.

[Hassibi, 2000] B. Hassibi. A fast square-root implementation for BLAST. *Proc. Asilomar Conf. on Signals, Systems and Computers*, **2**, 1255–1259, November 2000.

[Hassibi and Hochwald, 2000] B. Hassibi and B. Hochwald. Optimal training in space-time systems. *Proc. Asilomar Conf. on Signals, Systems and Computers*, **1**, 743–747, Pacific Grove, CA, October/November 2000.

[Hassibi and Hochwald, 2001] B. Hassibi and B. Hochwald. High-rate codes that are linear in space and time. *IEEE Trans. Inf. Theory*, April 2001. Submitted.

[Hassibi and Marzetta, 2002] B. Hassibi and T. Marzetta. Multiple antennas and isotropically random unitary inputs: The received signal density in closed form. *IEEE Trans. Inf. Theory*, **48(6)**, 1473–1484, June 2002.

[Hassibi and Vikalo, 2001] B. Hassibi and H. Vikalo. On the expected complexity of sphere decoding. *Proc. Asilomar Conf. on Signals, Systems and Computers*, **2**, 1051–1055, Pacific Grove, CA, 2001.

[Hata and Nagatsu, 1980] M. Hata and T. Nagatsu. Mobile location using signal strength measurements in cellular systems. *IEEE Trans. Veh. Tech.*, **29(2)**, 245–251, June 1980.

[Heath and Paulraj, 1999] R. Heath and A. Paulraj. Transmit diversity using decision directed antenna hopping. *Proc. Comm. Theory Mini-Conference*, 141–145, Vancouver, Canada, June 1999.

[Heath and Paulraj, 2001a] R. Heath and A. Paulraj. Antenna selection for spatial multiplexing systems based on minimum error rate. *Proc. IEEE ICC*, **7**, 2276–2280, Helsinki, Finland, June 2001.

[Heath and Paulraj, 2001b] R. Heath and A. Paulraj. Switching between multiplexing and diversity in MIMO communication links. *IEEE Trans. Comm.*, 2001. Submitted.

[Heath and Paulraj, 2002] R. Heath and A. Paulraj. Linear dispersion codes for MIMO systems based on frame theory. *IEEE Trans. Sig. Proc.*, **50(10)**, 2429–2441, October 2002.

[Heath et al., 2001] R. Heath, H. Bölcskei and A. Paulraj. Space-time signaling and frame theory. *Proc. IEEE ICASSP*, **4**, 2445–2448, Salt Lake City, UT, May 2001.

[Hiroike et al., 1992] A. Hiroike, F. Adachi and N. Nakajima. Combined effects of phase sweeping transmitter diversity and channel coding. *IEEE Trans. Veh. Tech.*, **41(2)**, 170–176, May 1992.

[Hochwald and ten Brink, 2001] B. Hochwald and S. ten Brink. Achieving near-capacity on a multiple-antenna channel, *IEEE Trans. Comm.*, 2001, Submitted.

[Hochwald et al., 2001] B. Hochwald, T. Marzetta and C. Papadias. A transmitter diversity scheme for wideband CDMA systems based on space-time spreading. *IEEE J. Sel. Areas Comm.*, **19(1)**, 48–60, January 2001.

[Huaiyu and Poor, 2001] D. Huaiyu and H. V. Poor. Sample-by-sample adaptive space-time processing for multiuser detection in multipath cdma systems. *Proc. IEEE VTC*, **3**, 1814–1818, Atlantic City, NJ, October 2001.

[Huang et al., 1999] H. Huang, H. Viswanathan and G. Foschini. Achieving high data rates in cdma systems using blast techniques. *Proc. IEEE GLOBECOM*, **5**, 2316–2320, Rio de Janeiro, Brazil, December 1999.

[Huang et al., 2002] H. Huang, H. Viswanathan and G. Foschini. Multiple antennas in cellular CDMA systems: transmission, detection, and spectral efficiency. *IEEE Trans. Wireless Comm.*, **1(3)**, 383–392, July 2002.

[Hwang et al., 2002] J. Hwang, C. Chen and M. Tsai. Design and analytical error performance of MMSE-DFE receiver with spatial diversity. *Proc. IEEE SPWAC*, 13–16, Taiwan, China, March 2002.

[Jafar et al., 2001] S. Jafar, S. Vishwanath and A. Goldsmith. Channel capacity and beamforming for multiple transmit and multiple receive antennas with covariance feedback. *Proc. IEEE ICC*, **7**, 2266–2270, Helsinki, Finland, 2001.

[Jafarkhani, 2001] H. Jafarkhani. A quasi-orthogonal space-time block code. *IEEE Trans. Comm.*, **49**, 1–4, January 2001.

[Jakes, 1974] W. Jakes. *Microwave Mobile Communications*. Wiley, New York, NY, 1974.

[Jindal *et al.*, 2001] N. Jindal, S. Vishwanath and A. Goldsmith. On the duality of multiple-access and broadcast channels. *Proc. Allerton Conf. on Communication, Control and Computing*, Monticello, IL, October 2001.

[Jöngren *et al.*, 2002] G. Jöngren, M. Skoglund and B. Ottersten. Combining beamforming with orthogonal space-time block coding. *IEEE Trans. Inf. Theory*, **48(3)**, 611–627, March 2002.

[Jorswieck and Boche, 2002] E. Jorswieck and H. Boche. On the optimality-range of beamforming for MIMO systems with covariance feedback. *Proc. Conf. on Information Sciences and Systems*, Princeton University, March 2002.

[Kannan, 1983] R. Kannan. Improved algorithms on integer programming and related lattice problems. *Proc. ACM Symp. on Theory of Comp.*, 193–206, Boston, MA, April 1983.

[Kapoor *et al.*, 1999] S. Kapoor, D. Marchok and Y. Huang. Adaptive interference suppression in multiuser wireless OFDM systems using antenna arrays. *IEEE Trans. Sig. Proc.*, **47(12)**, 3381–3391, December 1999.

[Kasami, 1966] T. Kasami. Weight distribution formula for some class of cyclic codes. Tech. Rep. R-285, Coordinated Science Laboratory, University of Illinois, Urbana, IL, April 1966.

[Kermoal *et al.*, 2000] J. Kermoal, L. Schumacher, P. Mogensen and K. Pedersen. Experimental investigation of correlation properties of MIMO radio channels for indoor picocell scenarios. *Proc. IEEE VTC*, **1**, 14–21, Boston, MA, September 2000.

[Kim and Bhargava, 2002] D. Kim and V. Bhargava. Combined multidimensional signaling and transmit diversity for high-rate wide-band CDMA. *IEEE Trans. Comm.*, **50(2)**, 262–275, February 2002.

[Kim *et al.*, 1998] J. Kim, L. Cimini and J. Chuang. Coding strategies for OFDM with antenna diversity high-bit-rate mobile data applications. *Proc. IEEE VTC*, **2**, 763–767, Ottawa, Canada, May 1998.

[Krenz and Wesolowski, 1997] R. Krenz and K. Wesolowski. Comparative study of space-diversity techniques for MLSE receivers in mobile radio. *IEEE Trans. Veh. Tech.*, **46(3)**, 653–663, August 1997.

[Kuo and Fitz, 1997] W. Kuo and M. Fitz. Design and analysis of transmitter diversity using intentional frequency offset for wireless communications. *IEEE Trans. Veh. Tech.*, **46(5)**, 871–881, November 1997.

[Kyritsi, 2002] P. Kyritsi. Capacity of multiple input-multiple output wireless systems in an indoor environment. PhD thesis, Stanford University, January 2002.

[Lang *et al.*, 1999] L. Lang, L. Cimini and J. Chuang. Turbo codes for OFDM with antenna diversity. *Proc. IEEE VTC*, **2**, 1664–1668, Houston, TX, May 1999.

[Larsson, 2001] E. Larsson. Ubiquitous signal processing: Applications to communications, spectral analysis and array processing. PhD thesis, Uppsala University, Uppsala, Sweden, 2001.

[Larsson *et al.*, 2001] E. Larsson, P. Stoica and J. Li. Space-time block codes: Ml detection for unknown channels and unstructured interference. *Proc. Asilomar Conf. on Signals, Systems and Computers*, **2**, 916–920, Pacific Grove, CA, November 2001.

[Larsson *et al.*, 2002] E. Larsson, P. Stoica, E. Lindskog and J. Li. Space-time block coding for frequency-selective channels. *Proc. IEEE ICASSP*, **3**, 2405–2408, Orlando, FL, May 2002.

[Lee, 1982] W. C. Y. Lee. *Mobile Communications Engineering*. McGraw-Hill, New York, 1982.

[Lee, 1995] W. Lee. *Mobile Cellular Telecommunications: Analog and Digital Systems*. McGraw-Hill Professional, 2nd edition, February 1995.

[Lee and Messerschmitt, 1993] E. Lee and D. Messerschmitt. *Digital Communications*. Kluwer, Norwell, MA, 2nd edition, 1993.

[Lempiäinen and Laiho-Steffens, 1999] J. Lempiäinen and J. Laiho-Steffens. The performance of polarization diversity schemes at a base station in small/micro cells at 1800 MHz. *IEEE Trans. Veh. Tech.*, **47(3)**, 1087–1092, August 1999.

[Leon-Garcia, 1994] A. Leon-Garcia. *Probability and Random Processes for Electrical Engineering.* Addison Wesley, New York, NY, 2nd edition, 1994.

[Li and Sollenberger, 1999] Y. Li and N. Sollenberger. Adaptive antenna arrays for OFDM systems with cochannel interference. *IEEE Trans. Comm.*, **47(2)**, 217–229, February 1999.

[Li et al., 1999] Y. Li, J. Chuang and N. Sollenberger. Transmitter diversity for OFDM systems and its impact on high-rate data wireless networks. *IEEE J. Sel. Areas Comm.*, **17(7)**, 1233–1243, July 1999.

[Li et al., 2000] X. Li, H. Huang, G. Foschini and R. Valenzuela. Effects of iterative detection and decoding on the performance of BLAST. *Proc. IEEE GLOBECOM*, **2**, 1061–1066, San Francisco, CA, November 2000.

[Li et al., 2001] J. Li, K. Letaief, R. Cheng and Z. Cao. Co-channel interference cancellation for space-time coded OFDM systems. *Proc. IEEE ICC*, **6**, 1638–1642, Helsinki, Finland, June 2001.

[Liang, 1998] J. Liang. Interference reduction and equalization with space-time processing in TDMA cellular networks. PhD thesis, Stanford University, Stanford, CA, June 1998.

[Liang and Paulraj, 1996] J. Liang and A. Paulraj. Two stage CCI/ISI reduction with space-time processing in TDMA cellular networks. *Proc. Asilomar Conf. on Signals, Systems and Computers*, **1**, 607–611, Pacific Grove, CA, November 1996.

[Liberti and Rappaport, 1999] J. Liberti and T. Rappaport. *Smart Antennas for Wireless Communications.* Prentice Hall, Upper Saddle River, NJ, April 1999.

[Lin et al., 2000] L. Lin, L. Cimini and J. Chuang. Comparison of convolutional and turbo codes for OFDM with antenna diversity in high-bit-rate wireless applications. *IEEE Comm. Letters*, **4(9)**, 277–279, September 2000.

[Lindskog, 1997] E. Lindskog. Multi-channel maximum likelihood sequence estimation. *Proc. IEEE VTC*, **2**, 715–719, Phoenix, AZ, May 1997.

[Lindskog, 1999] E. Lindskog. Space-time processing and equalization for wireless communications. PhD thesis, Uppsala University, 1999.

[Lindskog and Paulraj, 2000] E. Lindskog and A. Paulraj. A transmit diversity scheme for channels with intersymbol interference. *Proc. IEEE ICC*, **1**, 307–311, New Orleans, LA, 2000.

[Liu et al., 2000] Y. Liu, M. Fitz and O. Takeshita. QPSK space-time turbo codes. *Proc. IEEE ICC*, **1**, 292–296, New Orleans, LA, June 2000.

[Liu et al., 2001a] Y. Liu, M. Fitz and O. Takeshita. Outage probability and space-time code design criteria for frequency selective fading channels with fractional delay. *Proc. IEEE ISIT*, 80, Washington, DC, June 2001.

[Liu et al., 2001b] Y. Liu, M. Fitz and O. Takeshita. Space-time codes performance criteria and design for frequency selective fading channels. *Proc. IEEE ICC*, **9**, 2800–2804, Helsinki, Finland, June 2001.

[Liu et al., 2001c] Z. Liu, G. Giannakis, S. Barbarossa and A. Scaglione. Transmit antennae space-time block coding for generalized OFDM in the presence of unknown multipath. *IEEE J. Sel. Areas Comm.*, **19(7)**, 1352–1364, July 2001.

[Lo, 1999] T. Lo. Maximal ratio transmission. *IEEE Trans. Comm.*, **47(10)**, 1458–1461, October 1999.

[Lo et al., 1991] N. Lo, D. Falconer and A. Sheikh. Channel interpolation for digital mobile radio communications. *Proc. IEEE ICC*, **2**, 773–777, Denver, CO, June 1991.

[Lotter and van Rooyen, 1998] M. Lotter and P. van Rooyen. Space division multiple access for cellular CDMA. *Proc. IEEE Int. Symp. on Spread Spectrum Tech. and App.*, **3**, 959–964, Sun City, South Africa, September 1998.

[Lozano and Papadias, 2002] A. Lozano and C. Papadias. Layered space-time receivers for frequency-selective wireless channels. *IEEE Trans. Comm.*, **50(1)**, 65–73, January 2002.

[Lu et al., 2002] B. Lu, X. Wang and K. Narayanan. LDPC-based space-time coded OFDM systems over correlated fading channels: Performance analysis and receiver design. *IEEE Trans. Comm.*, **50(1)**, 74–88, January 2002.

[Lupas and Verdu, 1989] R. Lupas and S. Verdu. Linear multiuser detectors for synchronous code division multiple-access channels. *IEEE Trans. Inf. Theory*, **35(1)**, 123–136, November 1989.

[MacKay, 1995] D. MacKay. Good error-correcting codes based on very sparse matrices. *IEEE Trans. Inf. Theory*, **45(2)**, 399–430, May 1999.

[Madhow and Honig, 1994] U. Madhow and M. Honig. MMSE interference suppression for direct sequence spread spectrum CDMA. *IEEE Trans. Comm.*, **42(12)**, 3178–3188, December 1994.

[Madhow et al., 1999] U. Madhow, E. Visotsky and S. Warrier. Multiuser space-time communication. *IEEE Information Theory and Communications Workshop*, 15–17, Kruger National Park, South Africa, June 1999.

[Martin and Ottersten, 2002] C. Martin and B. Ottersten. Analytic approximations of eigenvalue moments and mean channel capacity for mimo channels. *Proc. IEEE ICASSP*, **3**, 2389–2392, Orlando, FL, May 2002.

[Marzetta, 1999] T. Marzetta. BLAST training: estimating channel characteristics for high-capacity space-time wireless. *Proc. Allerton Conf. on Communication, Control and Computing*, Monticello, IL, September 1999.

[Marzetta and Hochwald, 1999] T. Marzetta and B. Hochwald. Capacity of a mobile multiple-antenna communication link in Rayleigh flat fading. *IEEE Trans. Inf. Theory*, **45(1)**, 139–157, January 1999.

[Marzetta et al., 2001] T. Marzetta, B. Hassibi and B. Hochwald. Space-time autocoding constellations with pairwise-independent signals. *Proc. IEEE ISIT*, 326, Washington, DC, June 2001.

[Meyr et al., 1997] H. Meyr, M. Moeneclaey and S. A. Fletchel. *Digital Communication Receivers: Synchronization, Channel Estimation and Signal Processing*. Wiley, New York, NY, 1997.

[Mogensen, 1993] P. Mogensen. GSM base-station antenna diversity using soft-decision combining on up-link and delayed-signal transmission on down-link. *Proc. IEEE VTC*, 611–613, Secaucus, NL, May 1993.

[Molisch, 2002] A. Molisch. A generic model for MIMO wireless propagation channels. *Proc. IEEE ICC*, **1**, 277–282, New York, NY, April/May 2002.

[Molisch et al., 2001] A. Molisch, M. Win and J. Winters. Capacity of MIMO systems with antenna selection. *Proc. IEEE ICC*, **2**, 570–574, Helsinki, Finland, June 2001.

[Montalbano et al., 1998] G. Montalbano, I. Ghauri and D. Slock. Spatio-temporal array processing for cdma/sdma downlink transmission. *Proc. Asilomar Conf. on Signals, Systems and Computers*, **2**, 1337–1341, Pacific Grove, CA, October/November 1998.

[Mudulodu and Paulraj, 2000] S. Mudulodu and A. Paulraj. A simple multiplexing scheme for MIMO systems using multiple spreading codes. *Proc. Asilomar Conf. on Signals, Systems and Computers*, **1**, 769–774, Pacific Grove, CA, 2000.

[Muirhead, 1982] R. J. Muirhead. *Aspects of Multivariate Statistical Theory*. Wiley, New York, NY, 1982.

[Muller and Huber, 1997] S. Muller and J. Huber. A comparison of peak power reduction schemes for OFDM. *Proc. IEEE GLOBECOM*, **1**, 1–5, Phoenix, AZ, November 1997.

[Nabar *et al.*, 2001] R. Nabar, H. Bölcskei and A. Paulraj. Transmit optimization for spatial multiplexing in the presence of spatial fading correlation. *Proc. IEEE GLOBECOM*, **1**, 131–135, San Antonio, TX, November 2001.

[Nabar *et al.*, 2002a] R. Nabar, H. Bölcskei, V. Erceg, D. Gesbert and A. Paulraj. Performance of multi-antenna signaling techniques in the presence of polarization diversity. *IEEE Trans. Sig. Proc.*, **50(10)**, 2553–2562, October 2002.

[Nabar *et al.*, 2002b] R. Nabar, H. Bölcskei and A. Paulraj. Outage performance of space-time block codes for generalized MIMO channels. *IEEE Trans. Inf. Theory*, March 2002. Submitted.

[Nabar *et al.*, 2002c] R. U. Nabar, H. Bölcskei and A. J. Paulraj. Influence of propagation conditions on the outage capacity of space-time block codes. *Proc. EWC 2002*, Florence, Italy, February 2002.

[Nabar *et al.*, 2002d] R. U. Nabar, H. Bölcskei and A. J. Paulraj. Outage properties of space-time block codes in correlated Rayleigh or Ricean fading environments. *Proc. IEEE ICASSP*, **1**, 2381–2384, Orlando, FL, May 2002.

[Naguib, 1996] A. Naguib. Adaptive antennas for CDMA wireless networks. PhD thesis, Stanford University, August 1996.

[Naguib *et al.*, 1998a] A. Naguib, N. Seshadri and A. Calderbank. Applications of space-time block codes and interference suppression for high capacity and high data rate wireless systems. *Proc. Asilomar Conf. on Signals, Systems and Computers*, **2**, 1803–1810, Pacific Grove, CA, 1998.

[Naguib *et al.*, 1998b] A. Naguib, V. Tarokh, N. Seshadri and A. Calderbank. A space-time coding modem for high-data-rate wireless communications. *IEEE J. Sel. Areas Comm.*, **16(8)**, 1459–1478, October 1998.

[Nakagami, 1960] M. Nakagami. The *m* distribution: a general formula of intensity distribution of rapid fading. *Statistical Methods in Radio Wave Propagation*, pp. 3–36, Pergamon Press, 1960. Edited by W. G. Hoffman.

[Narula *et al.*, 1998] A. Narula, M. Lopez, M. Trott and G. Wornell. Efficient use of side information in multiple-antenna data transmission over fading channels. *IEEE J. Sel. Areas Comm.*, **16(8)**, 1423–1436, October 1998.

[Neeser and Massey, 1993] F. Neeser and J. Massey. Proper complex random processes with applications to information theory. *IEEE Trans. Inf. Theory*, **39(4)**, 1293–1302, July 1993.

[Neubauer and Eggers, 1999] T. Neubauer and P. Eggers. Simultaneous characterization of polarization diversity matrix components in pico cells. *Proc. IEEE VTC*, **3**, 1361–1365, Amsterdam, The Netherlands, 1999.

[Ng, 1998] B. Ng. Structured channel methods for wireless communications. PhD thesis, Stanford University, 1998.

[Oestges and Paulraj, 2003] C. Oestges and A. Paulraj. A physical scattering model for MIMO macrocellular wireless channels. *IEEE J. Sel. Areas Comm.*, 2003. Accepted for publication.

[Okumura *et al.*, 1968] Y. Okumura, E. Ohmuri, T. Kawano and K. Fukuda. Field strength and its variability in VHF and UHF land mobile radio service. *Rev. ECL*, **16**, 825–873, 1968.

[Olofsson *et al.*, 1997] H. Olofsson, M. Almgren and M. Hook. Transmitter diversity with antenna hopping for wireless communication systems. *Proc. IEEE VTC*, **3**, 1743–1747, Phoenix, AZ, May 1997.

[Omura, 1971] J. Omura. Optimal receiver design for convolutional codes and channels with memory via control theoretical concepts. *Inf. Sci.*, **3**, 243–266, 1971.

[O'Neill and Lopes, 1994] R. O'Neill and L. Lopes. Performance of amplitude limited multitone signals. *Proc. IEEE VTC*, **3**, 1675–1679, Stockholm, Sweden, June 1994.

[Ostling, 1993] P. Ostling. Performance of MMSE linear equalizer and decision feedback equalizer in single-frequency simulcast environment. *Proc. IEEE VTC*, 629–632, Secaucus, NJ, May 1993.

[Ottersten, 1996] B. Ottersten. Array processing for wireless communications. *Proc. Eighth IEEE Signal Processing Workshop on Statistical Signal and Array Processing*, 466–473, Corfu, Greece, June 1996.

[Oyman et al., 2002a] Ö. Oyman, D. Gore and A. Paulraj. Spectral efficiency of MIMO receivers. 2002. In preparation.

[Oyman et al., 2002b] Ö. Oyman, R. Nabar, H. Bölcskei and A. J. Paulraj. Tight lower bounds on the ergodic capacity of Rayleigh fading MIMO channels. *Proc. IEEE GLOBECOM*, Taipei, Taiwan, November 2002.

[Ozarow et al., 1994] L. Ozarow, S. Shamai and A. Wyner. Information theoretic considerations for cellular mobile radio. *IEEE Trans. Veh. Technol.*, **43(2)**, 359–378, May 1994.

[Papadias, 1999] C. Papadias. On the spectral efficiency of space-time spreading schemes for multiple antenna CDMA systems. *Proc. Asilomar Conf. on Signals, Systems and Computers*, **1**, 639–643, Pacific Grove, CA, October 1999.

[Papadias and Foschini, 2001] C. Papadias and G. Foschini. A space-time coding approach for systems employing four transmit antennas. *Proc. IEEE ICASSP*, **4**, 2481–2484, Salt Lake City, UT, 2001.

[Papadias and Foschini, 2002] C. Papadias and G. Foschini. On the capacity of certain space-time coding schemes. *EURASIP J. Appl. Signal Proc.*, **5**, 447–458, May 2002.

[Papadias et al., 1998] C. Papadias, H. Huang and L. Mailaender. Adaptive multi-user detection of fading cdma channels using antenna arrays. *Proc. Asilomar Conf. on Signals, Systems and Computers*, **2**, 1564–1568, Pacific Grove, CA, November 1998.

[Papoulis, 1984] Athanasios Papoulis. *Probability, Random Variables, and Stochastic Processes*. McGraw-Hill, New York, NY, 1984.

[Parsons, 1992] D. Parsons. *The Mobile Radio Propagation Channel*. Wiley, New York, NY, 1992.

[Paulraj, 2002] A. Paulraj. *Class Reader for EE492m – Space-time Wireless Communications*. Stanford University, Stanford, CA. Available online at http://www.stanford.edu/class/ee492, 2002.

[Paulraj and Kailath, 1994] A. Paulraj and T. Kailath. Increasing capacity in wireless broadcast systems using distributed transmission/directional reception. US Patent, 5 345 599, 1994.

[Paulraj and Papadias, 1997] A. Paulraj and C. Papadias. Space-time processing for wireless communications. *IEEE Signal Proc. Mag.*, **14(6)**, 49–83, November 1997.

[Paulraj et al., 1986] A. Paulraj, R. Roy and T. Kailath. ESPRIT – a subspace rotation approach to signal parameter estimation. *IEEE Proceedings*, **74(7)**, 1044–1045, July 1986.

[Paulraj et al., 1998] A. Paulraj, C. Papadias, V. Reddy and A. van der Veen. *Blind Space-Time Signal Processing: Wireless Communications*. Prentice Hall, Upper Saddle River, NJ, 1998. Edited by V. Poor and G. Wornell.

[Pickholtz et al., 1982] R. Pickholtz, D. Schilling and L. Milstein. Theory of spread-spectrum communications – a tutorial. *IEEE Trans. Comm.*, **COM-30**, 855–884, May 1982.

[Poor and Wornell, 1998] V. Poor and G. Wornell. *Wireless Communications: Signal Processing Perspectives*. Prentice Hall, Upper Saddle River, NJ, 1998.

[Proakis, 1995] J. Proakis. *Digital Communications*. McGraw-Hill, New York, NY, 3rd edition, 1995.

[Radon, 1922] J. Radon. Lineare scharen orthogonaler matrizen. *Abhandlungen aus dem Mathematischen Seminar der Hamburgishen Universität*, **1**, 1–14, 1922.

[Raleigh and Cioffi, 1998] G. Raleigh and J. Cioffi. Spatio-temporal coding for wireless communication. *IEEE Trans. Comm.*, **46(3)**, 357–366, March 1998.

[Ramos et al., 2000] J. Ramos, M. Zoltowski and H. Liu. Low-complexity space-time processor for ds-cdma communications. *IEEE Trans. Sig. Proc.*, **48(1)**, 39–52, January 2000.

[Rapajic and Vucetic, 1994] P. Rapajic and B. Vucetic. Adaptive receiver structures for asynchronous CDMA systems. *IEEE J. Sel. Areas Comm.*, **12(4)**, 685–697, May 1994.

[Rappaport, 1996] T. Rappaport. *Wireless Communications: Principles & Practice*. Prentice Hall, Upper Saddle River, NJ, 1996.

[Rashid-Farrokhi *et al.*, 1998] F. Rashid-Farrokhi, K. Liu and L. Tassiulas. Transmit beamforming and power control for cellular wireless systems. *IEEE J. Sel. Areas Comm.*, **16(8)**, 1437–1449, October 1998.

[Rashid-Farrokhi *et al.*, 2000] F. Rashid-Farrokhi, A. Lozano, G. Foschini and R. Valenzuela. Spectral efficiency of wireless systems with multiple transmit and receive antennas. *Proc. IEEE Int. Symp. on PIMRC*, **1**, 373–377, London, UK, September 2000.

[Rimoldi and Urbanke, 1996] B. Rimoldi and R. Urbanke. A rate-splitting approach to the gaussian multiple-access channel. *IEEE Trans. Inf. Theory*, **46(2)**, 364–375, March 1996.

[Robertson *et al.*, 1974] P. Robertson, E. Villebrun and P. Hoeher. A comparison of optimal and sub-optimal MAP decoding algorithms operating in the log domain. *IEEE Trans. Inf. Theory*, **20**, 284–287, March 1974.

[Rooyen *et al.*, 2000] P. Rooyen, M. Lötter and D. Wyk. *Space-Time Processing for CDMA Mobile Communications*. Kluwer, Norwell, MA, 2000.

[Roy, 1997] R. Roy. Spatial division multiple access technology and its application to wireless communication systems. *Proc. IEEE VTC*, **2**, 730–734, Phoenix, AZ, May 1997.

[Roy and Ottersten, 1996] R. Roy and B. Ottersten. Spatial division multiple access wireless communication systems. US Patent, 5 515 378, May 1996.

[Roy *et al.*, 1986] R. Roy, A. Paulraj and T. Kailath. ESPRIT – A subspace rotation approach to estimation of parameters of cisoids in noise. *IEEE Trans. Acoust., Speech, Sig. Proc.*, **34(5)**, 1340–1342, October 1986.

[Salvekar *et al.*, 2001] A. Salvekar, J. Tellado and J. Cioffi. Peak-to-average power ratio reduction for block transmission systems in the presence of transmit filtering. *Proc. IEEE ICC*, **1**, 175–178, Helsinki, Finland, June 2001.

[Sampath, 2001] H. Sampath. Linear precoding and decoding for multiple input multiple output (MIMO) wireless channels. PhD thesis, Stanford University, May 2001.

[Sampath and Paulraj, 2001] H. Sampath and A. Paulraj. Linear precoding for space-time coded systems with known fading correlations. *Proc. Asilomar Conf. on Signals, Systems and Computers*, **1**, 246–251, Pacific Grove, CA, November 2001.

[Sampath *et al.*, 2001] H. Sampath, P. Stoica and A. Paulraj. Generalized linear precoder and decoder design for MIMO channels using the weighted MMSE criterion. *IEEE Trans. Comm.*, **49(12)**, 2198–2206, December 2001.

[Sampath *et al.*, 2002] H. Sampath, S. Talwar, J. Tellado, V. Erceg and A. Paulraj. A fourth-generation mimo-ofdm broadband wireless system: Design, performance, and field test results. *IEEE Comm. Mag.*, **40(9)**, 143–149, September 2002.

[Sandhu, 2002] S. Sandhu. Signal design for multiple-input multiple-output (MIMO) wireless: a unified perspective. PhD thesis, Stanford University, 2002.

[Sandhu and Paulraj, 2001] S. Sandhu and A. Paulraj. Union bound on error probability of linear space-time block codes. *Proc. IEEE ICASSP*, **4**, 2473–2476, Salt Lake City, UT, May 2001.

[Sandhu *et al.*, 2001] S. Sandhu, R. Heath and A. Paulraj. Space-time block codes versus space-time trellis codes. *Proc. IEEE ICC*, **4**, 1132–1136, Helsinki, Finland, June 2001.

[Saraydar *et al.*, 2002] C. Saraydar, N. Mandayam and D. Goodman. Efficient power control via pricing in wireless data networks. *IEEE Trans. Comm.*, **50(2)**, 291–303, February 2002.

[Sayeed and Veeravalli, 2002] A. Sayeed and V. Veeravalli. The essential degrees of freedom in space-time fading channels. *Proc. IEEE PIMRC*, Lisbon, Portugal, September 2002.

[Scaglione, 2002] A. Scaglione. Statistical analysis of the capacity of mimo frequency selective Rayleigh fading channels with arbitrary number of inputs and outputs. *Proc. IEEE ISIT*, 278, Lausanne, Switzerland, June/July 2002.

[Scaglione *et al.*, 2002] A. Scaglione, P. Stoica, S. Barbarossa, G. Giannakis and H. Sampath. Optimal designs for space-time linear precoders and decoders. *IEEE Trans. Sig. Proc.*, **50(5)**, 1051–1064, May 2002.

[Schlegel and Grant, 2001] C. Schlegel and A. Grant. Concatenated space-time coding. *Proc. IEEE Int. Symp. on PIMRC*, **1**, C139–C143, San Diego, CA, September/October 2001.

[Schmidt, 1981] R. Schmidt. A signal subspace approach to multiple emitter locations and spectral estimation. PhD thesis, Stanford University, Stanford, CA, 1981.

[Schubert and Boche, 2002] M. Schubert and H. Boche. A unifying theory for uplink and downlink multi-user beamforming. *Proc. Int. Zurich Seminar on Broadband Communications*, 27(1)–27(6), Zurich, Switzerland, February 2002.

[Sellathurai and Haykin, 2000] M. Sellathurai and S. Haykin. Turbo-BLAST for high-speed wireless communications. *Proc. IEEE WCNC*, 315–320, Chicago, IL, September 2000.

[Seshadri and Winters, 1994] N. Seshadri and J. Winters. Two signaling schemes for improving the error performance of frequency-division-duplex (FDD) transmission systems using transmitter antenna diversity. *Int. J. Wireless Information Networks*, **1**, 49–60, January 1994.

[Shamai and Zaidel, 2001] S. Shamai and B. Zaidel. Enhancing the cellular downlink capacity via co-processing at the transmitting end. *Proc. IEEE VTC*, **3**, 1745–1749, 2001.

[Shannon, 1948] C. Shannon. A mathematical theory of communication. *Bell Labs Tech. J.*, **27**, 379–423, 623–656, July and October 1948.

[Sharma and Papadias, 2002] N. Sharma and C. Papadias. Improved quasi-orthogonal codes through constellation rotation. *Proc. IEEE ICASSP*, **4**, 3968–3971, Orlando, FL, May 2002.

[Shiu and Kahn, 1998] D. Shiu and J. Kahn. Power allocation strategies for wireless systems with multiple antennas. available at http://www.eecs.berkeley.edu/ jmk/pubs/poweralloc.pdf, 1998.

[Simon, 2002] M. Simon. *Probability Distributions Involving Gaussian Random Variables: A Handbook for Scientists and Engineers*. Kluwer, Norwell, MA, 2002.

[Simon *et al.*, 1994] M. Simon, J. Omura, R. Scholtz and B. Levitt. *Spread Spectrum Communications Handbook*. McGraw-Hill, New York, NY, 1994.

[Sokal and Rohlf, 1995] R. Sokal and F. Rohlf. *Biometry: The Principles and Practice of Statistics in Biological Research*. Freeman Co., New York, NY, 3rd edition, 1995.

[Soma *et al.*, 2002] P. Soma, D. Baum, V. Erceg, R. Krishnamoorthy and A. Paulraj. Analysis and modeling of multiple-input multiple-output (MIMO) radio channel based on outdoor measurements conducted at 2.5 GHz for fixed BWA applications. *Proc. IEEE ICC*, **1**, 396–400, New York, NY, April/May 2002.

[Stamoulis *et al.*, 2002] A. Stamoulis, L. Zhiqiang and G. Giannakis. Space-time block-coded OFDMA with linear precoding for multirate services. *IEEE Trans. Sig. Proc.*, **50(1)**, 119–129, January 2002.

[Stefanov and Duman, 2001] A. Stefanov and T. Duman. Turbo-coded modulation for systems with transmit and receive antenna diversity over block fading channels: system model, decoding approaches and practical considerations. *IEEE J. Sel. Areas Comm.*, **19**, 958–968, May 2001.

[Stoica and Lindskog, 2001] P. Stoica and E. Lindskog. Space-time block coding for channels with intersymbol interference. *Proc. Asilomar Conf. on Signals, Systems and Computers*, **1**, 252–256, Pacific Grove, CA, November 2001.

[Stoica and Nehorai, 1991] P. Stoica and A. Nehorai. Performance comparison of subspace rotation and MUSIC methods for direction estimation. *IEEE Trans. Sig. Proc.*, **39(2)**, 446–453, February 1991.

[Stridh *et al.*, 2000] R. Stridh, B. Ottersten and P. Karlsson. MIMO channel capacity of a measured indoor radio channel at 5.8 GHz. *Proc. Asilomar Conf. on Signals, Systems and Computers*, **1**, 733–737, Pacific Grove, CA, November 2000.

[Stüber, 1996] G. Stüber. *Principles of Mobile Communication*. Kluwer, Norwell, MA, 1996.

[Su and Geraniotis, 2001] H. Su and E. Geraniotis. Space-time turbo codes with full antenna diversity. *IEEE Trans. Comm.*, **49(1)**, 47–57, January 2001.

[Suard *et al.*, 1998] B. Suard, G. Xu and T. Kailath. Uplink channel capacity of space-division-multiple-access schemes. *IEEE Trans. Inf. Theory*, **44(4)**, 1468–1476, July 1998.

[Swindlehurst *et al.*, 2001] A. Swindlehurst, G. German, J. Wallace and M. Jensen. Experimental measurements of capacity for MIMO indoor wireless channels. *Proc. IEEE Signal Proc. Workshop on Signal Processing Advances in Wireless Communications*, 30–30, Taoyuan, Taiwan, March 2001.

[Tarokh *et al.*, 1998] V. Tarokh, N. Seshadri and A. R. Calderbank. Space-time codes for high data rate wireless communication: Performance criterion and code construction. *IEEE Trans. Inf. Theory*, **44(2)**, 744–765, March 1998.

[Tarokh *et al.*, 1999a] V. Tarokh, H. Jafarkhani and A. Calderbank. Space-time block coding for wireless communications: Performance results. *IEEE J. Sel. Areas Comm.*, **17(3)**, 451–460, March 1999.

[Tarokh *et al.*, 1999b] V. Tarokh, H. Jafarkhani and A. R. Calderbank. Space-time block codes from orthogonal designs. *IEEE Trans. Inf. Theory*, **45(5)**, 1456–1467, July 1999.

[Tarokh *et al.*, 1999c] V. Tarokh, A. Naguib, N. Seshadri and A. Calderbank. Combined array processing and space-time coding. *IEEE Trans. Inf. Theory*, **45(4)**, 1121–1128, May 1999.

[Tarokh *et al.*, 1999d] V. Tarokh, A. Naguib, N. Seshadri and A. Calderbank. Space-time codes for high data rate wireless communication: Performance criteria in the presence of channel estimation errors, mobility, and multiple paths. *IEEE Trans. Comm.*, **47(2)**, 199–207, February 1999.

[Tehrani *et al.*, 1999] A. Tehrani, R. Negi and J. Cioffi. Space-time coding over a code division multiple access system. *Proc. IEEE WCNC*, **1**, 134–138, New Orleans, LA, September 1999.

[Telatar, 1995] I. Telatar. Capacity of multi-antenna Gaussian channels. Technical Report #BL0112170-950615-07TM, AT & T Bell Laboratories, 1995.

[Telatar, 1999a] I. Telatar. Capacity of multi-antenna Gaussian channels. *European Trans. Tel.*, **10(6)**, 585–595, November/December 1999.

[Telatar, 1999b] I. E. Telatar. Capacity of multi-antenna Gaussian channels. *European Trans. Tel.*, **10(6)**, 585–595, November/December 1999.

[Tellado, 1998] J. Tellado. PAR reduction in multicarrier transmission systems. PhD thesis, Stanford University, 1998.

[ten Brink, 1999] S. ten Brink. On the convergence of iterative coding. *Electron. Lett.*, **35(10)**, 1459–1460, May 1999.

[ten Brink *et al.*, 1998] S. ten Brink, J. Speidel and R.-H. Yan. Iterative demapping for QPSK modulation. *Electron. Lett.*, **34(15)**, 1459–1460, July 1998.

[Tomlinson, 1971] M. Tomlinson. New automatic equalizer using modulo arithmetic. *Electronic Lett.*, **7**, 138–139, March 1971.

[Tonello, 2000] A. Tonello. Space-time bit interleaved coded modulation with an iterative decoding strategy. *Proc. IEEE VTC*, **1**, 473–478, Boston, MA, September 2000.

[Tong *et al.*, 1994] L. Tong, G. Xu and T. Kailath. Blind identification and equalization based on second-order statistics: A time domain approach. *IEEE Trans. Inf. Theory*, **40(2)**, 340–349, March 1994.

[Tong, 2001] L. Tong. Channel estimation for space-time orthogonal block codes. *Proc. IEEE ICC*, **4**, 1127–1131, Helsinki, Finland, June 2001.

[Turin, 1960] G. Turin. The characteristic function of hermitian quadratic forms in complex normal variables. *Biometrika*, **47**, 199–201, 1960.

[Turin, 1980] G. Turin. Introduction to spread spectrum antimultipath techniques and their application to urban digital radio. *IEEE Proceedings*, **68(3)**, 328–353, March 1980.

[Turkmani, 1995] A. Turkmani. An experimental evaluation of the performance of two-branch space and polarization diversity schemes at 1800 MHz. *IEEE Trans. Veh. Tech.*, **44(2)**, 318–326, May 1995.

[Uysal and Georghiades, 2001] M. Uysal and C. N. Georghiades. Effect of spatial fading correlation on performance of space-time codes. *Electronics Lett.*, **37(3)**, 181–183, February 2001.

[van der Veen et al., 1995] A. van der Veen, S. Talwar and A. Paulraj. Blind identification of FIR channels carrying multiple finite alphabet signals. *Proc. IEEE ICASSP*, **2**, 1213–1216, Detroit, MI, 1995.

[van Zelst et al., 2001] A. van Zelst, R. van Nee and G. Awater. Turbo-BLAST and its performance. *Proc. IEEE VTC*, **2**, 1282–1286, Rhodes, Greece, May 2001.

[Vandenameele, 2001] P. Vandenameele. *Space Division Multiple Access for Wireless Local Area Networks*. Kluwer, Norwell, MA, 2001.

[Vanmarcke, 1983] E. Vanmarcke. *Random Fields: Analysis and Synthesis*. MIT Press, Cambridge, MA, 1983.

[Varanasi and Guess, 1997] M. Varanasi and T. Guess. Optimum decision feedback multiuser equalization with successive decoding achieves the total capacity of the Gaussian multiple access channel. *Proc. Asilomar Conf. on Signals, Systems and Computers*, **2**, 1405–1409, Pacific Grove, CA, November 1997.

[Vaughan, 1990] R. Vaughan. Polarization diversity in mobile communications. *IEEE Trans. Veh. Tech.*, **39(3)**, 177–186, August 1990.

[Vaughan and Andersen, 2001] R. Vaughan and J. Andersen. *Channels, Propagation and Antennas for Mobile Communications*. IEE Press, 2001.

[Venkatesh et al., 2002] S. Venkatesh, Ö. Oyman and A. Paulraj. On the capacity and error performance in degenerate channels. 2002. In preparation.

[Ventura-Travest et al., 1997] J. Ventura-Travest, G. Caire, E. Biglieri and G. Taricco. Impact of diversity reception on fading channels with coded modulation – part i: Coherent detection. *IEEE Trans. Comm.*, **45(5)**, 676–686, May 1997.

[Verdu, 1993] S. Verdu. Multi-user detection. In V. Poor, editor, *Advances in Statistical Signal Processing*, pp. 369–409. JAI Press, Greenwich, CT, 1993.

[Verdu, 1998] S. Verdu. *Multiuser Detection*. Cambridge University Press, Cambridge, UK, 1998.

[Vishwanath et al., 2002] S. Vishwanath, N. Jindal and A. Goldsmith. On the capacity of multiple input multiple output broadcast channels. *Proc. IEEE ICC*, **3**, 1444–1450, New York, NY, April 2002.

[Visotsky and Madhow, 2001] E. Visotsky and U. Madhow. Space-time transmit precoding with imperfect feedloack. *IEEE IT*, **47(6)**, 2362–2369, September 2001.

[Viterbi, 1995] A. Viterbi. *CDMA Principles of Spread Spectrum Communications*. Addison Wesley, New York, NY, 1995.

[Viterbo and Boutros, 1999] E. Viterbo and J. Boutros. A universal lattice code decoder for fading channels. *IEEE Trans. Inf. Theory*, **45**, 1639–1642, July 1999.

[Wang and Poor, 1999] X. Wang and V. Poor. Iterative (turbo) soft interference cancellation and decoding for coded CDMA. *IEEE Trans. Comm.*, **47(7)**, 1046–1067, July 1999.

[Wang and Xia, 2002] H. Wang and X. Xia. Upper bounds on rates of complex orthogonal space-time block codes. *IEEE Trans. Inf. Theory*, 2002. Submitted.

[Wesel and Cioffi, 1995] R. Wesel and J. Cioffi. Fundamentals of coding for broadcast OFDM. *Proc. Asilomar Conf. on Signals, Systems and Computers*, **1**, 2–6, Pacific Grove, CA, October/November 1995.

[Win and Winters, 2001] M. Win and J. Winters. Virtual branch analysis of symbol error probability for hybrid selection/maximal-ratio combining in Rayleigh fading. *IEEE Trans. Comm.*, **49(11)**, 1926–1934, November 2001.

[Winters, 1998] J. Winters. The diversity gain of transmit diversity in wireless systems with Rayleigh fading. *IEEE Trans. Vehicular Technology*, **47(1)**, 119–123, February 1998.

[Winters et al., 1994] J. Winters, J. Salz and R. Gitlin. The impact of antenna diversity on the capacity of wireless communications systems. *IEEE Trans. Comm.*, **42(2)**, 1740–1751, February 1994.

[Wittneben, 1991] A. Wittneben. Basestation modulation diversity for digital SIMULCAST. *Proc. IEEE VTC*, 848–853, St. Louis, MO, May 1991.

[Wolniansky et al., 1998] P. Wolniansky, G. Foschini, G. Golden and R. Valenzuela. V-BLAST: An architecture for realizing very high data rates over the rich-scattering wireless channel. *Proc. URSI ISSSE*, 295–300, September 1998.

[Wozencraft and Jacobs, 1965] J. M. Wozencraft and I. M. Jacobs. *Principles of Communication Engineering*. Wiley, New York, NY, 1965.

[Yang and Roy, 1994] J. Yang and S. Roy. On joint transmitter and receiver optimization for multiple-input-multiple-output (MIMO) transmission systems. *IEEE Trans. Comm.*, **42(12)**, 3221–3231, December 1994.

[Yu and Cioffi, 2001a] W. Yu and J. Cioffi. Sum capacity of a gaussian vector broadcast channel. *IEEE Trans. Inf. Theory*, 2001. Submitted.

[Yu and Cioffi, 2001b] W. Yu and J. Cioffi. Trellis precoding for the broadcast channel. *Proc. IEEE GLOBECOM*, **2**, 1338–1344, San Antonio, TX, November 2001.

[Yu et al., 2001a] K. Yu, M. Bengtsson, B. Ottersten, D. McNamara, P. Karlsson and M. Beach. Second order statistics of NLOS indoor MIMO channels based on 5.2 GHz measurements. *Proc. IEEE GLOBECOM*, **1**, 156–160, San Antonio, TX, November 2001.

[Yu et al., 2001b] W. Yu, W. Rhee and J. Cioffi. Optimal power control in multiple access fading channels with multiple antennas. *Proc. IEEE ICC*, **2**, 575–579, Lisbon, Portugal, September 2001.

[Zheng and Tse, 2001] L. Zheng and D. Tse. Optimal diversity-multiplexing trade-off in multi-antenna channels. *Proc. Allerton Conf. on Communication, Control and Computing*, Monticello, IL, October 2001.

[Zheng and Tse, 2002] L. Zheng and D. Tse. Communicating on the Grassmanian manifold: A geometric approach to the noncoherent multiple-antenna channel. *IEEE Trans. Inf. Theory*, **48(2)**, 359–383, February 2002.

[Zhou and Giannakis, 2001] S. Zhou and G. Giannakis. Space-time coding with maximum diversity gains over frequency-selective fading channels. *IEEE Sig. Proc. Letters*, **8(10)**, 269–272, October 2001.

[Zhu and Murch, 2001] X. Zhu and R. D. Murch. MIMO-DFE based BLAST over frequency selective channels. *IEEE GLOBECOM*, **1**, 499–503, San Antonio, TX, November 2001.

[Zoltowski and Stavrinides, 1989] M. Zoltowski and D. Stavrinides. Sensor array signal processing via a Procrustes rotations based eigenanalysis of the ESPRIT data pencil. *IEEE Trans. Sig. Proc.*, **37(6)**, 832–861, June 1989.

Index of common variables

Subject index

Printed in the United Kingdom by
Lightning Source UK Ltd., Milton Keynes
139043UK00001B/37/P